The Science of Land Subsidence

The Science of Land Subsidence explains the current science underpinning natural and human-made land subsidence events, and provides students and interested readers with sufficient background on the basics of geology, natural science, chemical, and environmental engineering. Moreover, it presents a wide-ranging discussion presented in the author's comprehensible conversational style describing the impact of land subsidence events on health, sustenance, and society in general, and provides various case studies covering catastrophic land subsidence events. This book is directed at undergraduate and graduate students, professionals, scientists, and the general reading public who would like to gain a broad multidisciplinary view of one of the greatest challenges of our generation.

- Describes the impact of land subsidence events on health, sustenance, and society in general.
- Provides various case studies covering catastrophic land subsidence events.

The Science of Land Subsidence

Frank R. Spellman

CRC Press
Taylor & Francis Group
Boca Raton London New York

CRC Press is an imprint of the
Taylor & Francis Group, an **informa** business

Designed cover image: Shutterstock

First edition published 2025
by CRC Press
6000 Broken Sound Parkway NW, Suite 300, Boca Raton, FL 33487-2742

and by CRC Press
4 Park Square, Milton Park, Abingdon, Oxon, OX14 4RN

CRC Press is an imprint of Taylor & Francis Group, LLC

Library of Congress Cataloging-in-Publication Data
Names: Spellman, Frank R., author.
Title: The science of land subsidence / Frank R. Spellman.
Description: First edition. | Boca Raton : CRC Press, 2025. |
Includes bibliographical references and index. | Identifiers: LCCN 2023057618 |
ISBN 9781032609560 (hardback) | ISBN 9781032609577 (paperback) |
ISBN 9781003461265 (ebook)
Subjects: LCSH: Subsidences (Earth movements) | Subsidences (Earth movements)–Prevention. | Subsidences (Earth movements)–Case studies.
Classification: LCC QE600.2 .S64 2025 | DDC 551.3/07–dc23/eng/20240324
LC record available at https://lccn.loc.gov/2023057618

ISBN: 978-1-032-60956-0 (hbk)
ISBN: 978-1-032-60957-7 (pbk)
ISBN: 978-1-003-46126-5 (ebk)

DOI: 10.1201/9781003461265

Typeset in Times
by codeMantra

Contents

SECTION II   That Sinking Feeling

SECTION III   Land Subsidence Mitigation

Preface

This book, *The Science of Land Subsidence*, is the 13th volume in the acclaimed series that includes *The Science of Green Energy* (in production), *The Science of Carbon Capture and Storage (CCS)* (in production), *The Science of Ocean Pollution*, *The Science of Lithium (Li)*, *The Science of Electric Vehicles (EVs): Concepts and Applications*, *The Science of Rare Earth Elements: Concepts and Applications*, *The Science of Water*, *The Science of Air*, *The Science of Environmental Pollution*, *The Science of Renewable Energy*, *The Science of Waste*, and *The Science of Wind Power*, all of which bring this highly successful series fully into the 21st century. This book continues the series mantra based on good science and not feel-good science. It also continues to be presented in the author's trademark conversational style—making sure communication is certain—not a failure ... and, maybe because I just want to converse with the reader. My aim, my goal, my objective is to be comprehensive in coverage and comprehensible in what I deliver.

Following the successful format of the other editions in this series, the aim of this no-holds-barred book is to provide an understanding of the current science underpinning natural and human-made land subsidence events and to provide students and interested readers with sufficient background on the basics of geology, natural science, and chemical and environmental engineering so that they can understand the current state and art of the field. Moreover, this book provides a wide-ranging discussion and again presented in the author's comprehensible conversational style describing the impact of land subsidence events on health, sustenance, and society in general, and provides various case studies covering catastrophic land subsidence events. This book is directed at undergraduate and graduate students, professionals, scientists, and the general reading public who would like to gain a broad multidisciplinary view of one of the greatest challenges of our generation and others.

Frank Spellman,
Norfolk, VA

Author

Frank R. Spellman, PhD, is a retired assistant professor of environmental health at Old Dominion University, Norfolk, Virginia, and the author of more than 160 books covering topics ranging from concentrated animal feeding operations to all areas of environmental science and occupational health. Many of his texts are readily available online, and several have been adopted for classroom use at major universities throughout the United States, Canada, China, Europe, and Russia; two have been translated into Spanish for South American markets. Dr. Spellman has been cited in more than 1150 publications. He serves as a professional expert witness for three law groups and as an incident/accident investigator for the US Department of Justice and a northern Virginia law firm. In addition, he consults on homeland security vulnerability assessments for critical infrastructures including water/wastewater facilities nationwide and conducts pre-Occupational Safety and Health Administration/Environmental Protection Agency audits throughout the country. Dr. Spellman receives frequent requests to co-author with well-recognized experts in several scientific fields; for example, he is a contributing author of the prestigious text *The Engineering Handbook*, 2nd ed. (CRC Press). Dr. Spellman lectures on sewage treatment, water treatment, and homeland security and safety topics throughout the country and teaches water/wastewater operator short courses at Virginia Tech (Blacksburg, Virginia). He holds a BA in public administration, a BS in business management, an MBA, and an MS and PhD in environmental engineering.

Section I

Identifying Land Subsidence

1 Introduction

IT'S AN OCCURRENCE

Land subsidence is not an isolated problem. From the lower reaches of the Chesapeake Bay Region in the Hampton Roads and Tidewater Region to the San Francisco Bay/ Delta to the Florida Everglades, to the New Orleans and from upstate New York to Houston, inhabitants (including wildlife) are coming to terms with (well, sort of) a common problem in these varied locations—land subsidence due to various activities such as withdrawal of groundwater, application of water at land surface, building high-rise buildings, drainage of organic soils, natural compaction (e.g., thawing of permafrost), and in some locations, underground mining (e.g., solid and liquid hydrocarbon removal). Again, land subsidence is not limited to the US regions just mention; at least 45 states (approximately 15,000 square miles) are experiencing this mostly anthropogenic (human-caused) event. In this book, case studies are presented using these locations with a special focus on the Lower Chesapeake Bay Region, which the author has spent more than 50 years studying, sampling, testing, and monitoring. This book focuses on four principal processes causing land subsidence: the formation of aquifer systems by bolide (i.e., that is the natural occurrence of the exceptionally bright explosive fireball that formed the crater that is now known as the Lower Chesapeake Bay Region), the compaction of aquifer systems, the oxidation of organic soils, and the collapse of cavities in carbonate and evaporate rocks (resulting in sinkholes and more).

An important aspect of this book is to provide information (the 411) that describes the Earth and Earth processes, its resources, and the procedures and processes that govern the availability and quality of those resources. This book seeks to broaden and enhance public understanding of land subsidence as an Earth process and the serious impacts that land subsidence can cause if those effects are not known and understood, expected, and correctly managed. It's all about science—the application of scientific understanding and engineering approaches to problems of land subsidence can mitigate and eliminate many of the negative issues/impacts of subsidence while allowing beneficial water and purified (drinkable) wastewater. It is the intention and hope of the author that the information provided in this book will reach a wide range of citizens, general readers, water users, water/wastewater managers, students, and officials responsible for public properties and investments along with regulation of land and water/wastewater use.

This book will be an end to itself in providing an understanding of the phenomena of land subsidence that satisfies individuals' need to act as informed decision-makers and citizens. Moreover, the information presented herein will satisfy the curiosity of readers about an important Earth process. There will be some readers who read this book to provide several references for interested individuals.

DOI: 10.1201/9781003461265-2

The bottom line: This book is a science book because it is critical to the formulation of sensible decisions about management of land and water resources, and wastewater reuse.

IT'S A SUBTLE OCCURRENCE

Land subsidence is a subtle, not-so-obvious occurrence—in these cases, subsidence can only be observed or gauged by measuring it. Well, this is the case, of course, unless a catastrophic sinkhole appears in your sight or takes out your house or swimming pool or sends you on an unexpected journey, so to speak, like those in Retsof Salt Mine in Genesee Valley, New York, or the one in Winter Park, Florida. Now keep in mind that catastrophic disasters like sinkholes are only discoverable in local regions and not in the woods, uninhabited terrain, or any other remote locations. Again, where the occurrence of land subsidence is not indicated by protruding wells, failed well casings, broken pipelines, and drainage reversal, measurement is required to detect land subsidence. Subsidence that occurs from groundwater mining or drainage of organic soils is typically gradual and widespread like in San Joaquin Valley, California, where more than 30 feet of subsidence has occurred, but one would hardly know that without measurements made over time.

Note that using measurements to gauge land subsidence is not always easy to accomplish. The detection of such occurrences is complicated by the large area-based scale of elevation changes and the requirement for vertical stable benchmarks (reference marks) located outside the area affected by subsidence. Note that the determination of subsidence trends in time and space is limited in part by the inherently sparse distribution of available benchmarks with which comparisons can be made. However, when stable benchmarks are available and repeat surveys are made, subsidence can be easily measured by professional surveyors. Truth be told, this is one of the principal ways in which experts first detect subsidence. It is interesting to note that time and again, public agencies and private contractors discover that key local benchmarks have moved only after repeat surveys that span decades or longer. Before such discovery, when the rate of cumulative subsidence is low, apparent errors in surveys may be adjusted throughout the network under the assumption that the discrepancies reflect random errors committed by the survey party team: the transit operator and/or his/her rod-and-chain person. Any subsidence may then go undetected until later surveys when steps are taken to confirm the current elevations of the affected benchmarks (USGS, 2013).

DID YOU KNOW?

A **benchmark** is a type of survey marker that indicates vertical position (elevation); it is a point of reference. **Benchmarking** is the practice (hobby) of hunting for these marks.

As mentioned earlier, sometimes subsidence is obvious. A protruding well-casing event is commonly found in agricultural areas and also in some urban areas where groundwater has been extracted from an alluvial (sedimentary) aquifer system.

In such cases, the land surface and aquifer system are displaced downward relative to the well casing. Stressed well casings are subject to failure through collapse and dislocation. Where the frequency of well-casing failures is high, land subsidence is generally thought of as the cause.

Another indication of land subsidence and compaction is the formation of fissures in alluvial aquifer systems. There are several other indicators of subsidence, including changes in flood-inundation frequency and distribution; stagnation or reversals of streams, aqueducts, storm drainages, or sewer lines; failure, overtopping, or reduction in freeboard along reaches of levees, canals, and flood conveyance structures; and, more generally, cracks and/or changes in the gradient of linear engineered structures (envision a bundle of spaghetti) such as pipelines and roadways.

DID YOU KNOW?

During the summer of 1977, many irrigation wells that penetrated valley-fill deposits were damaged. Most of the damaged wells occurred in the southwestern part of the San Joaquin Valley. The damage seems to have been caused by compaction of the aquifer system, which resulted in the vertical compression and rupture of well casings (USGS, 1999).

So, to summarize, the physical indicators of land subsidence include the following:

*   Ground surface cracks
*   Ground surface sinking or settling
*   Leaning or tilting of structures
*   Flooding in low-lying areas
*   Sinkholes becoming more common and deeper
*   Changes in vegetation patterns
*   Increased sedimentation in streams and rivers

The bottom line: Note that subsidence detection is needed to understand and manage current and future land and water resources in areas where subsidence is or could be a problem.

DIFFERENTIAL SURVEYS

In differential surveys, relative changes in the position of the land surface are measured. Note that the observable position is typically a geodetic mark (aka survey mark, survey monument) or one that has been established to some depth (usually higher than 10 feet when in soil), so that any measured movement can be attributed to deep-seated ground movement and not to surficial effects such as first heave (i.e., upward heaving of the soil). Occasionally, geodetic marks, especially those used to measure benchmarks (vertical movement), are established in massive artificial foundations, such as wells coupled to Earth bridge abutments. Any vertical or horizontal movement of a geodetic mark is measured in relation to other observation points.

When a stable benchmark is the control point, the absolute position of the observation point can be determined. This method has been used for decades to measure land subsidence using repeat surveys of benchmarks referenced to some known, or presumed stable, reference frame. In order to map land subsidence, access to a stable reference frame is essential. It is important to point out that in many areas where subsidence has been recognized and other areas where subsidence has not yet been well documented, accurate assessment has been hindered or delayed by the lack of a sufficiently stable benchmark (vertical reference frame—control) (USGS, 1999).

DID YOU KNOW?

A somewhat relative term, "sufficiently stable" has meaning in the context of a particular time frame of interest and magnitude of differential movement.

It Takes a Network

Because of the continuous and episodic crustal motions caused mostly by postglacial rebound, tectonism, volcanism, and anthropogenic alteration of the Earth's surface, it is sometimes necessary to remeasure geodetic control on a national scale. Thus, networks of geodetic control consist of known positions that are determined relative to a horizontal or vertical datum or both. To establish a national network and partnerships with various public and private parties, the National Geodetic Survey (NGS) has implemented high-accuracy reference networks (HARNs) in every state, which resulted in the establishment of approximately 16,000 survey stations. This enhancement of the network required not only to replace thousands of historic benchmarks and horizontal control marks lost to development, natural causes, and vandalism but also to provide geodetic monuments easily accessible by roadways. Early detection and measurement of land subsidence has facilitated its accurate detection and measurement.

From the Old to the New

Before the advent of the Global Positioning System (GPS)—a result of great advancements in space technology (satellites and digital advancement)—the most common means of conducting land surveys used the theodolite (i.e., a rotating telescope for measuring horizontal and vertical angles) or an electronic distance measuring (EDM) device that has been around since the 1950s. Before the invention of these new measuring devices, the spirit level (i.e., an instrument that uses a bubble level) was a common method for determining elevation. Note that the spirit level might be the instrument of choice if only vertical position is being sought. The technique of differential leveling (i.e., a surveying technique used to determine the elevation between two points) allows the surveyor to calculate the elevation from a known reference point to other points using a precisely leveled telescope and graduated vertical rods (aka rod, pole, or perch in US units and length of 16.5 feet). This method may seem simple but, when conducted with care, has proven to be highly accurate.

Moreover, when surveying to meet the standards set for even lower orders of accuracy in geodetic leveling, 0.05-foot changes in elevation can be routinely measured over distances of miles (USGS, 1999). However, leveling and EDM device errors increase at large-scale measurements. The big advantage of the spirit level is its low cost and accuracy when the scale is small (i.e., less than 5 miles); thus, it is still commonly used. Now, for sizeable regional networks, the more efficient GPS surveying is recommended for assorted surveys. So, let's discuss GPS, which utilizes Earth-orbiting satellites.

The invention of GPS is a revolution in the surveying and measurement of crustal motion that occurred in the 1980s. Tests of the satellite-based NAVSTAR GPS show that it is possible to achieve a precision of 1 part in 1 million between points that are 5 to more than 25 miles apart.

Two GPS receivers can determine the relative position of two points at each observation point when receiving signals simultaneously from the same set of four or more satellites. Note that when the same points are reoccupied at another time interval, any relative motion between the points that occurred during the time interval can be measured.

SIDEBAR 1.1 ANTELOPE VALLEY, MOJAVE DESERT, CALIFORNIA

A geodetic network was used to measure historical subsidence in Antelope Valley, Mojave Desert, California. This project used 85 stations in Antelope Valley and required about 150 days of observation in 1992. Results from GPS surveys and conventional leveling surveys spanning more than 60 years showed a maximum subsidence of 6.6 feet; more than 200 square miles had subsided more than 2 feet since about 1930 (Ikehara and Phillips, 1994). Moreover, investigators found that much of the area in the Lancaster groundwater subbasin has subsided at least 2 feet since about 1930. Subsidence of more than 5 feet occurred between about 1930 and 1992 in three areas: near Avenue I and Division Street, near Avenue G-8 and 90th Street East, and near Avenue I and 45th Street East. Subsidence at these and other locations occurred mostly after 1957; however, near Avenue I and 60th Street West, most of the subsidence (2–3 feet) occurred prior to 1960 (Ikehara and Phillips, 1994). The bottom line: subsidence in the Lancaster subbasin coincides with large declines in groundwater levels. This points to the need to measure land subsidence while monitoring corresponding groundwater levels.

THE SENTINEL

Another method of measuring land subsidence is using a borehole extensometer, which measures both subsidence and horizontal displacement. An *extensometer* is a device that is used to measure changes in the length of an object. Being used in applications related to land subsidence measurement, a borehole extensometer measures the compaction or expansion of an aquifer system independently of other vertical movements, such as crustal and tectonic motions (Galloway et al., 1999). An extensometer measures changes in the thickness of aquifer systems by recording changes in the distance between two points in a well (Figure 1.1). Usually, the two measurement points

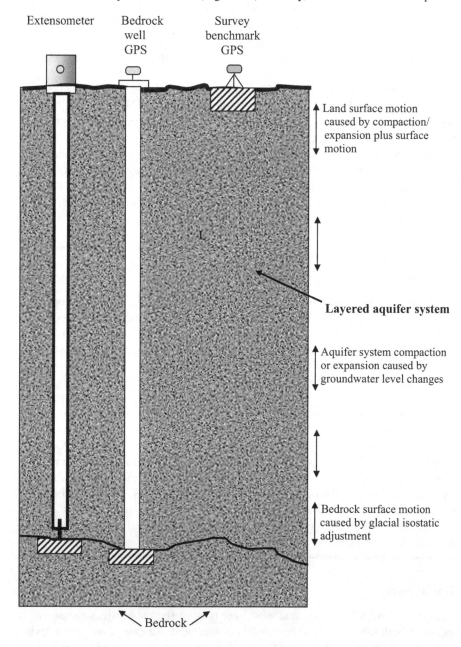

FIGURE 1.1 Subsidence monitoring methods. Survey Benchmark GPS to measure land surface motion. Bedrock Well GPS to measure bedrock surface motion and extensometer to measure aquifer-system compaction or expansion. Source: USGS (2016).

are established at the top and bottom of the well to measure the total compaction between the land surface and the bottom of the aquifer system. Alternatively, specific intervals within a well can be measured, for example, to determine the compaction of just one aquifer within a layered aquifer system. Extensometer measurements are often combined with surface monitoring techniques to determine the contribution of aquifer-system compaction to total land subsidence (Poland, 1984).

The borehole extensometer generates a continuous record of changes in vertical distance between the land surface and the reference point or subsurface benchmark at the bottom of a deep borehole (Riley, 1986). It is especially effective when used in areas undergoing aquifer-system compaction because it is the most effective means of determining continuous deformation at a point. If the subsurface benchmark is established below the base of the compacting aquifer system, the extensometer can be used as the stable reference of starting point for local geodetic surveys. Note that the current practice utilizes incorporating a multiple-stage extensometer in a single instrument to measure aquifer-system compaction simultaneously in different depth intervals (USGS, 1999).

The borehole extensometer used as a stand-alone instrument may be regarded simply as a sentinel against the undetected onset of unacceptable rates of aquifer-system compaction. Riley (1969) pointed out, however, that when used in conjunction with good well logs and water-level data from an adjacent observation well, the deformation history generated by an extensometer can provide the basis for stress–strain analysis and inverse modeling that defines the average compressibility and vertical hydraulic conductivity of aquitards (Helm, 1975). This is attributable to the fact that the extensometer is directly related to the water volume of aquitards. Modern, thoroughly upgraded extensometers record minute elastic compression and expansion that inevitably accompany even very small fluctuations in groundwater levels in unconsolidated alluvial aquifer systems, as well as larger deformations typical of the irreversible compaction of aquitards. Note that reliable estimates of aquitard properties are necessary for predictive modeling, regardless of whether the objective is the prevention or mitigation of land subsidence or simply the optimal use of the storage capacity of the aquifer system (USGS, 1999).

There exist several types of extensometers (Figure 1.2, which shows a type of tensiometer not used in land subsidence monitoring), but in this book, the extensometer that measures differential horizontal ground motion as Earth fissures caused by changes in groundwater levels is discussed (Carpenter, 1993). Buried horizontal extensometers constructed of quartz tubes or invar wires (i.e., nickel–iron low-expansion alloy wire) are useful when precise, continuous measurements are required on a scale of 10–100 feet. Tape extensometers measure changes across intermonument distances up to approximately 100 feet with a repeatability of 0.01 inches. They are used in conjunction with geodetic monuments specially equipped with ball-bearing instrument mounts, which can serve as both horizontal and vertical control points. Collections or lines of monuments can be extended for arbitrary distances, usually in the range of 200–600 feet.

Radar Technique

InSAR (Interferometric Synthetic Aperture Radar) is a new tool for measuring and mapping subsidence; it is a powerful tool that uses radar signals to measure deformation of the Earth's crust at an unprecedented level of spatial detail and high

FIGURE 1.2 A type of tensiometer. This meter does not measure land subsidence and is shown here only to point out that there are other types of tensiometers. Photograph by F. Spellman.

degree of measurement resolution. A huge advantage of using InSAR is that radar waves penetrate clouds and are equally effective in darkness. InSAR has been used to investigate surface deformation resulting from land subsidence (Galloway and Hoffman, 2007). With InSAR, as little as 5 mm of elevation change can be measured over hundreds or thousands of square kilometers with a horizontal spatial resolution down to 20 m (Pritchard, 2006). Interferograms (maps) show land-surface elevation changes that are produced by combining two synthetic aperture rate (SAR) images acquired by multiple satellite or airborne passes over the same area at different times. InSAR analysis has the advantage of measuring subsidence over a large area, whereas traditional geodetic leveling and GPS surveying are performed at only one or a handful of locations, during a survey (Sneed et al., 2002; Stork and Sneed, 2002).

Using InSAR in the Chesapeake Bay Region has potential limitations. Subsidence rates determined by InSAR might have errors that are larger than the subsidence rates observed in the region (1.1–4.8 mm/yr). The region's high humidity and dense vegetation would generate spurious radar signals, require persistent scatter techniques, and result in lower measurement resolution than that is found in more arid regions (Rancoules et al., 2009). Also, available satellite data cover only a relatively short time span. The best available SAR satellite data for the southern Chesapeake Bay Region cover from 1992 to 2000, so the time of accumulated subsidence determined from these data would be no more than 8 years. Despite these limitations, InSAR could be used to identify hotspot areas of subsidence. Such mapping could be useful for identifying unexpected areas of subsidence, focusing attention on important areas, and choosing locations for other ground-based subsidence monitoring techniques (USGS, 2013).

InSAR takes two radar images of the same area at different times from similar vantage points in space and compares them. Any movement of the ground surface toward or away from the satellite can be measured and portrayed as a "picture"—not of the surface itself but of how much the surface deformed (moved) during the time between the images. Note that geophysical applications of radar interferometry take advantage of the phase component of the reflected radar signals to measure apparent changes in the range of distance of the land surface (USGS, 1999). On a typical Earth-orbiting satellite, ordinary radar has a very poor ground resolution of about 3–4 miles because of the restricted size of the antenna on the satellite. SAR is a type of active data collection where a sensor produces its own energy and then records the amount of energy reflected back after interacting with the Earth.

SAR was invented by the mathematician Carl A. Wiley in 1951 while working on a correlations guidance system for the Atlas ICBM program. A simplified conceptual illustration of Wiley's concept is shown in Figure 1.3, where the gray rectangle indicates a radar antenna of a reasonably short length that is moving at a velocity V along its flight path from the right to the left while also constantly transmitting short radar pulses and receiving echoes returned from objects on the ground. Each radar pulse illuminates an instantaneous footprint of size S on the Earth surface (NASA, 2019).

In short, SAR takes advantage of the motion of the spacecraft along its orbital path to mathematically synthesize (reconstruct) an operationally larger antenna and yield high-spatial-resolution imaging capability on the order of tens of feet. Note that although optical imagery is similar to interpreting a photograph, SAR data require a different approach in that the signal is instead responsive to surface characteristics like structure and moisture (NASA, 2019).

For landscapes with more or less stable radar reflectors—buildings or other engineered structures, or undisturbed rocks and ground surfaces—over a period of time, it's possible to make high-precision measurements of the change in the position of the reflectors by subtracting or "interfering" two radar scans made of the same area at different times. This is the principle behind InSAR (USGS, 1999).

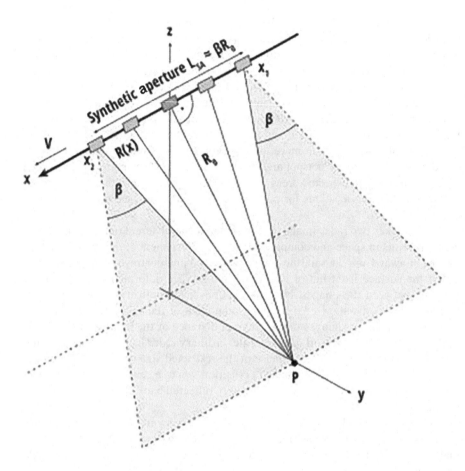

FIGURE 1.3 Geometry of observations used to form the synthetic aperture for target P at along-track position $x = 0$. Source: NASA SAR Handbook (2019)—Public Domain. Accessed 8/27/23 @ https://ntrs.nasa.gov/citation/20190002563.

Mapping Potential for Subsidence

Note that the areal and vertical distribution of subsidence-prone materials, their current state of stress, and the stress history determine the potential for subsidence. In the case of organic soil subsidence (oxidation), subsidence-prone material is generally surficial and both thickness and areal extent are easily mapped; the subsidence rate is primarily determined by the degree of drainage. Subsidence-prone (fine-grained) material in an aquifer-system compaction is buried and must be mapped indirectly by drilling, sampling, assembling drilling logs of subsurface lithology, and various borehole and surface geophysical techniques. Acoustic profiling has been successfully used to map the possible locations of buried cavities in west-central Florida.

SIDEBAR 1.2 PRECONSOLIDATION STRESS (CRITICAL HEAD)

The maximum level of past stressing of a skeletal element is termed "preconsolidation stress" (aka preconsolidation pressure, pre-compressions stress, pre-compaction stress, and preload stress (Dawidowski and Koolen, 1994)). In other words, preconsolidation stress is the maximum effective vertical overburden stress that a particular soil has sustained in the past. When the load on the aquitard skeleton exceeds the preconsolidation stress, the aquitard skeleton may undergo significant, permanent rearrangement, resulting in irreversible compaction. Because the skeleton defines the pore structure of the aquitard, this results in a permanent reduction in pore volume as the pore fluid is "squeezed" out of the aquitard into aquifers. In confined aquifer systems subject to large-scale overdraft, the volume of water derived from irreversible aquitard compaction is essentially equal to the volume of subsidence and can typically range from 10% to 30% of the total volume of water pumped. This represents a one-time mining of stored groundwater and a small permanent reduction in the storage capacity of the aquifer system.

Preconsolidation is the result of various factors that cause soil to approach its preconsolidation pressure. These factors include:

- *Change in total stress due to the removal of overburden can cause preconsolidation pressure in a soil*—A good example is the melting of glaciers.
- *Change in pore water pressure*—Basically, a change in water table elevation, Artesian pressure, deep pumping or flow into tunnels, and desiccation due to surface drying, or plant life can bring soil to its preconsolidation pressure (Holtz and Kovacs, 2010; Tomas et al., 2007).
- *Change in soil structure due to aging (secondary compression)*—Over time, soil will consolidate under high pressures from loading and poor water pressure have been depleted (Holtz and Kovacs, 2010; Tomas et al., 2007).
- *Environmental changes*—Changes in pH, temperature, and salt concentration can cause soil to approach its preconsolidation pressure (Holtz and Kovacs, 2010; Tomas et al., 2007).
- *Chemical weathering*—Different types of chemical weathering will cause preconsolidation pressure. Precipitation, cementing agents, and ion exchange are a few examples (Holtz and Kovacs, 2010; Tomas et al., 2007).

Keep in mind that determining new preconsolidation stress thresholds in aquifer systems is as difficult as determining the native preconsolidation stresses in undeveloped aquifer systems. Additionally, the problem is exacerbated when developed aquifer systems contain thick aquitards affected by hydrodynamic lag.

For successful evaluation of the historical compaction of aquifer systems, an accurate initial estimate of preconsolidation stress (critical head) is important (Hanson, 1989; Hanson and Benedict, 1994). Note that land subsidence becomes obvious only after substantial water-level drawdowns have caused increased intergranular stresses and initiated inelastic compaction. Holzer (1981) identified a variety of natural mechanisms that can cause such an overconsolidated condition in alluvial basins, including removal of overburden by erosion, prehistoric groundwater-level declines, desiccation, and diagenesis (i.e., the process of changing chemical and physical properties of sediments). It is interesting to note that a few investigations have examined the elastic responses of aquifer systems to changes in effective stress under natural conditions before large-scale groundwater withdrawal has begun to cause irreversible subsidence. Consequently, information on critical aquifer hydraulic head (i.e., a measure of fluid potential), representing the native preconsolidation stress of the system, is usually deduced from paired time series of groundwater levels and land subsidence (Holzer, 1981; Anderson, 1988, 1989) measured at wells and nearby benchmarks, or inferred from groundwater flow models (Hanson et al., 1990; Hanson and Benedict, 1994).

Related uncertainties exist for systems that have undergone lowered groundwater levels and land subsidence for some period, followed by groundwater-level recharge and slowing or cessation of subsidence.

THE ROLE OF AQUITARDS IN COMPACTION

In recent decades, increasing recognition has been given to the critical role of aquitards in the intermediate- and long-term response of alluvial systems to groundwater pumpage. Aquitard systems play an important role in compaction. In many confined aquifer systems, interbedded layers of silt sand clays, once dismissed as non-water yielding, comprise the bulk of the groundwater storage capacity. This is based on their substantially higher porosity and compressibility and, in many cases, their higher aggregate thickness compared to the more transmissive, coarser-grained sand and gravel layers (USGS, 1999).

Aquitards are less permeable than aquifers. Thus, the vertical drainage of aquitards into adjacent pumped aquifers may proceed very slowly and thus lag far behind the changing water levels in adjacent aquifers. The delayed response within the inner portions of a thick aquitard may be largely isolated from the higher-frequency seasonal fluctuations and more influenced by lower-frequency, longer-term trends in groundwater levels. Because the migration of the increased internal stress into the aquitard accompanies its drainage, as more fluid is squeezed from the interior of the aquitard, larger and larger intern stresses from the interior of the aquitard, larger and larger internal stresses propagate farther into the aquitard.

DID YOU KNOW?

Responses to changing water levels following several decades of groundwater development suggest that stresses directly driving much of the compaction are somewhat insulated from the changing stresses caused by short-term water-level variations in aquifers.

When the preconsolidation stress is exceeded by internal stresses, compressibility increases dramatically, typically by a factor of 20–100 times, and the resulting compaction is largely nonrecoverable. At stresses greater than the preconsolidation stress, the lag in aquitard drainage increases by comparable factors, and concomitant compaction may require decades or centuries to approach completion. The theory of hydrodynamic consolidation (Terzaghi, 1925)—an essential element of the *aquitard drainage model*—describes the delay involved in draining aquitards when heads are lowered in adjacent aquifers, as well as the residual compaction that may continue long after drawdowns in aquifers have essentially stabilized. Numerical modeling based on Terzaghi's theory has successfully simulated complex histories of compaction observed in response to water-level fluctuations (Helm, 1975).

DID YOU KNOW?

Hydrocompaction—compaction due to wetting—is a near-surface phenomenon that produces land-surface subsidence through a mechanism entirely different from the compaction of deep, overpumped aquifer systems.

OUT OF SIGHT OUT OF MIND

Despite our strong reliance on groundwater, groundwater has for many years been one of the most neglected and ignored natural resources. It has been ignored because it is less visible than other environmental resources—rivers or lakes, for example. The public do not worry about—or even think about—things that they cannot see or observe—. However, recent publicity about events concerning groundwater contamination is making the public more aware of the problem, and regulators have also noticed this. In addition, the ongoing occurrence of land subsidence caused by groundwater depletion via over-withdrawal has started to get noticed—in some locations, a lot of notice—e.g., New Orleans and Hampton Roads, Virginia, and other locations that will be discussed later in this book.

THE BOTTOM LINE

Based on years of observation and measurement, it is obvious that somewhere in the neighborhood, 75%–80% of land subsidence is the consequence of groundwater removal; the focus of this book is on groundwater withdrawal as the principal driver of land subsidence. Other mechanisms of land subsidence are also covered. One thing is certain: when describing land subsidence, we are, obviously and to the point, talking about land, ground, and *terra firma*, all of which are composed of soil—soil, loam, dirt, mud, rock, and Earth. To get to the margin of understanding land subsidence, it is necessary to obtain a fundamental (very basic) understanding of soil. Therefore, Chapter 2 covers soil(s) in its/their most basic forms and substances. Chapter 3 presents a discussion on basic hydraulics simply because we can't get a complete understanding of groundwater movement and its effect on land subsidence unless we have some knowledge of the flow of groundwater.

REFERENCES

Anderson, S.R. (1988). Potential for aquifer compaction, land, subsidence, and earth fissures in Tucson Basin, Pima County, Arizona: U.S. Geological Survey Hydrologic Investigations Atlas HA-713, 3 sheets, scale 1:250,000.

Anderson, S.R. (1989). Potential for aquifer compaction, land, subsidence, and earth fissures in AVRa Valley, Pima and Pinal Counties, Arizona: U.S. Geological Survey Hydrologic Investigations Atlas HA-718, 3 sheets, scale 1:250,000.

Carpenter, M.C. (1993). Earth-fissure movements associated with fluctuations in ground-water levels near the Picacho Mountains, South Central Arizona, 1980-84: U.S. Geological Survey Professional Paper 497-H, 49 p.

Dawidowski, J.B. and Koolen, J.J. (1994). Computerized determination of the preconsolidation stress in compaction texting of field core samples. *Soil and Tillage Research*, v. 31, no. 2, pp. 277–282.

Galloway, D.L., Jones, D.R. & Ingebritsen, S.E. (1999). Land Subsidence in the U.S. United States Global Survey. https://doi.org/10.3133/cir/182

Galloway, D.L. and Hoffmann, J. (2007). The application of satellite differential SAR interferometry-derived ground displacements in hydrogeology. *Hydrogeology Journal*, v. 15, no. 1, pp. 133–154.

Hanson, R.T. (1989). Aquifer-system compaction, Tucson Basin and Avra Valley, Arizona: U.S. Geological Survey Open-File Report 99-4172, 69 p.

Hanson, R.T., Anderson, S.R., and Pool, D.R. (1990). Simulation of groundwater flow and potential land subsidence, Avra Calley, Arizona: U.S. Geological Survey Water-Resources Investigation Report 90-4178, 41 p.

Hanson, R.R. and Benedict, J.F. (1994). Simulation of groundwater flow and potential land subsidence, upper Santa Cruz Basin, Arizona: U.S Geological Survey Water-Resources Investigations Report 93-4196, 47 p.

Helm, D.C. (1975). One-dimensional simulation of aquifer system compaction ear Pixley, Calif., part 1. *Constant Parameters: Water Resource Research*, v. 11, pp. 465–478.

Holtz, R.D. and Kovacs, W.D. (2010). *An Introduction to Geotechnical Engineering*, 2nd ed. New York: Pearson Education India.

Holzer, D.C. (1981). Preconsolidation stress of aquifer systems in areas of induced land subsidence. *Water Resources Research*, v. 17, pp. 693–704.

Ikehara, M.E. and Phillips, S.P. (1994). Determination of land subsidence related groundwater level declines using global positioning system and leveling surveys in Antelope Valley, Los Angeles and Kern Counties, California, 1992: U.S. Geological Survey Water-Resources Investigations Report 94-4184, 101 p.

NASA. (2019). SAR Handbook. Accessed 8/27/23 @ https://ntrs.nasa.gov/citation/20190002563.

Poland, J (1984) Guidebook to studies of land subsidence due to groundwater withdrawal. New York: United Nations.

Pritchard, M.E. (2006). InSAR, a tool for measuring Earth's surface deformation. *Physics Today*, v. 59, no. 7, pp. 68–69.

Rancoules, D., Bourgine, B., de Michele, M., Le Cozannet, G., Closset, L., Bremmer, C., Veldkamp, H., Tragheim, D., Bateson, I., Crosetto, M., Agudo, M., and Engdahl, M. (2009). Validation and intecomparison of persistent scatterrers interferometry-PSIC4 project results. *Journal of Applied Geophysics*, v. 68, no. 3, pp. 335–347.

Riley, F.S. (1969). Analysis of borehole extensometer data from central California: International Association of Scientific Hydrology Publication 89, pp. 423–431.

Riley, F.S. (1986). Developments in borehole extensometry: International Symposium on Land Subside, 3rd, Venice, March 1–25, 1984, [Proceedings, Johnson, I.A., Carbognin, L., and Ubertini, L., eds.], International Association of Scientific Hydrology Publication 151, pp. 169–186.

Sneed, M., Stork, S.V., and Ikehara, M.E. (2002). Detection and measurement of land subsidence using global position system and interferometric synthetic aperture radar, Coachella Valley, California, 1998-2000: U.S. Geological Survey Water-Resources Investigations Report 02-4239, 29 p.

Stork, S.V. and Sneed, M. (2002). Houston-Galveston Bay area, Texas, from space-A new tool for mapping land subsidence: U.S. Geological Survey Fact Sheet 2002-110, 6 p. https://pubs.usgs.gov/fs/fs-110-02/.

Terzaghi, K. (1925). Principles of soil mechanics, IV-Settlement and consolidation of clay. *Engineering New-Record*, v. 95, no. 3, pp. 874–878.

Tomas, R., Domenech, C., Mira, A., Cuenca, A., and Delgado, J. (2007). Preconsolidation stress in the Vega Baja and Media areas of the River Segura (SE Spain). Causes and relationship with piezometric level changes. *Engineering Geology*, v. 91, no. 2–4, pp. 135–151.

USGS. (1999). *Land Subsidence in the United* States. Washington, DC: U.S. Department of the Interior, U.S. Geological Survey.

USGS. (2013). *Land Subsidence in the United States. Circular 1182*. Washington, DC: U.S. Department of the Interior, U.S. Geological Survey.

USGS. (2016). Las Vegas Valley: Land Subsidence and Fissuring due to Ground-Water Withdrawal. Accessed from https://geochange.er.usgs.gov/sw/impacts/hydroogoly/vegas_gov.

2 About Soil

SOIL IS NOT DIRT

A chapter about soil? Yes. Why? Why not? But this book is about land subsidence, you say. Yes it is, for sure. But what is land, I ask? Land can be an estate, an area, a country, grounds, a nation, a parcel, a region, terra firma, and it can be soil. What can soil be, you ask? Well, it can be clay, dust, ground, grime, loam, soot, terra firma, organic material, gases, minerals, and finally it can be land. The point is that when we are talking about land subsidence, we are essentially talking about soil; subsidence is a function of what happens to soil; therefore, we need to have a clear understanding of soil. With regard to land decline, land decrease, and land reduction, we are talking about land subsidence, which is the same as soil subsidence.

Weekend gardeners tend to think of soil as the first few inches below the Earth's surface—the thin layer that needs to be weeded and that provides a firm foundation for plants. But the soil actually extends from the surface down to the Earth's hard rocky crust. It is a zone of transition, and, as in many of nature's transition zones, the soil is the site of important chemical and physical processes. In addition, because plants need soil to grow, it is arguably the most valuable of all the mineral resources on Earth (Beazley, 1992). By the way, many call soil dirt and dirt soil. It is important to point out that soil is not dirt and dirt is not soil. Dirt is displaced soil.

A NEVER-ENDING PROCESS[1]

We take soil for granted. It's always been there, with the implied corollary that it will always be there. But where does soil come from? Of course, soil was formed and, in a never-ending process, it is still being formed; however, soil formation is a slow process—one at work over the course of millennia as mountains are worn away to dust through bare rock succession. Any activity, human or natural, that exposes rock to air begins the process. Through the agents of physical and chemical weathering, through extremes of heat and cold, through storms and earthquake and entropy, bare rock is gradually broken, pulverized, reduced, and worn away. As its exterior structures are exposed and weakened, plant life appears to speed the process along.

Lichens cover the bare rock first, growing on the rock's surface, etching it with mild acids and collecting a thin film of soil that is trapped against the rock and clings. This changes the conditions of growth so much that the lichens can no longer survive and are replaced by mosses. The mosses establish themselves in the soil trapped and enriched by the lichens and collect even more soil. They hold moisture to the surface of the rock, setting up another change in environmental conditions. Well-established mosses hold enough soil to allow herbaceous plant seeds to invade the rock. Grasses and small flowering plants move in, sending out fine root systems that hold more soil

DOI: 10.1201/9781003461265-3

and moisture and work their way into minute fissures in the rock's surface. More and more organisms join the increasingly complex community. Weedy shrubs are the next invaders, with heavier root systems that find their way into every crevice.

Each stage of succession affects the decay of the rock's surface and adds its own organic material to the mix. Over the course of time, mountains are worn away, eaten away to soil, as time, plants, weather, and extremes of weather work on them. The parent material, the rock, becomes smaller and weaker as the years, decades, centuries, and millennia go by, creating the rich, varied, and valuable mineral resource we call soil.

SOIL: WHAT IS IT?[2]

Perhaps no term causes more confusion in communication between various groups of average persons, soil scientists, soil engineers, and Earth scientists than the word *soil* itself. In simple terms, *soil* can be defined as the topmost layer of decomposed rock and organic matter which usually contains air, moisture, and nutrients and can therefore support life. Most people would have little difficulty in understanding and accepting this simple definition. Then why are various groups confused about the exact meaning of the word *soil*? Quite simply, confusion reigns because soil is not simple—it is quite complex. In addition, the term soil has different meanings to different groups (like pollution, the exact definition of soil is a personal judgment call). Let's take a look at how some of these different groups view soil.

Typical people seldom give soil a thought because it usually doesn't directly impact their lives. They seldom think about soil as soil, but they might think of soil in terms of dirt. First of all, soil is not dirt. Dirt is misplaced soil—soil where we don't want it, such as on our hands, clothes, automobiles, or floors. Dirt we try to clean up and to keep out of our living environments.

Second, soil is too special to be called dirt, because soil is mysterious and, whether we realize it or not, essential to our existence. Because we think of it as common, we relegate soil to an ignoble position. As our usual course of action, we degrade it, abuse it, throw it away, contaminate it, and ignore it. We treat it like dirt; only feces hold a lowlier status. Soil deserves better. Why? Because soil is not dirt (how could it be?); moreover, it is not filth, or grime, or squalor. Soil is composed of clay, air, water, sand, loam, and organic detritus of former life forms. If water is Earth's blood and air is Earth's breath, then soil is Earth's flesh, bone, and marrow. Simply, typical people with much exposure to soil would know that soil is a natural, three-dimensional body at the Earth's surface. It is capable of supporting plants and has properties resulting from the integrated effect of climate and living matter acting on earthy parent material, as conditioned by relief and by the passage of time.

Soil scientists (or pedologists) are people interested in soil as a medium for plant growth. Their focus is on the upper meter or so beneath the land surface (this is known as the weathering zone, which contains the organic-rich material that supports plant growth) directly above the unconsolidated *parent material*. Soil scientists have developed a classification system for soils based on the physical, chemical, and biological properties that can be observed and measured in the soil.

Soils engineers are typically soiling specialists who look at soil as a medium that can be excavated using tools. Soils engineers are not concerned with the plant-growing potential of a particular soil, but rather are concerned with a particular soil's ability to support a load. They attempt to determine (through examination and testing) a soil's particle size, particle-size distribution, and the plasticity of the soil.

Earth scientists (or geologists) have a view that typically falls between pedologists and soils engineers—they are interested in soils and the weathering processes as past indicators of climatic conditions and in relation to the geological formation of useful materials ranging from clay deposits to metallic ores.

To gain a new understanding of soil, go out to a plowed farm field, pick up a handful of soil, and look at it—very closely. What are you holding in your hand? Read the two descriptions that follow to gain a better understanding of what soil actually is and why it is critically important to us all (Spellman and Whiting, 2006):

1. A handful of soil is alive, a delicate living organism—as lively as an army of migrating caribou and as fascinating as a flock of egrets. Literally teeming with life of incomparable forms, soil deserves to be classified as an independent ecosystem, or more correctly stated as many ecosystems as possible.
2. When we reach down and pick up a handful of soil, it should remind us (and maybe startle some of us) to the realization that without its thin living soil layer Earth would be a planet as lifeless as our own moon.

KEY TERMS DEFINED[3]

Every branch of science, including soil science, has its own language. To work even at the edge of soil science, soil pollution, and soil pollution remediation, you must acquire a familiarity with the vocabulary used in this text. Some might ask, "Why are these terms not included in a glossary at the end of the text?" This text does have a glossary, but there are terms that need to be defined up front to gain a better understanding as we proceed. Thus, important soil terms are provided herein.

- *Ablation till*—A superglacial coarse-grained sediment or till, accumulating as the subadjacent ice melts and drains away, finally deposited on the exhumed subglacial surface.
- *Absorption*—Movement of ions and water into the plant roots as a result of either metabolic processes by the root (active absorption) or diffusion along a gradient (passive absorption).
- *Acid rain*—Atmospheric precipitation with pH values less than about 5.6, the acidity being due to inorganic acids such as nitric and sulfuric acids that are formed when oxides of nitrogen and sulfur are emitted into the atmosphere.
- *Acid soil*—A soil with a pH value of <7.0 or neutral. Soils may be naturally acid from their rocky origin, by leaching, or may become acid from decaying leaves or from soil additives such as aluminum sulfate (alum). Acid soils can be neutralized by the addition of lime products.

- *Actinomycetes*—A group of organisms intermediate between the bacteria and the true fungi that usually produce a characteristic branched mycelium. Includes many (but not all) organisms belonging to the order of Actinomycetales.
- *Adhesion*—Molecular attraction that holds the surfaces of two substances (e.g., water and sand particles) in contact.
- *Adsorption*—The attraction of ions or compounds to the surface of a solid.
- *Aeration, soil*—The process by which air in the soil is replaced by air from the atmosphere. In a well-aerated soil, the soil air is similar in composition to the atmosphere above the soil. Poorly aerated soil usually contains more carbon dioxide and correspondingly less oxygen than the atmosphere above the soil.
- *Aerobic*—Growing only in the presence of molecular oxygen, as aerobic organisms.
- *Aggregates, soil*—Soil structural units of various shapes, composed of mineral and organic material, formed by natural processes, and having a range of stabilities.
- *Agronomy*—A specialization of agriculture concerned with the theory and practice of field crop production and soil management. The scientific management of land.
- *Air capacity*—Percentage of soil volume occupied by air spaces or pores.
- *Air porosity*—The proportion of the bulk volume of soil that is filled with air at any given time or under a given condition, such as a specified moisture potential; usually the large pores.
- *Alkali*—A substance capable of liberating hydroxide ions in water, measured by a pH of more than 7.0, and possessing caustic properties; it can neutralize hydrogen ions, with which it reacts to form a salt and water, and is an important agent in rock weathering.
- *Alluvium*—A general term for unconsolidated, granular sediments deposited by rivers.
- *Amendment, soil*—Any substance other than fertilizers (such as compost, sulfur, gypsum, lime, and sawdust) used to alter the chemical or physical properties of a soil, generally to make it more productive.
- *Ammonification*—The production of ammonia and ammonium-nitrogen through the decomposition of organic nitrogen compounds in soil organic matter.
- *Anaerobic*—Without molecular oxygen.
- *Anion*—An atom that has gained one or more negatively charged electrons and is thus itself negatively charged.
- *Aspect (of slopes)*—The direction that a slope faces with respect to the sun.
- *Assimilation*—The taking up of plant nutrients and their transformation into actual plant tissues.
- *Atterberg limits*—Water contents of fine-grained soils at different states of consistency.
- *Autotrophs*—Plants and microorganisms capable of synthesizing organic compounds from inorganic materials by either photosynthesis or oxidation reactions.

- *Available water*—The portion of water in a soil that can be readily absorbed by plant roots. The amount of water released between the field capacity and the permanent wilting point.
- *Bedrock*—The solid rock underlying soils and the regolith in depths ranging from zero (where exposed by erosion) to several hundred feet.
- *Biological function*—The role played by a chemical compound or a system of chemical compounds in living organisms.
- *Biomass*—The total weight of living biological organisms within a specified unit (area, community, population).
- *Biome*—A major ecological community extending over large areas.
- *Blow-out*—A deflation depression, eroded by wind from the face of a vegetated dune.
- *Breccia*—A rock composed of coarse angular fragments cemented together.
- *Calcareous soil*—Containing sufficient calcium carbonate (often with magnesium carbonate) to effervesce visibly when treated with hydrochloric acid.
- *Caliche*—A layer near the surface, more or less cemented by secondary carbonates of calcium or magnesium precipitated from the soil solution. It may occur as a soft, thin soil horizon, as a hard, thick bed just beneath the solum, or as a surface layer exposed to erosion.
- *Capillary water*—Held within the capillary pores of soils; mostly available to plants.
- *Catena*—The sequence of soils which occupy a slope transect, from the topographic divide to the bottom of the adjacent valley.
- *Cation*—An atom that has lost one or more negatively charged electrons and is thus itself positively charged.
- *Chelate*—(from Greek *chele* for "claw") a complex organic compound containing a central metallic ion surrounded by organic chemical groups.
- *Class, soil*—A group of soils having a definite range in a particular property such as acidity, degree of slope, texture, structure, land-use capability, degree of erosion, or drainage.
- *Clay*—A soil separate consisting of particles <0.0002 mm in equivalent diameter.
- *Cohesion*—Holding together: force holding a solid or liquid together, owing to attraction between like molecules. Decreases with a rise in temperature.
- *Colloidal*—Matter of very fine particle size.
- *Convection*—A process of heat transfer in a fluid involving the movement of substantial volumes of the fluid concerned. Convection is very important in the atmosphere and, to a lesser extent, in the oceans.
- *Denitrification*—The biochemical reduction of nitrate or nitrite to gaseous nitrogen, either as molecular nitrogen or as an oxide of nitrogen.
- *Detritus*—Debris from dead plants and animals.
- *Diffusion*—The movement of atoms in a gaseous mixture, or ions in a solution, primarily as a result of their own random motion.
- *Drainage*—The removal of excess water, both surface and subsurface, from plants. All plants (except aquatics) will die if exposed to an excess of water.
- *Duff*—The matted, partly decomposed organic surface layer of forest soils.

- *Erosion*—Wearing away of the land surface by running water, wind, ice, or other geological agents, including such processes as gravitational creep.
- *Eutrophication*—A process of lake aging whereby aquatic plants are abundant and waters are deficient in oxygen. The process is usually accelerated by enrichment of waters with surface runoff containing nitrogen and phosphorus.
- *Evapotranspiration*—The combined loss of water from a given area, during a specified period of time, by evaporation from the soil surface and by transpiration from plants.
- *Exfoliation*—Mechanical or physical weathering that involves the disintegration and removal of successive layers of rock mass.
- *Fertility, soil*—The quality of a soil that enables it to provide essential chemical elements in quantities and proportions for the growth of specified plants.
- *Fixation*—The transformation in soil of a plant nutrient from an available to an unavailable state.
- *Fluvial*—Deposits of parent materials laid down by rivers or streams.
- *Friable*—A soil consistency term pertaining to the ease of crumbling of soils.
- *Heaving*—The partial lifting of plants, buildings, roadways, fence posts, etc., out of the ground, as a result of freezing and thawing of the surface soil during the winter.
- *Heterotroph*—An organism capable of deriving energy for life processes only from the decomposition of organic compounds, and incapable of using inorganic compounds as sole sources of energy or for organic synthesis.
- *Horizon, soil*—A layer of soil, approximately parallel to the soil surface, differing in properties and characteristics from adjacent layers below or above it.
- *Humus*—More or less stable fraction of the soil organic matter (usually dark in color) remaining after the major portions of added plant and animal residues have decomposed.
- *Hydration*—The incorporation of water into the chemical composition of a mineral, converting it from an anhydrous to a hydrous form; the term is also applied to a form of weathering in which hydration swelling creates tensile stress within a rock mass.
- *Hydraulic conductivity*—The rate at which water is able to move through a soil.
- *Hydrolysis*—The reaction between water and a compound (commonly a salt). The hydroxyl from the water combines with the anion from the compound undergoing hydrolysis to form a base; the hydrogen ion from the water combines with the cation from the compound to form an acid.
- *Hygroscopic coefficient*—The amount of moisture in a dry soil when it is in equilibrium with some standard relative humidity near a saturated atmosphere (about 98%), expressed in terms of percentage on the basis of oven-dried soil.
- *Infiltration*—The downward entry of water into the soil.

- *Ions*—Atoms that have lost or gained one or more negatively charged electrons.
- *Land classification*—The arrangement of land units into various categories based upon the properties of the land and its suitability for some particular purpose.
- *Leaching*—The removal of materials in solution from the soil by percolating waters.
- *Liebig's law*—The growth and reproduction of an organism are determined by the nutrient substance (oxygen, carbon dioxide, calcium, etc.) that is available in minimum quantity with respect to organic needs, the limiting factor.
- *Loam*—The textural class name for soil having moderate amounts of sand, silt, and clay.
- *Loess*—An accumulation of windblown dust (silt) that may have undergone mild digenesis.
- *Marl*—An earthy deposit consisting mainly of calcium carbonate, usually mixed with clay. Marl is used for liming acid soils. It is slower-acting than most lime products used for this purpose.
- *Mineralization*—The conversion of an element from an organic form to an inorganic state as a result of microbial decomposition.
- *Nitrogen fixation*—The biological conversion of elemental nitrogen (N_2) to organic combinations, or to forms readily utilized in biological processes.
- *Osmosis*—The movement of a liquid across a membrane from a region of high concentration to a region of low concentration. Water and nutrients move into roots independently.
- *Oxidation*—The loss of electrons by a substance.
- *Parent material*—The unconsolidated and more or less chemically weathered mineral or organic matter from which the solum of soils is developed by pedogenic processes.
- *Ped*—A unit of soil structure such as an aggregate, crumb, prism, block, or granule, formed by natural processes.
- *Pedogenic/pedological process*—Any process associated with the formation and development of soil.
- *pH*—The degree of acidity or alkalinity of the soil. Also referred to as soil reaction, this measurement is based on the pH scale where 7.0 is neutral—values from 0.0 to 7.0 are acid, and values from 7.0 to 14.0 are alkaline. The pH of soil is determined by a simple chemical test where a sensitive indicator solution is added directly to a soil sample in a test tube.
- *Photosynthesis*—The process by which green leaves of plants, in the presence of sunlight, manufacture their own needed materials from carbon dioxide in the air, and water and minerals taken from the soil.
- *Porosity, soil*—The volume percentage of the total bulk not occupied by solid particles.
- *Profile, soil*—A vertical section of the soil through all its horizons and extending into the parent material.
- *Reduction*—The gain of electrons and therefore the loss of positive valence charge by a substance.

- *Regolith*—The unconsolidated mantle of weathered rock and soil material on the Earth's surface; loose Earth materials above solid rock.
- *Rock*—The material that forms the essential part of the Earth's solid crust, including loose incoherent masses such as sand and gravel, as well as solid masses of granite and limestone.
- *Rock cycle*—The global geological cycling of lithospheric and crustal rocks from their igneous origins through all or any stages of alteration, deformation, resorption, and reformation.
- *Runoff*—The portion of the precipitation in an area that is discharged from the area through stream channels.
- *Salinization*—The process of accumulation of salts in soil.
- *Sand*—A soil particle between 0.05 and 2.0 mm in diameter; a soil textural class.
- *Silt*—A soil separate consisting of particles between 0.05 and 0.002 mm in equivalent diameter; a soil textural class.
- *Slope*—The degree of deviation of a surface from horizontal, measured in a numerical ratio, percent, or degrees.
- *Soil*—An assemblage of loose and normally stratified granular minerogenic and biogenic debris at the land surface; it is the supporting medium for the growth of plants.
- *Soil air*—The soil atmosphere; the gaseous phase of the soil, being that volume not occupied by soil or liquid.
- *Soil horizon*—A layer of soil, approximately parallel to the soil surface, with distinct characteristics produced by soil-forming processes. These characteristics form the basis for systematic classification of soils.
- *Soil profile*—A vertical section of soil from the surface through all its horizons, including C horizon.
- *Soil structure*—The combination or arrangement of primary soil particles into secondary particles, units, or peds. These secondary units may be, but usually are not, arranged in the profile in such a manner as to give a distinctive characteristic pattern. The secondary units are characterized and classified on the basis of size, shape, and degree of distinctness into classes, types, and grades, respectively.
- *Soil texture*—The relative proportion of the various soil separates in a soil.
- *Soluble*—Will dissolve easily in water.
- *Solum*—(pl. *sola*) The upper and most weathered part of the soil profile; the A, E, and B horizons.
- *Subsoil*—That part of the soil below the plow layer.
- *Till*—Unstratified glacial drift deposited directly by the ice and consisting of clay, sand, gravel, and boulders intermingled in any proportion.
- *Tilth*—The physical condition of soil as related to its ease of tillage, fitness as a seedbed, and its impedance to seedling emergence and root penetration.
- *Topsoil*—The layer of soil moved in cultivation.
- *Weathering*—All physical and chemical changes produced in rocks, at or near the Earth's surface, by atmospheric agents.

FUNCTIONS OF SOIL

Before we begin a journey that takes us through the territory that is soil and examine soil from micro- to macrolevels, we need to stop for a moment and discuss why, beyond the obvious reason, soil is so important to us, to our environment, and to our very survival.

We normally relate soil to our backyards, to farms, to forests, or to a regional watershed. We think of soil as the substance upon which plants grow. Soils play other roles, though. They have six main functions important to us as shown in Figure 2.1: (1) soil is a medium for plant growth; (2) soils regulate our water supplies; (3) soils are recyclers of raw materials; (4) soils provide a habitat for organisms; (5) soils are used as an engineering medium; and (6) soils provide materials. Let's take a closer look at each of the functions of soils.

SOIL AS A PLANT GROWTH MEDIUM

We are all aware of the primary function of soil: soil serves as a plant growth medium, a function that becomes more important with each passing day as Earth's population continues to grow. However, while it is true that soil is a medium for plant growth, soil is actually alive as well. Soil exists in paradox: we depend on soil for life, and at the same time, soil depends on life. Its very origin, its maintenance, and its true nature are intimately tied to living plants and animals. What does this mean? Let's take a look at how the elegant prose of renowned environmental writer Rachel Carson (1962) explained this paradox.

> The soil community ... consists of a web of interwoven lives, each in some way related to the others—the living creatures depending on the soil, but the soil in turn is a vital element of the earth only so long as this community within it flourishes.

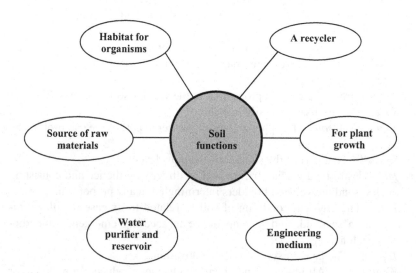

FIGURE 2.1 Functions of soil.

The soil might say to us if it could, "Don't kill off the life within me and I will do the best I can to provide life that will help to sustain your life." What we have here is a trade-off—one vitally important to both soil and ourselves. Remember that most of Earth's people are tillers of the soil—and the soil is their source of livelihood—and those soil tillers provide food for us all.

As a plant growth medium, soil provides vital resources and performs important functions for the plant. To grow in soil, plants must have water and nutrients. Soil provides both of these. To grow and sustain growth, a plant must have a root system. Soil provides pore spaces for roots. To grow and maintain growth, a plant's roots must have oxygen for respiration and carbon dioxide exchange and ultimate diffusion out of the soil. Soil provides the air and pore spaces (the soil's ventilation system) for this. To continue to grow, a plant must have support. Soil provides this support.

If a plant seed is planted in a soil and is exposed to the proper amount of sunlight, for growth to occur, the soil must provide nutrients through a root system that has space to grow, a continuous stream of water (about 500 g of water is required to produce 1 g of dry plant material) for root nutrient transport and plant cooling, and a pathway for both oxygen and carbon dioxide transfer. Just as important, soil water provides the plant with its normal fullness or tension (turgor) that it needs to stand— the structural support it needs to face the sun for photosynthesis to occur.

As well as the functions stated above, soil is also an important moderator of temperature fluctuations. If you have ever dug in a garden on a hot summer day, you probably noticed that the soil was warmer (even hot) on the surface but much cooler just a few inches below the surface.

SOIL AS A REGULATOR OF WATER SUPPLIES

When we walk on land, few of us probably realize that we are actually walking across a bridge. This bridge (in many areas) transports us across a veritable ocean of water below us, deep—or not so deep—under the surface of the Earth. Consider what happens to rain. Where does the rainwater go? Some, falling directly over water bodies, become part of the water body again, but an enormous amount falls on land. Some of the water, obviously, runs off—always following the path of least resistance. In modern communities, stormwater runoff is a hot topic. Cities have taken giant steps to try and control runoff—to send it where it can be properly handled, to prevent flooding. Let's take a closer look at precipitation and the "sinks" it "pours" into, and then relate this usually natural operation to soil water. We begin with surface water and then move onto that ocean of freshwater below the soil's surface: groundwater.

Surface water (water on the Earth's surface as opposed to subsurface water, or groundwater) is mostly a product of precipitation: rain, snow, sleet, or hail. Surface water is exposed or open to the atmosphere and results from the movement of water on and just under the Earth's surface (overland flow). This overland flow is the same thing as surface runoff, which is the amount of rainfall that passes over the Earth's surface. Specific sources of surface water include rivers, streams, lakes, impoundments, shallow wells, rain catchments, and tundra ponds or muskegs (peat bogs).

Most surface water is the result of surface runoff. The amount and flow rate of surface runoff are highly variable. This variability stems from two main factors:

(1) human interference (influences) and (2) natural conditions. In some cases, surface water runs quickly off land. Generally this is undesirable (from a water resources standpoint) because it does not provide enough time for water to infiltrate into the ground and recharge groundwater aquifers. Other problems associated with quick surface water runoff are erosion and flooding. Probably the only good thing that can be said about surface water that quickly runs off land is that it does not have enough time (normally) to become contaminated with high mineral content. Surface water running slowly off land may be expected to have all the opposite effects.

Surface water travels over the land to what amounts to a predetermined destination. What factors influence how surface water moves? Surface water's journey over the face of the Earth typically begins at its drainage basin, sometimes referred to as its *drainage area*, *catchment*, or watershed. For a groundwater source, this is known as the *recharge area*, the area from which precipitation flows into an underground water source.

A surface water drainage basin is usually an area measured in square miles, acres, or sections, and if a city takes water from a surface water source, how large (and what lies within) the drainage basin is essential information for the assessment of water quality.

We know that water doesn't run uphill; instead, surface water runoff (like the flow of electricity) follows along the path of least resistance. Generally speaking, water within a drainage basin will naturally (by the geological formation of the area) be shunted toward one primary watercourse (a river, stream, creek, and brook) unless some man-made distribution system diverts the flow. Various factors directly influence the surface water's flow over land. The principal factors are the following:

- *Rainfall duration*—Length of the rainstorm affects the amount of runoff. Even a light, gentle rain will eventually saturate the soil if it lasts long enough. Once the saturated soil can absorb no more water, rainfall builds up on the surface and begins to flow as runoff.
- *Rainfall intensity*—The harder and faster it rains, the more quickly soil becomes saturated. With hard rains, the surface inches of soil quickly become inundated, and with short, hard storms, most of the rainfall may end up as surface runoff, because the moisture is carried away before significant amounts of water are absorbed into the Earth.
- *Soil moisture*—Obviously, if the soil is already laden with water from previous rains, the saturation point will be reached sooner than if the soil were dry. Frozen soil also inhibits water absorption: up to 100% of snow melts or rainfall on frozen soil will end up as runoff because frozen ground is impervious.
- *Soil composition*—Runoff amount is directly affected by soil composition. Hard rock surfaces will shed all rainfall, obviously, but so will soils with heavy clay composition. Clay soils possess small void spaces that swell when wet. When the void spaces close, they form a barrier that does not allow additional absorption or infiltration. On the opposite end of the spectrum, coarse sand allows easy water flow-through, even in a torrential downpour.

- *Vegetation cover*—Runoff is limited by ground cover. Roots of vegetation and pine needles, pinecones, leaves, and branches create a porous layer (sheet of decaying natural organic substances) above the soil. This porous "organic" sheet (ground cover) readily allows water into the soil. Vegetation and organic waste also act as a cover to protect the soil from hard, driving rains. Hard rains can compact bare soils, close off void spaces, and increase runoff. Vegetation and ground cover work to maintain the soil's infiltration and water-holding capacity. Note that vegetation and ground cover also reduce evaporation of soil moisture as well.
- *Ground slope*—Flat land water flow is usually so slow that large amounts of rainfall can infiltrate the ground. Gravity works against infiltration on steeply sloping ground where up to 80% of rainfall may become surface runoff.
- *Human influences*—Various human activities have a definite impact on surface water runoff. Most human activities tend to increase the rate of water flow. For example, canals and ditches are usually constructed to provide steady flow, and agricultural activities generally remove ground cover that would work to retard the runoff rate. On the opposite extreme, man-made dams are generally built to retard the flow of runoff.

Human habitations, with their paved streets, tarmac, paved parking lots, and buildings create surface runoff potential, since so many surfaces are impervious to infiltration. All these surfaces hasten the flow of water, and they also increase the possibility of flooding, often with devastating results. Because of urban increases in runoff, a whole new field (industry) has developed: stormwater management.

Paving over natural surface acreage has another serious side effect. Without enough area available for water to infiltrate the ground and percolate through the soil to eventually reach and replenish (recharge) groundwater sources, those sources may eventually fail, with a devastating impact on local water supply.

Now let's shift gears and take a look at groundwater. Water falling to the ground as precipitation normally follows three courses. Some runs off directly to rivers and streams, some infiltrate to ground reservoirs, and the rest evaporates or transpires through vegetation. The water in the ground (groundwater) is "invisible" and may be thought of as a temporary natural reservoir (ASTM, 1969; Spellman, 2008). Almost all groundwater is in constant motion toward rivers or other surface water bodies.

Groundwater is defined as water below the Earth's crust, but above a depth of 2500 feet. Thus, if water is located between the Earth's crust and the 2500-foot level, it is considered usable (potable) fresh water. In the United States, it is estimated "that at least 50% of total available freshwater storage is in underground aquifers" (Kemmer, 1979).

In this text, we are concerned with the amount of water retained in the soil for two principal reasons. First, we are concerned with the amount of water retained to ensure plant life and growth. Recall that earlier we stated that producing 1 g of dry plant material requires about 500 g of water. Note that about 5 g of this water becomes an integral part of the plant. Unless rainfall is frequent, you don't have to be a rocket scientist to figure out that the ability of soil to hold water against the force of gravity is very important. Thus, one of the vital functions of soil is to regulate the water supply to plants. Second, we are concerned with soil's ability to hold and retain water in aquifers to enable wells to be dug to provide a potable water supply where it is needed and also to maintain a groundwater level that will prevent or retard land subsidence.

SOIL AS A RECYCLER OF RAW MATERIALS

Imagine what it would be like to step out into the open air and be hit by a stench that would not only offend your olfactory sense but could almost reach out and grab you (like the situation we had in the cave earlier, but worse). Imagine looking out upon the cluttered fields in front of your domicile and seeing nothing but stack upon stack upon stack of the sources of horrible, putrefied, foul, decaying, gagging, choking, retching stench. We are talking about plant and animal remains and waste (mountains of it), reaching toward the sky and surrounded by colonies of landing and spiting flies of all varieties. "Ugh," you say. Well, thankfully (in most cases) you are right. However, if it were not for the power of the soil to recycle waste products, then this scene or something like it is imaginable and even possible, building a mountain toward the moon. Of course, this scenario is impossible because under these described conditions there would be no life to die and to stack up.

Soil is a recycler—probably the premier recycler on Earth. The simple fact is that if it were not for soil's incredible recycling ability, plants and animals would have run out of nourishment long ago. Soil recycles in other ways. For example, consider the geochemical cycles (i.e., the chemical interactions between soil, water, air, and life on Earth) in which soil plays a major role.

Soil possesses the incomparable ability and capacity to assimilate great quantities of organic wastes and turn them into beneficial organic matter (humus) and then to convert the nutrients in the wastes to forms that can be utilized by plants and animals. In turn, the soil returns carbon to the atmosphere as carbon dioxide, where it again will eventually become part of living organisms through photosynthesis. Soil performs several different recycling functions—most of them good, some of them not so good.

Consider one recycling function of soil that may not be so good. Soils have the capacity to accumulate large amounts of carbon as soil organic matter, which can have a major impact on global change such as greenhouse effect. Moreover, it is important that waste be applied in appropriate amounts and not include toxic and environmentally harmful elements or compounds that could poison soils, wastes, and plants.

SOIL AS A HABITAT FOR SOIL ORGANISMS

Life not only formed the soil, but other living things of incredible abundance and diversity now exist within it; if this were not so the soil would be a dead and sterile thing.

—*Carson (1962)*

One thing is certain; most soils are not dead and sterile things. The fact is that a handful of soil is an ecosystem. It may contain up to billions of organisms, belonging to thousands of species. Table 2.1 lists a few (very few) of these organisms. Obviously, communities of living organisms inhabit the soil. What is not so obvious is that they are as complex and intrinsically valuable as are those organisms that roam the land surface and waters of Earth.

SOIL AS AN ENGINEERING MEDIUM

We usually think of soil as being firm and solid (solid ground, *terra firma*). As solid ground, soil is usually a good substrate upon which to build highways and structures. However, not all soils are firm and solid—some are not as stable as others. While construction of buildings and highways may be suitable in one location on one type of soil, it may be unsuitable in another location with different soil. To construct structurally sound, stable highways and buildings, construction on soils and with soil materials requires knowledge of the diversity of soil properties. Note that working with manufactured building materials that have been "engineered" to withstand certain stresses and forces is much different than working with natural soil materials, even though engineers have the same concerns about soils as they do with man-made building materials (concrete and steel). It is much more difficult to make

TABLE 2.1
Soil Organisms (a Representative Sample)

Microorganisms (Protists)

Bacteria
Fungi
Actinomycetes
Algae
Protozoa

Non-arthropod Animals

Nematodes
Earthworms and potworms

Arthropod Animals

Springtails
Mites
Millipedes and centipedes
Harvestman
Ants
Diplopoda
Diptera
Crustacea

Vertebrates

Mice, moles, voles
Rabbits, gophers, squirrels

these predictions or determinations for soil's ability to resist compression, to remain in place, its bearing strength, shear strength, and stability, than it is to make the same determinations for manufactured building materials.

CONCURRENT SOIL FUNCTIONS

Soils perform specific critical functions no matter where they are located (USDA, 2009), and they perform more than one function at the same time as shown in Figure 2.2 and described below.

- Soils act like sponges, soaking up rainwater and limiting runoff. Soil also impacts groundwater recharge and flood-control potentials in urban areas.
- Soils act like faucets, storing and releasing water and air for plants and animals to use.
- Soils act like supermarkets, providing valuable nutrients and air and water to plants and animals. Soil also stores carbon and prevents its loss into the atmosphere.
- Soils act like strainers or filters, filtering and purifying water and air that flow through them.
- Soils buffer, degrade, immobilize, detoxify, and trap pollutants, such as oil, pesticides, herbicides, and heavy metals, and keep them from entering groundwater supplies. Soils also store nutrients for future use by plants and animals above ground and by microbes within soils.

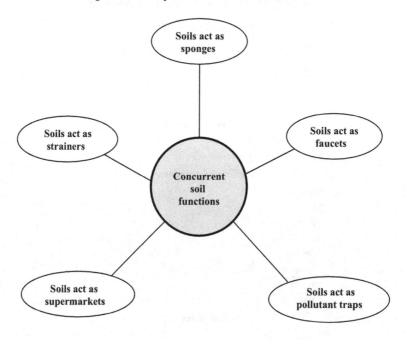

FIGURE 2.2   Concurrent soil functions.

Okay, after the preceding introduction to soil, let's talk about soil basics. Any fundamental discussion about soil should begin with a definition of what soil is. The word soil is derived through Old French from the Latin *solum*, which means floor or ground. Steinbeck referenced soil by its scars, crusts, and crusting. The Swiss writer Charles-Ferdinand Ramuz referred to soil as that soft stuff under the feet. A student of Hippocrates talks about soil as an immense quantity of forces. A more current and concise definition is made difficult by the great diversity of soils throughout the globe. However, here is a generalized definition from the Soil Science Society of America (SSSA, 2008):

> Soil is unconsolidated mineral matter on the surface of the earth that has been subjected to and influenced by genetic and environmental factors of parent material, climate, macro- and microorganisms, and topography, all acting over a period of time and producing a product—soil—that differs from the material from which it is derived in many physical, chemical, and biological properties, and characteristics.

DID YOU KNOW?

SSSA (2008) points out that five major factors interact to create different types of soil: **climate, organisms, relief, parent material, and time = CLORPT—for short.**

Engineers might define soil by saying that soil occupies the unconsolidated mantle of weathered rock making up the loose materials on the Earth's surface, commonly known as the regolith (see Figure 2.3). Soil can be described as a three-phase system, composed of a solid, liquid, and gaseous phase (see Figure 2.4a). (*Note*: This phase relationship is important in dealing with soil pollution because each of the three phases of soil is in equilibrium with the atmosphere, and with rivers, lakes, and the oceans. Thus, the fate and transport of pollutants are influenced by each of these components.)

Soil is also commonly described (see Figure 2.4b) as a mixture of air, water, mineral matter, and organic matter; the relative proportions of these four components greatly influence the productivity of soil. The interface (where the regolith meets the atmosphere) of these materials that make up soil is what concerns us here.

Keep in mind that the four major ingredients that make up soil are not mixed or blended like cake batter. Instead, a major and critically important constituent of soil is pore spaces, which are vital to air and water circulation, providing space for roots to grow and microscopic organisms to live. Without sufficient pore space, soil would be too compacted to be productive. Ideally, the pore space will be divided roughly equally between water and air, with about one-quarter of the soil volume consisting of air and one-quarter consisting of water. The relative proportions of air and water in a soil typically fluctuate significantly as water is added and lost. Compared to surface soils, subsoils tend to contain less total pore space, less organic matter, and a larger proportion of micropores, which tend to be filled with water.

Let's take a closer look at the four major components that make up soil.

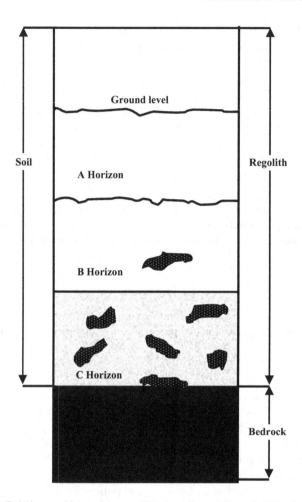

FIGURE 2.3   Relative positions of the regolith, its soil, and the underlying bedrock.

Soil air circulates through soil pores in the same way air circulates through a ventilation system. Only when the pores (the ventilation ducts) become blocked by water or other substances does the air fail to circulate. Though soil pores normally connect to interface with the atmosphere, soil air is different from atmospheric air. It differs in composition from place to place. Soil air also normally has a higher moisture content than the atmosphere. The content of carbon dioxide (CO_2) is usually higher as well and that of oxygen (O_2) lower than accumulations of these gases found in the atmosphere.

Earlier, we stated that only when soil pores are occupied by water or other substances does air fail to circulate in the soil. For proper plant growth, this is of particular importance, because in soil pore spaces that are water-dominated, air oxygen content is low and carbon dioxide levels are high, which restricts plant growth.

The presence of water in soil (often reflective of climatic factors) is essential for the survival and growth of plants and other soil organisms. Soil moisture is a major determinant of the productivity of terrestrial ecosystems and agricultural systems. Water moving through soil materials is a major force behind soil formation. Along

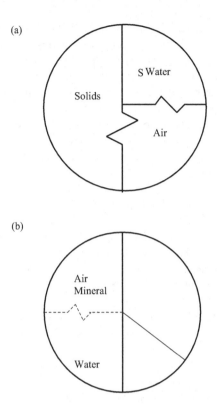

FIGURE 2.4   (a) Three phases of soil: solids, water, and air. Broken lines indicate that these phases are not constant but change with conditions; (b) another view of soil (a loam surface soil). Makeup: air, water, and solids in mineral and organic content.

with air, water, and dissolved nutrients, soil moisture is critical to the quality and quantity of local and regional water resources.

Mineral matter varies in size and is a major constituent of nonorganic soils. Mineral matter consists of large particles (rock fragments) including stones, gravel, and coarse sand. Many of the smaller mineral matter components are made of a single mineral. Minerals in the soil (for plant life) are the primary source of most of the chemical elements essential for plant growth.

Soil organic matter consists primarily of living organisms and the remains of plants, animals, and microorganisms that are continuously broken down (biodegraded) in the soil into new substances that are synthesized by other microorganisms. These other microorganisms continually use this organic matter and reduce it to carbon dioxide via respiration until it is depleted, making repeated additions of new plant and animal residues necessary to maintain soil organic matter (Brady and Weil, 2007).

Now that we have defined soil, let's take a closer look at a few of the basics pertaining to soil, and some of the common terms used in any discussion related to soil basics. Soil is the layer of bonded particles of sand, silt, and clay that covers the land surface of the Earth. Most soils develop in multiple layers. The topmost layer, topsoil, is the layer of soil moved in cultivation and in which plants grow. This topmost layer is actually an ecosystem composed of both biotic and abiotic components—inorganic chemicals, air, water, decaying organic material that provides vital nutrients for plant photosynthesis, and living organisms. Below the topmost layer is the subsoil, the part of the soil below the plow level, usually no more than a meter in thickness. Subsoil is much less productive, partly because it contains much less organic matter.

Below that is the parent material, the unconsolidated (and more or less chemically weathered) bedrock or other geological material from which the soil is ultimately formed. The general rule of thumb is that it takes about 30 years to form 1 inch of topsoil from subsoil; it takes much longer than that for subsoil to be formed from parent material, with the length of time depending on the nature of the underlying matter (Franck and Brownstone, 1992).

PHYSICAL PROPERTIES OF SOIL

From the soil pollution technologist's point of view (regarding land conservation and methodologies for contaminated soil remediation through reuse and recycling), five major physical properties of soil are of interest. They are soil texture, slope, structure, organic matter, and soil color. Soil texture or the relative proportion of the various soil separates in a soil (see Figure 2.5) is a given and cannot be easily or practically changed significantly. It is determined by the size of the rock particles (sand,

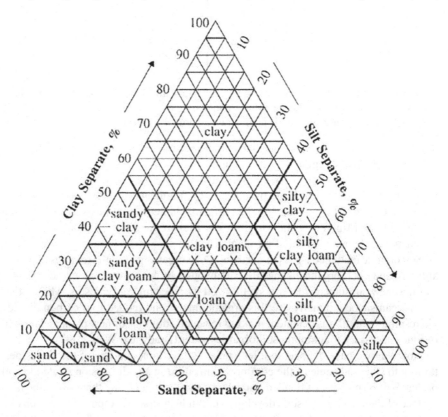

FIGURE 2.5   Textural triangle. Adapted from USDA (2009).

silt, and clay particles) or the soil separates within the soil. The largest soil particles are gravel, which consists of fragments larger than 2.0 mm in diameter.

Particles between 0.05 and 2.0 mm are classified as sand. Silt particles range from 0.002 to 0.05 mm in diameter, and the smallest particles (clay particles) are less than 0.002 mm in diameter. Though clays are composed of the smallest particles, those particles have stronger bonds than silt or sand, though once broken apart, they erode more readily. Particle size has a direct impact on erodibility. Rarely does a soil consist of only one single size of particle—most are a mixture of various sizes.

The slope (or steepness of the soil layer) is another given, important because the erosive power of runoff increases with the steepness of the slope. Slope also allows runoff to exert increased force on soil particles, which breaks them apart more readily and carries them farther away.

Soil structure (tilth) should not be confused with soil texture—they are different. In fact, in the field, the properties determined by soil texture may be considerably modified by soil structure. Soil structure refers to the combination or arrangement of primary soil particles into secondary particles (units or peds). Simply stated, soil structure refers to the way various soil particles clump together. The size, shape, and arrangement of clusters of soil particles called aggregates form naturally formed larger clumps called peds. Sand particles do not clump because sandy soils lack structure. Clay soils tend to stick together in large clumps. Good soil develops small friable (easily crumbled) clumps. Soil develops a unique, fairly stable structure in undisturbed landscapes, but agricultural practices break down the aggregates and peds, lessening erosion resistance.

The presence of decomposed or decomposing remains of plants and animals (organic matter) in soil helps not only fertility but also soil structure—especially the soil's ability to store water. Live organisms such as protozoa, nematodes, earthworms, insects, fungi, and bacteria are typical inhabitants of soil. These organisms work either to control the population of organisms in the soil or to aid in the recycling of dead organic matter. All soil organisms, in one way or another, work to release nutrients from the organic matter, changing complex organic materials into products that can be used by plants.

Just about anyone who has looked at soil has probably noticed that soil color is often different from one location to another. Soil colors range from very bright to dull grays, to a wide range of reds, browns, blacks, whites, yellows, and even greens. Soil color is dependent primarily on the quantity of humus and the chemical form of iron oxides present.

Soil scientists use a set of standardized color charts (the *Munsell—Soil Color Book*) to describe soil colors. They consider three properties of color—hue, value, and chroma—in combination to produce a large number of color chips to which soil scientists can compare the color of the soil being investigated.

Soil Separates

As pointed out in the previous section, soil particles have been divided into groups based on their size termed (*soil separates*) based on their size (sand, silt, and clay; see Figure 2.6) by the International Soil Science Society System, the United States

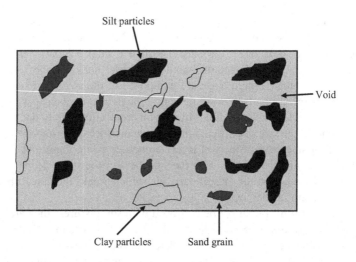

FIGURE 2.6   Enlarged view: cross section of sandy soil.

TABLE 2.2
Characteristics of Soil Separates (USDA)

Separate	Diameter (mm)	Number of Particles/Gram
Very coarse sand	2.00–1.00	90
Coarse sand	1.00–0.50	720
Medium sand	0.50–0.25	5700
Fine sand	0.25–0.10	46,000
Very fine sand	0.10–0.05	722,000
Silt	0.05–0.002	5,776,000
Clay	Below 0.002	90,260,853,000

Public Roads Administration, and the United States Department of Agriculture (USDA). In this text, we use the classification established by the USDA. The size ranges in these separates reflect major changes in how the particles behave, and in the physical properties they impart to soils.

In Table 2.2, the names of the separates are given, together with their diameters, and the number of particles in 1 gram of soil (according to USDA, 1999).

Sand ranges in diameter from 2 to 0.05 mm and is divided into five classes (see Table 2.2). Sand grains are more or less spherical (rounded) in shape, with variable angularity, depending on the extent to which they have been worn down by abrasive processes such as rolling around by flowing water during soil formation.

Sand forms the framework of soil and gives it stability when in a mixture of finer particles. Sand particles are relatively large, which allows voids that form between each grain to also be relatively large. This promotes free drainage of water and the entry of air into the soil. Sand is usually composed of a high percentage of quartz, because it is most resistant to weathering, and its breakdown is extremely slow. Many other minerals are found in sand, depending upon the rocks from which the sand was derived. In the short term (on an annual basis), sand contributes little to plant nutrition in the soil. However, in the long term (thousands of years of soil formation), soils with a lot of weatherable minerals in their sand fraction develop a higher state of fertility.

Silt (essentially microsand), though spherically and mineralogically similar to sand, is smaller—too small to be seen with the naked eye (see Figure 2.6). It weathers faster and releases soluble nutrients for plant growth faster than sand. Too fine to be gritty, silt imparts a smooth feel (like flour) without stickiness. The pores between silt particles are much smaller than those in sand (sand and silt are just progressively finer and finer pieces of the original crystals in the parent rocks). In flowing water, silt is suspended until it drops out when flow is reduced. On the land surface, silt, if disturbed by strong winds, can be carried great distances and is deposited as loess.

The clay soil separate is (for the most part) much different from sand and silt (see Figure 2.6). Clay is composed of secondary minerals that were formed by the drastic alteration of the original forms or by the recrystallization of the products of their weathering. Because clay crystals are plate-like (sheeted) in shape, they have a tremendous surface area-to-volume ratio, giving clay a tremendous capacity to adsorb water and other substances on its surfaces. Clay actually acts as a storage reservoir for both water and nutrients. There are many kinds of clay, each with different internal arrangements of chemical elements, which give them individual characteristics.

SOIL FORMATION

Everywhere on Earth's land surface is either rock formation or exposed soil. When rocks formed deep in the Earth are thrust upward and exposed to the Earth's atmosphere, the rocks adjust to the new environment, and soil formation begins. Soil is formed as a result of physical, chemical, and biological interactions in specific locations. Just as vegetation varies among biomes, so do the soil types that support that vegetation. The vegetation of the tundra and that of the rainforest differ vastly from each other and from vegetation of the prairie and coniferous forest; soils differ in similar ways.

In the soil-forming process, two related, but fundamentally different, processes are occurring simultaneously. The first is the formation of soil parent materials by weathering of rocks, rock fragments, and sediments. This set of processes is carried out in the zone of weathering. The end point is to produce parent material for the soil to develop in and is referred to as C horizon material (see Figure 2.7). It applies in the same way for glacial deposits as for rocks. The second set of processes is the formation of the soil profile by soil-forming processes, which gradually changes the C horizon material into A, E, and B horizons. Figure 2.7 illustrates two soil profiles: one on hard granite and the other on a glacial deposit.

WEATHERING

Soil development takes time and is the result of two major processes: weathering and morphogenesis (morphogenesis was described earlier as bare rock succession). Weathering, the breaking down of bedrock and other sediments that have been

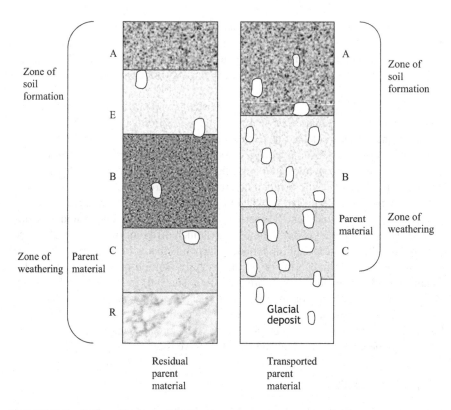

FIGURE 2.7 Soil profiles on residual and transported parent materials.

deposited on the bedrock by wind, water, volcanic eruptions, or melting glaciers, happens physically, chemically, or a combination of both. Weathering is the first step in the erosion process; again, it causes the breakdown of rocks, either to form new minerals that are stable on the surface of Earth or to break the rocks down to smaller particles. Simply, weathering (which projects itself on all surface material above the water table) is the general term used for all the ways in which a rock may be broken down. The factors that influence weathering include the following:

- *Rock type and structure*—Each mineral contained in rocks has a different susceptibility to weathering. A rock with bedding planes, joints, and fractures provides pathways for the entry of water, leading to more rapid weathering. Differential weathering (rocks erode at differing rates) can occur when rock combinations consist of rocks that weather faster than more resistant rocks.
- *Slope*—On steep slopes, weathering products may be quickly washed away by rains. Wherever the force of gravity is greater than the force of friction holding particles upon a slope, these tend to slide downhill.
- *Climate*—Higher temperatures and high amounts of water generally cause chemical reactions to run faster. Rates of weathering are higher in warmer than in colder dry climates.

- *Animals*—Rodents, earthworms, and ants that burrow into soil bring material to the surface where it can be exposed to the agents of weathering.
- *Time*—Depends on slope, animals, and climate.

Although weathering processes are separated, it is important to recognize that these processes work in tandem to break down rocks and minerals into smaller fragments. Geologists recognize two categories of weathering processes:

- *Physical (or mechanical) weathering*—Disintegration of rocks and minerals by a physical or mechanical process.
- *Chemical weathering*—Decomposition of rock by chemical changes or solution.

Physical Weathering

Physical weathering involves the disintegration of a rock by physical processes. These include freezing and thawing of water in rock crevices, disruption by plant roots or burrowing animals, and the changes in volume that result from chemical weathering with the rock. These and other physical weathering processes are discussed below.

- *Development of joints*—Joints are another way that rocks yield to stress. Joints are fractures or cracks in which the rocks on either side of the fracture have not undergone relative movement. Joints form as a result of expansion due to cooling or relief of pressure as overlying rocks are removed by erosion. They form free space in rock by which other agents of chemical or physical weathering can enter (unlike faults that show offset across the fracture). They play an important part in rock weathering as zones of weakness and water movement.
- *Crystal growth*—As water percolates through fractures and pore spaces, it may contain ions that precipitate to form crystals. When crystals grow, they can cause the necessary stresses needed for mechanical rupturing of rocks and minerals.
- *Heat*—It was once thought that daily heating and cooling of rocks was a major contributor to the weathering process. This view is no longer shared by most practicing geologists. However, it should be pointed out that sudden heating of rocks from forest fires may cause expansion and eventual breakage of rock.
- *Biological activities*—Plant and animal activities are important contributors to rock weathering. Plants contribute to the weathering process by extending their root systems into fractures and growing, causing expansion of the fracture. Growth of plants and their effects are evident in many places where they are planted near cement work (streets, brickwork, and sidewalks). Animal burrowing in rock cracks can break rock.
- *Frost wedging*—Frost wedging is often produced by alternate freezing and thawing of water in rock pores and fissures. Expansion of water during freezing causes the rock to fracture. Frost wedgings are more prevalent at high altitudes where there may be many freeze–thaw cycles. One classic and striking

example of weathering of Earth's surface rocks by frost wedging is illustrated by the formation of hoodoos in Bryce Canyon National Park, Utah. "Although Bryce Canyon receives a meager 18 inches of precipitation annually, it's amazing what this little bit of water can do under the right circumstances!" (NPS, 2008). Approximately 200 freeze–thaw cycles occur annually in Bryce. During these periods, snow and ice melt in the afternoon and water seeps into the joints of the Bryce or Claron Formation. When the sun sets, temperatures plummet and the water refreezes, expanding up to 9% as it becomes ice. This frost wedging process exerts tremendous pressure or force on the adjacent rock and shatters and pries the weak rock apart. The assault from frost wedging is a powerful force but, at the same time, rainwater (the universal solvent), which is naturally acidic, slowly dissolves away the limestone, rounding off the edges of these fractured rocks and washing away the debris. Small rivulets of water round down Bryce's rim, forming gullies. As gullies are cut deeper, narrow walls of rock known as fins begin to emerge. Fins eventually develop holes known as windows. Windows grow larger until their roofs collapse, creating hoodoos. As old hoodoos age and collapse, new ones are born.

DID YOU KNOW?

Bryce Canyon National Park lies along the high eastern escarpment of the Paunsaugunt Plateau in the Colorado Plateau region of southern Utah. Its extraordinary geological character is expressed by thousands of rock chimneys (hoodoos) that occupy amphitheater-like alcoves in the Pink Cliffs, whose bedrock host is Claron Formation of Eocene age (Davis and Pollock, 2003).

DID YOU KNOW?

Hoodoo Pronunciation: 'hu-du
  Noun:
  Etymology: West African; from voodoo
  A natural column of rock in western North America often in fantastic form.

—Merriam-Webster Online (www.m-w.com)

Chemical Weathering

Chemical weathering involves the decomposition of rock by chemical changes or solutions. Rocks that are formed under conditions present deep within the Earth are exposed to conditions quite different (i.e., surface temperatures and pressures are lower on the surface and copious amounts of free water and oxygen are available) when uplifted onto the surface. The chief chemical weathering processes are oxidation, carbonation and hydration, and solution in water.

The Persistent Hand of Water

Because of its unprecedented impact on shaping and reshaping Earth, at this point in the text, it is important to point out that given time, nothing, absolutely nothing, on Earth is safe from the heavy hand of water. The effects of water sculpting by virtue of movement and accompanying friction will be covered later in the text. For now, in regard to water exposure and chemical weathering, the main agent responsible for chemical weathering reactions is not water movement but instead is water and weak acids formed in water. The acids formed in water are solutions that have abundant free hydrogen$^+$ ions. The most common weak acid that occurs in surface waters is carbonic acid. Carbonic acid (H_2CO_3) is produced when atmospheric carbon dioxide dissolves in water; it exists only in solution. Hydrogen ions are quite small and can easily enter crystal structures, releasing other ions into the water.

$$\underset{\text{water}}{H_2O} + \underset{\text{carbon dioxide}}{CO_2} \rightarrow \underset{\text{carbonic acid}}{H_2CO_3} \rightarrow \underset{\text{hydrogen ion}}{H^+} + \underset{\text{bicarbonate ion}}{HCO_3^-}$$

Types of Chemical Weathering Reactions

As mentioned, chemical weathering breaks rocks down chemically adding or removing chemical elements and changes them into other materials. Again, as stated, chemical weathering consists of chemical reactions, most of which involve water. Types of chemical weathering include the following:

- *Hydrolysis*—This is a water–rock reaction that occurs when an ion in the mineral is replaced by H^+ or OH^-.
- *Leaching*—Ions are removed by dissolution into water.
- *Oxidation*—Oxygen is plentiful near Earth's surface; thus, it may react with minerals to change the oxidation state of an ion.
- *Dehydration*—Occurs when water or a hydroxide ion is removed from a mineral.
- *Complete dissolution*

Physical and chemical weathering does not always (if ever) occur independently of each other. Instead, they normally work in combination. A classic example of the effect—the power of their simultaneous actions—can be seen in an ecological process known as bare rock succession, described in the chapter opening.

DID YOU KNOW?

As described earlier, plants such as mosses and lichens also penetrate rock and loosen particles. Bare rocks are also subjected to chemical weathering, which involves chemical attack and dissolution of rock. Accomplished primarily through oxidation via exposure to oxygen gas in the atmosphere, acidic precipitation (after having dissolved small amounts of carbon dioxide gas from the atmosphere) and acidic secretions of microorganisms (bacteria, fungi, and lichens), chemical weathering speeds up in warm climates and slows down in cold ones.

The final stages of soil formation consist of the processes of morphogenesis, or the production of a distinctive soil profile with its constituent layers or horizons. The soil profile (the vertical section of the soil from the surface through all its horizons, including C horizons) gives the environmental scientist critical information. When properly interpreted, soil horizons can provide a warning on potential problems in using the land and tell much about the environment and history of a region. Soil profiles allow us to describe, sample, and map soils.

Soil horizons are distinct layers, roughly parallel to the surface, which differ in color, texture, structure, and content of organic matter. The clarity with which horizons can be recognized depends upon the relative balance of the migration, stratification, aggregation, and mixing processes that take place in the soil during morphogenesis. In podzol-type soils (formed mainly in cool, humid climates), striking horizonation is quite apparent; in vertisol-type soils (soils high in swelling clays), the horizons are less distinct. When horizons are studied, they are given a letter symbol to reflect the genesis of the horizon.

Certain processes work to create or destroy clear soil horizons. Processes that tend to create clear horizons by vertical redistribution of soil materials include the leaching of ions in soil solutions, movement of clay-sized particles, upward movement of water by capillary action, and surface deposition of dust and aerosols. Clear soil horizons are destroyed by mixing processes that occur because of organisms, cultivation practices, creep processes on slopes, frost heave, and swelling and shrinkage of clays—all part of the natural soil formation process.

SOIL CHARACTERIZATION

Classification schemes of natural objects seek to organize knowledge so that the properties and relationships of the objects may be most easily remembered and understood for some specific purpose. The ultimate purpose of soil classification is maximum satisfaction of human wants that depend on the use of the soil. This requires grouping soils with similar properties so that lands can be efficiently managed for crop production. Furthermore, soils that are suitable or unsuitable for pipelines, roads, recreation, forestry, agriculture, wildlife, building sites, and so forth can be identified (Foth, 1978).

When people become ill, they may go to a doctor to seek a diagnosis of what is causing the illness and, hopefully, a prognosis of how long before they feel well again. What do diagnosis and prognosis have to do with soil? Actually, quite a lot. The diagnostic techniques used by a physician to identify the causative factors leading to a particular illness are analogous to the soil practitioner using diagnostic techniques to identify a particular soil. Sound far-fetched? It shouldn't because it isn't. Soil scientists must be able to determine the type of soil they study or work with.

Determining the type of soil makes sense, but what does prognosis have to do with all this? Soil practitioners not only need to be able to identify or classify a soil type, but this information allows them to correctly predict how a particular pollutant will react or respond when spilled in that type of soil. The fate of the pollutant is important in determining the possible damage incurred to the environment—soil, groundwater, and air—because, ultimately, a spill could easily affect all three. Thus, the soil practitioner must not only use diagnostic tools in determining soil type, but also be familiar with the soil type, to judge how a particular pollutant or contaminant will respond when spilled in the soil type.

Let's take a closer look at the genesis of soil classification. From the time humans first advanced from hunter-gatherer status to cultivators of crops, they noticed differences in productive soils and unproductive soils. The ancient Chinese, Egyptians, Romans, and Greeks all recognized and acknowledged the differences in soils as media for plant growth. These early soil classification practices were based primarily upon texture, color, and wetness.

Soil classification as a scientific practice did not gain a foothold until the later 18th and early 19th centuries when the science of geology was born. Such terms (with an obvious geological connotation) as limestone soils and lake-laid soils as well as clayey and sandy soils came into being. The Russian scientist Dokuchaev was the first to suggest a generic classification of soils—that soils were natural bodies. Dokuchaev's classification work was then further developed by Europeans and Americans. The system is based on the theory that each soil has a definite form and structure (morphology), related to a particular combination of soil-forming factors. This system was used until 1960, when the U.S. Department of Agriculture published *Soil Classification, A Comprehensive System* (USDA, Soil Survey Staff, 1960). This classification system places major emphasis on soil morphology and gives less emphasis to genesis or the soil-forming factors as compared to previous systems. In 1975, *Soil Classification, A Comprehensive System* was replaced by *Soil Taxonomy* (USDA, 1999), which classifies objects according to their natural relationships. Soils are classified based on measurable properties of soil profiles.

Note that no clear delineation or line of demarcation can be drawn between the properties of one soil and those of another. Instead, a gradation (sometimes quite subtle—like from one shade of white to another) occurs in soil properties as one moves from one soil to another. Brady and Weil (2007) noted that "The gradation in soil properties can be compared to the gradation in the wavelengths of light as you move from one color to another. The changing is gradual, and yet we identify a boundary that differentiates what we call 'green' from what we call 'blue'" (p. 58).

To properly characterize the primary characteristics of a soil, a soil must be identified down to the smallest three-dimensional characteristic sample possible. However, to accurately perform a particular soil sample characterization, a sampling unit must be large enough so that the nature of its horizons can be studied and the range of its properties identified. The *pedon* (rhymes with head-on) is this unit. The pedon is roughly polygonal in shape and designates the smallest characteristic unit that can still be called a soil.

Because pedons occupy a very small space (from approximately 1 to $10\,m^2$), they cannot, obviously, be used as the basic unit for a workable field soil classification system. To solve this problem, a group of pedons, termed a *polypedon*, is of sufficient size to serve as a basic classification unit (or commonly called a *soil individual*). In the United States, these groupings have been called a *soil series*.

There is a difference between *a* soil and *the* soil. This difference is important in the soil classification scheme. A soil is characterized by a sampling unit (pedon), which as a group (polypedons) forms a soil individual. The soil, on the other hand, is a collection of all these natural ingredients and is distinguishable from other bodies such as water, air, solid rock, and other parts of the Earth's crust. By incorporating the difference between a soil and the soil, a classification system has been developed that is effective and widely used.

DIAGNOSTIC HORIZONS, TEMPERATURE, AND MOISTURE REGIMES

Soil taxonomy uses a strict definition of soil horizons called diagnostic horizons, which are used to define most of the orders. Two kinds of diagnostic horizons are recognized: surface and subsurface. The surface diagnostic horizons are called epipedons (Greek *epi* over; *pedon* soil). The epipedon includes the dark (organic-rich) upper part of the soil, of the upper eluvial horizons, and/or sometimes both. Those soils beneath the epipedons are called subsurface diagnostic horizons. Each of these layers is used to characterize different soils in soil taxonomy.

In addition to using diagnostic horizons to strictly define soil horizons, soil moisture regime classes can also be used. A soil moisture regime refers to the presence of plant-available water or groundwater at a sufficiently high level. The control section of the soil (ranging from 10 to 30 cm for clay and from 30 to 90 cm for sandy soils) designates that section of the soil where water is present or absent during given periods in a year. The control section is divided into sections: upper and lower portions. The upper portion is defined as the depth to which $2.5 \, m^3$ of water will penetrate within 24 hours. The lower portion is the depth that $7.5 \, m^3$ of water will penetrate.

Six soil moisture regimes are identified:

- *Aridic*—Characteristic of soils in arid regions
- *Xeric*—Characteristic of having long periods of drought in the summer
- *Ustic*—Soil moisture is generally high enough to meet plant needs during the growing season
- *Udic*—Common soil in humid climatic regions
- *Perudic*—An extremely wet moisture regime annually
- *Aquic*—Soil saturated with water and free of gaseous oxygen

Table 2.3 lists the moisture regime classes and the percentage distribution of areas with different soil moisture regimes.

TABLE 2.3
Soil Moisture Regimes (Percent of Global Area Occupied by Each)

Moisture Regime	Percent of Soils
Aridic	35.9
Xeric	3.5
Ustic	18.0
Udic	33.1
Perudic	1.0
Aquic	8.3

Source: Adaptation from Eswaran (1993).

In soil taxonomy, several soil temperature regimes are also used to define classes of soils. Based on mean annual soil temperature, mean summer temperature, and the difference between mean summer and winter temperatures, soil temperature regimes are shown in Table 2.4. The diagnostic horizons and moisture/temperature regimes just discussed are the main criteria used to define the various categories in soil taxonomy.

SOIL TAXONOMY

The U.S. Soil Conservation Service's soil classification system, Soil Taxonomy (which is based on measurable properties of soil profiles), places soils in categories (see Table 2.5). Let's take a closer look at each one of these categories.

- *Order*—Soils not too dissimilar in their genesis. There are 11 soil orders in soil taxonomy. The names and major characteristics of each soil order are shown in Table 2.6.
- *Suborder*—Fifty-five subdivisions of order that emphasize properties that suggest some common features of soil genesis.
- *Great group*—Diagnostic horizons are the major bases for differentiating approximately 230 great groups.
- *Subgroup*—Approximately 1200 subdivisions of the great groups.
- *Family*—Approximately 7500 soils with subgroups having similar physical and chemical properties.
- *Series*—A subdivision of the family and the most specific unit of the classification system. More than 18,000 soil series are recognized in the United States.

TABLE 2.4
Soil Temperature Regimes (Percent of Global Areas Occupied by Each)

Soil Temperature Regimes/ Mean Annual Temperature (°C)	Percent
Pergelic (0)	10.9
Cryic (0–8)	13.5
Frigid (0–8)	1.2
Mesic (8–15)	12.5
Thermic (15–22)	11.4
Hyperthermic (>22)	18.5
Isofrigid (0–8)	0.1
Isomesic (8–15)	0.3
Isothermic (15–22)	2.4
Isohyperthermic (>22)	26.0
Water (NA)	1.2
Ice (NA)	1.4

Source: Adaptation from Eswaran (1993).

TABLE 2.5

Subdivision of Soil Taxonomy Classification System (in Hierarchical Order)

Category	Number of Taxa
Order	11
Suborder	55
Great group	Approximately 230
Subgroup	Approximately 1200
Family	Approximately 7500
Series	Approximately 18,500 in U.S.

TABLE 2.6

Soil Orders (with Simplified Definitions)

Alfisol	Mild forest soil with gray to brown surface horizon, medium to high base supply (refers to amount of interchangeable cations that remain in soil), and a subsurface horizon of clay accumulation
Andisol	Formed on volcanic ash and cinders and lightly weathered
Aridisol	Dry soil with pedogenic (soil-forming) horizon, low in organic matter
Entisol	Recent soil without pedogenic horizons
Histosol	Organic (peat or bog) soil
Inceptisol	Soil at the beginning of the weathering process with weakly differentiated horizons
Mollisol	Soft soil with a nearly black, organic-rich surface horizon and high base supply
Oxisol	Oxide-rich soil principally a mixture of kaolin, hydrated oxides, and quartz
Spodosol	Soil that has an accumulation of amorphous materials in the subsurface horizons
Ultisol	Soil with a horizon of silicate clay accumulation and low base supply
Vertisol	Soil with high activity clays (cracking clay soil)

Source: USDA Soil Survey Staff (1960).

SOIL ORDERS

As stated earlier, 11 soil orders are recognized; they constitute the first category of the classification (see Table 2.6).

SOIL SUBORDERS

Soil orders are further divided into 64 suborders, based primarily on the chemical and physical properties that reflect either the presence or absence of water logging or genetic differences caused by climate and vegetation—to give the class the greatest genetic homogeneity. Thus, the aqualfs (formed under wet conditions) are "wet" (aqu for *aqua*); alfisols become saturated with water sometime during the year. The suborder names all have two syllables, with the first syllable indicating the order, such as *alf* alfisol and *oll* for Mollisol.

SOIL GREAT GROUPS AND SUBGROUPS

Suborders are divided into great groups that are defined largely by the presence or absence of diagnostic horizons and the arrangements of those horizons. Great group names are coined by prefixing one or more additional formative elements to the appropriate suborder name. More than 230 great groups are identified. Subgroups are subdivisions of great groups. Subgroup names indicate to what extent the central concept of the great group is expressed. A Typic Fragiaqualf is a soil that is typical for the Fragiaqualf great group.

SOIL FAMILIES AND SERIES

The family category of classification is based on features that are important to plant growth such as texture, particle size, mineralogical class, and depth. Terms such as clayey, sandy, loamy, and others are used to identify textural classes. Terms used to describe mineralogical classes include *mixed, oxidic, carbonatic*, and others. For temperature classes, terms such as *hypothermic, frigid, cryic*, and others are used. The soil series (subdivided from soil family) gets down to the individual soil, and the name is that of a natural feature or place near where the soil was first recognized. Familiar series names include Amarillo (Texas), Carlsbad (New Mexico), and Fresno (California). In the United States, there are more than 18,000 soil series.

THE BOTTOM LINE

Land subsidence is all about soil. More specifically, land subsidence is related to what is in or not in the soil. That is, dry soil without any moisture (desert areas in particular) does not experience subsidence because of lack of water; there is no water and therefore the pore spaces within the soil normally occupied by water and air collapse, leading to land subsidence. This is important because much of the uninformed literature and uneducated opinions have formed around the idea that it is the soil itself that compacts in land subsidence events. No. It is the vacant pore spaces that collapse or compact, and not necessarily the soil mineral components. To understand land subsidence, we must understand not only soil but also soil mechanics, which is discussed in the next chapter.

NOTES

1 Much of the information in the following sections is from F.R. Spellman (2017) *The Science of Environmental Pollution*, 3rd ed. Boca Raton, FL: CRC Press.

2 Adaptation from F. Spellman (2017). *Land Subsidence Mitigation: Aquifer Recharge using Treated Wastewater Injection*. Boca Raton, FL: CRC Press.

3 This section was compiled and adapted from several sources including *Soil Taxonomy*, Washington, D.C.: USDA, 1976; *Resource Conservation Glossary*, Ankeny, IA: Soil Conservation Society of America, 1982; *Glossary of Soil Science Terms*, Madison, WI: Soil Science Society of America, 1987.

REFERENCES

ASTM. (1969). *Manual on Water*. Philadelphia, PA: American Society for Testing and Materials.

Beazley, J. (1992). *The Way Nature Works*. New York: Macmillan.

Brady, N.L. & Weil, R.R. (2007). *The Nature and Properties of Soils*. College Park, Maryland: Pearson Press.

Carson, R. (1962). *Silent Spring*, Boston, MA: Houghton Mifflin Company.

Davis, G.H. and Pollock, G.L. (2003). Geology of Bryce Canyon National Park, Utah. In *Geology of Utah's Parks and Monuments*, 2nd ed. Eds. Sprinkel, D.A., et al. Salt Lake City: Utah Geological Association.

Eswaran, H. (1993). Assessment of global resources: Current status and future needs. *Pedologie*, v. XL111, pp. 19–39.

Foth, H.D. (1978). *Fundamentals of Soil Science*, 6th ed. New York: John Wiley and Sons.

Franck, I. and Brownstone, D. (1992). *The Green Encyclopedia*. New York: Prentice-Hall.

Kemmer, F.N. (1979). *Water: The Universal Solvent*. Oak Ridge, IL: NALCO Chemical Company.

NPS. (2008). *The Hoodoo*. Washington, DC: National Park Service.

Spellman, F.R. (2008). *The Science of Water*. Boca Raton, FL: CRC Press.

Spellman, F.R. (2017). *The Science of Environmental Pollution*, 3rd ed. Boca Raton, FL: CRC Press.

Spellman, F.R. and Whiting, N.E. (2006). *Environmental Science and Technology*, 2nd ed. Rockville, MD: Government Institutes.

SSSA. (2008). Soil Science of America. Accessed 11/13/23 @ www.soils.org/files/about-soils/soil-overview.pdf.

USDA. (1999). *Soil Taxonomy: A Basic System of Soil Classification for Making and Interpreting Soil Surveys*, 2nd ed. Washington, DC: USDA Natural Resources Conservation Service.

USDA. (2009). *Urban Soil Primer*. Washington, DC: Untied States Department of Agriculture.

USDA Soil Survey Staff (1960). Soil Classification: A Comprehensive System. Washington, DC: United States Department of Agriculture.

3 Soil Mechanics

INTRODUCTION[1]

Soil mechanics is the application of laws of mechanics and hydraulics to engineering problems dealing with sediments and other unconsolidated accumulations of solid particles produced by the mechanical and chemical disintegration of rocks regardless of whether or not they contain an admixture of organic constituents (Terzaghi, 1942)

Why does the Leaning Tower of Pisa lean? The tower leans because it was built on a non-uniform consolidation of a clay layer beneath the structure. This process is ongoing (by about 1/25 of an inch per year) and may eventually lead to failure of the tower. The factors that caused the Leaning Tower of Pisa to lean (and affect using soil as foundational and building materials) are what this chapter is all about. The mechanics of soil are important factors in determining as to whether a particular building site is viable for building. Simply put, these factors can help to answer the question: will the soils present support buildings? This concerns us because wherever humans build, the opportunity for land subsidence may follow.

The *mechanics* of soil are physical factors important to engineers because their focus is on the soil's suitability as a construction material. Simply put, the engineer must determine the response of a particular volume of soil to internal and external mechanical forces. Obviously, this is important in determining the soil's suitability to withstand the load applied by structures of various types. By studying soil survey maps and reports, checking with soil scientists and other engineers familiar with the region, and the soil types of that region, an engineer can determine the suitability of a particular soil for whatever purpose. Conducting field sampling to ensure that the soil product possesses the soil characteristics for its intended purpose is also essential. The soil characteristics important for engineering purposes include soil texture, kinds of clay present, depth to bedrock, soil density, erodibility, corrosivity, surface geology, content of organic matter, salinity, depth to seasonal water table, plasticity, and consistency limits (aka Atterberg Limits). Engineers will also want to know the soil's space and volume (weight–volume relationships), stress-strain, slope stability, and compaction. Because these concepts are also of paramount importance to determining the fate of materials that are carried through the soil and to subsidence occurrence, these concepts are presented in this section.

WEIGHT-VOLUME OR SPACE AND VOLUME RELATIONSHIPS

As we mentioned earlier, all-natural soil consists of at least three primary components (or phases), solid (mineral) particles, water, and/or air (within void spaces between the solid particles). Examining the physical relationships (for soils in particular) between these phases is essential (see Figure 3.1). For convenience and clarity,

DOI: 10.1201/9781003461265-4

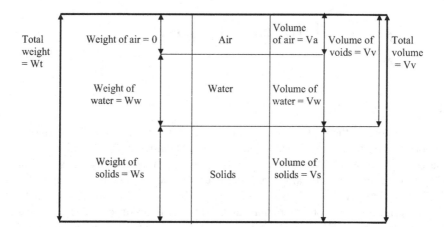

FIGURE 3.1 Weight/volume relationship of soil.

in Figure 3.1, the mass of soil is represented as a block diagram. Each phase shown in the diagram is a separate block, and each major component has been reduced to a concentrated commodity within a unit volume.

Note that the proportions of the components shown in Figure 3.1 will vary (sometimes widely) between and within various soil types. Remember that all water that is not chemically attached acts as a void filler. The relationship between free water and void spaces depends on available water (moisture). *Note:* This is an important point in regard to land subsidence. Remember, it is not the soil that compacts because of itself, but instead, it is the lack of or withdrawal of pore water that enables the soil to collapse or compact.

The volume of the soil mass (V_T) is the sum of the volumes of the three components, or

$$V_T = V_a + V_w + V_s \tag{3.1}$$

The volume of the voids is the sum of V_a (void air) and V_w (void water). However, the weight of the voids is only W_w, the weight of the water. Because weighing air in soil voids must be done within the Earth's atmosphere, the weight of air in the soil is factored in at zero. The total weight is expressed as the sum of the weights of the soil solids and water

$$W_T = W_s + W_w \tag{3.2}$$

The relationship between weight and volume can be expressed as

$$W_m = V_m G_m Y_w \tag{3.3}$$

where
  W_m = weight of the material (solid, liquid, or gas)
  V_m = volume of the material
  G_m = specific gravity of the material (dimensionless)
  Y_w = unit weight of water

We can solve a few useful problems with the relationships described above. More importantly, this information about a particular location's soil allows engineers to mechanically adjust the proportions of the three major components, by reorienting the mineral grains by compaction or tilling. In remediation, a decision to blend soil types to alter the proportions (such as increasing or decreasing the percentage of void space) may be part of a site cleanup process.

Relationships between volumes of soil and voids are described by the *void ratio*, e, and *porosity*, η. We must first determine the void ratio, which is the ratio of the void volume (V_v) to the volume of solids (V_s)

$$e = V_v/V_s \tag{3.4}$$

The first step is to determine the ratio of the volume of void spaces to the total volume. We do this by determining the porosity (η) of the soil, which is the ratio of void volume to total volume. Porosity is usually expressed as a percentage.

$$\eta = V_v/V_t \times 100\% \tag{3.5}$$

The terms *moisture content*, (w) and *degree of saturation*, (S) relate the water content of the soil and the volume of the water in the void space to the total void volume

$$w = W_w/W_s \times 100\% \tag{3.6}$$

and

$$S = V_w/V_v \times 100\% \tag{3.7}$$

SOIL PARTICLE CHARACTERISTICS

Size and shape of particles in the soil, along with density and other characteristics, provide information to the engineer on shear strength, compressibility, and other aspects of soil behavior. These index properties are used to create engineering classifications of soil. Simple classification tests are used to measure index properties (see Table 3.1) in the lab or the field.

From the engineering point of view, the separation of the *cohesive* (fine-grained) from the *incohesive* (coarse-grained) soils is an important distinction. Let's take a closer look at these two terms.

The level of cohesion of a soil describes the tendency of the soil particles to stick together. Cohesive soils contain silt and clay, which, along with water content, make these soils hold together through the attractive forces between individual clay and water particles. Because the clay particles so strongly influence cohesion the index properties of cohesive soils are more complicated than the index properties of cohesionless soils. The soil's *consistency—the arrangement of clay particles—*describes the resistance of soil at various moisture contents to mechanical stresses or manipulations and is the most important characteristic of cohesive soils.

TABLE 3.1
Index Property of Soils

Soil Type	Index Property
Cohesive (fine-grained)	Water content
	Sensitivity
	Type and amount of clay
	Consistency
	Atterberg limits
Incohesive (coarse-grained)	Relative density
	In-place density
	Particle-size distribution
	Clay content
	Shape of particles

Source: Adaptation from Kehew (2021).

Sensitivity (the ratio of unconfined compressive strength in the undisturbed state to strength in the remolded state (see Equation 3.8)) is another important index property of cohesive soils. Soils with high sensitivity are highly unstable.

$$S_t = \frac{\text{strength in undisturbed condition}}{\text{strength in remolded condition}} \tag{3.8}$$

As described earlier, soil water content also influences soil behavior. Water content values of soil—the *Atterberg limits*, a collective designation of the so-called limits of consistency of fine-grained soils determined with simple laboratory tests—are usually presented as the liquid limit (LL), plastic limit (PL) [*note* Plasticity is exhibited over a range of moisture contents referred to as *plasticity limits*], and the plasticity index (PI). The plastic limit is the lower water level at which soil begins to be malleable in the semi-solid state, but while the molded pieces still crumble easily with a little applied pressure. When the volume of the soil becomes nearly constant with further decreases in water content, the soil reaches the shrinkage state. The upper plasticity limit (or *liquid limit*) is reached when the water content in a soil-water mixture changes from a liquid to a semi-fluid or plastic state and tends to flow when jolted. A soil that tends to flow when wet presents special problems for both engineering purposes and remediation of contamination. The range of water content over which the soil is plastic called the *plasticity index*, provides the difference between the liquid limit and the plastic limit. Soils with the highest plasticity indices are unstable in bearing loads—a key point to remember.

The Bottom line: as water is added to dry plastic soil, the remolded mixture will eventually have the characteristics of a liquid. Soil consistency limits are important parameters (Atterberg Limits) related to land subsidence due to groundwater withdrawal; therefore, Figure 3.2 along with key defined parameters is provided.

ATTERBURG LIMITS

(Consistency of fine-grained soil varies in proportion to the water content)

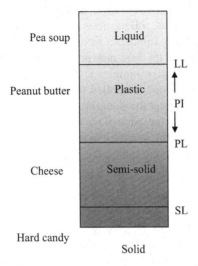

Liquid Limit (LL): is the minimum moisture content at which the soil will flow under its own weight.

Plastic Limit (PL): is the moisture content (in %) at which the soil when rolled into threads of 3.2mm in diameter, crumbles.

Shrinkage Limit (SL): is the moisture content (in %) at which the volume change of the soil mass ceases.

Plasticity Index (PI): is a measure of the range of the moisture content over which a soil is plastic.

$$PI = LL - PL$$

Liquidity Index (LI): the relative consistency of a cohesive soil in a natural state can be defined by the ration called LI.

$$LI = (w - PL)/ (LL - PL)$$

Activity: is the ratio of PI to the clay fraction (% by dry weight of particles < 2μm)

$$A = PI/ (Clay fraction \%)$$

FIGURE 3.2   Atterberg limits and key definitions.

The best-known and probably the most useful system of the several systems designed for classifying the stability of soil materials is called the *Unified System of Classification*. This classification gives each soil type (14 classes) a two-letter designation, based on particle-size distribution, liquid limit, and plasticity index.

Cohesionless coarse-grained soils are classified by index properties including the size and distribution of particles in the soil. Other index properties (including particle shape, in-place density, and relative density) are important in describing cohesionless soils because they relate to how closely particles can be packed together.

Soil Stress

Just as water pressure increases as you go deep into the water, the pressure within the soil increases as the depth increases. A soil with a unit weight of 75 pounds per cubic foot exerts a pressure of 75 psi at 1-foot depth and 225 psi at 3 feet. Of course, as the pressure on a soil unit increases, soil particles reorient themselves to support the cumulative load. This is critically important information to remember, because a soil sample retrieved from beneath the load may not be truly representative, once delivered to the surface. Representative samples are essential. The response of a soil to pressure (*stress*), as when a load is applied to a solid object is transmitted throughout the material. The load puts the material under pressure, which equals the amount of load divided by the surface area of the external face of the object over which it is applied. The response to this pressure or stress is called *displacement* or *strain*. Stress (like pressure) at any point within the object can be defined as force per unit area.

Soil Compressibility

Compressibility, the tendency of soil to decrease in volume under load, is most significant in clay soils because of inherently high porosity. The actual evaluation process for these properties is accomplished in the *consolidation test*. This test subjects a soil sample to an increasing load. The change in thickness is measured after the application of each load increment.

Soil Compaction

Compaction reduces the void ratio and increases the soil density, which affects how materials (including pollutants) travel through the soil. In sedimentology, compaction refers to the process by which sediment progressively loses its porosity due to the effects of loading. Compaction is accomplished by working the soil to reorient the soil grains into a more compact state. Water content sufficient to lubricate particle movement is critical to obtaining efficient compaction.

Soil Failure

Soil structural implications are involved with natural processes such as frost heave (which could damage a septic system and disturb improperly set footings or shift soils under an improperly seated underground storage tank and its piping) as well as changes applied to soils during remediation efforts (e.g., when excavating to mitigate a hazardous materials spill). When soil cannot support a load, *soil failure* occurs, which can include events as diverse as foundation overload, collapse of the sides of an excavation, or slope failure on the sides of a dike or hill. Because of the safety factors involved, a soil's structural stability is critically important.

Classifying the type of soil before an excavation can be accomplished involves determining the soil type. Finding a combination of soil types at an excavation site is common.

Soil types include the following:

- *Stable rock* is generally stable (may lose stability when excavated). Natural solid mineral material that can be excavated with vertical sides and will remain intact while exposed
- *Type A soil* is the most stable soil; it includes clay, silty clay, sandy clay, clay loam, and sometimes silty clay loam and sandy clay loam.
- *Type B soil* is moderately stable soil that includes silt, silt loam, sandy loam, and sometimes silty clay loam and sandy clay loam.
- *Type C soil* is the least stable soil; it includes granular soils like gravel, sand, loamy sand, submerged soil, as well as soil from which water is freely seeping, and unstable submerged rock.

Both visual and manual tests are used to classify soil for excavation. Visual soil testing concerns soil particle size and type. In a mixture of soils, if the soil clumps when dug, it could be clay or silt. The presence of cracks in walls and spalling (breaks up in chips or fragments—smaller pieces) may indicate Type B or C soil. Standing water or water seeping through trench walls automatically classifies the soil as Type C.

Manual soil testing includes the sedimentation test, wet shaking test, thread test, and ribbon test. A sample taken from the soil should be tested as soon as possible to preserve its natural moisture. Soil can be tested either on-site or off-site. A *sedimentation test* determines how much silt and clay are in sandy soil. Saturated sandy soil is placed in a straight-sided jar with about 5 inches of water. After the sample is thoroughly mixed (by shaking it) and allowed to settle, the percentage of sand is visible. Samples containing 80% sand, for example, will be classified as Type C.

The *wet shaking test* is another way to determine the amount of sand versus clay and silt in a soil sample. This test is accomplished by shaking a saturated sample by hand to gauge soil permeability based on the following facts: (1) shaken clay resists water movement through it, and (2) water flows freely through sand and less freely through silt.

The *thread test* is used to determine cohesion (remember, cohesion relates to stability—how well the grains hold together). A representative soil sample is rolled between the palms of the hands to about 1/8 inch in diameter and several inches in length. The rolled piece is placed on a flat surface and then picked up. If the sample holds together for 2 inches, it's considered cohesive.

The *ribbon test* is used as a backup for the thread test. It also determines cohesion. A representative soil sample is rolled out (using the palms of your hands) to 3/4 inch in diameter, and several inches in length. The sample is then squeezed between the thumb and the eforefinger into a flat unbroken ribbon 1/8–1/4 inch thick, which is allowed to fall freely over the fingers. If the ribbon does not break off before several inches are squeezed out, the soil is considered cohesive.

Once soil has been properly classified, the necessary measures for safe excavation can be chosen, based on both soil classification and site restrictions. The two standard protective systems include sloping or benching and shoring or shielding.

SOIL WATER

As a dynamic, heterogeneous body, soil is non-isotropic—it does not have the same properties in all directions. Because soil properties vary directionally, various physical processes are always active in the soil, as Winegardner (1996) makes clear: "all of the factors acting on a particular soil, in an established environment, at a specified time, are working from some state of imbalance to achieve a balance" (p. 63). Soil practitioners must understand the factors involved in the physical processes that are active in soil. These include physical interactions related to soil water, soil grains, organic matter, soil gases, and soil temperature. Because the main focus of this text is land subsidence due to groundwater withdrawal our focus here is on soil water.

WATER AND SOIL

Water is not only a vital component of every living being, but also essential to plant growth, is essential to the microorganisms that live in the soil, and is important in the weathering process, which involves the breakdown of rocks and minerals to form soil and release plant nutrients. In this section, we focus on soil water and its importance in soil. But first, we need to take a closer look at water—what it is and its physical properties. Water exists as a liquid between 0°C and 100°C (32°F and 212°F); as a solid at or below 0°C (32°F); and as a gas at or above 100°C (212°F). One gallon of water weighs 8.33 pounds (3.778 kg). One gallon of water equals 3.785 L. One cubic foot of water equals 7.5 gallons (28.35 L). One ton of water equals 240 gallons. One acre foot of water equals 43,560 cubic feet (325,900 gallons). Earth's rate of rainfall equals 340 cubic miles per day (16 million tons per second). Finally, water is dynamic (constantly in motion), evaporating from sea, lakes, and the soil; being transported through the atmosphere; falling to earth; running across the land; and filtering downward into and through the soil to flow along rock strata.

WATER: WHAT IS IT?[2]

Water is often assumed to be one of the simplest compounds known on Earth. But water is not simple—nowhere in nature is absolutely simple (pure) water to be found. Here on Earth, with a geologic origin dating back over three to five billion years, water found in even its purest form is composed of many constituents. Along with H_2O molecules, hydrogen (H^+), hydroxyl (OH^-), sodium, potassium and magnesium, other ions, and elements are present. Water contains additional dissolved compounds, including various carbonates, sulfates, silicates, and chlorides. Rainwater (often assumed to be the equivalent of distilled water) is not immune to contamination, which it collects as it descends through the atmosphere. The movement of water across the face of land contributes to its contamination, taking up dissolved gases such as carbon dioxide and oxygen, and a multitude of organic substances and minerals leached from the soil.

In soil physics, the physical properties (which are also a function of water's chemical structure) of water that concern us are density, viscosity, surface tension, and capillary action. Let's take a closer look at each of these physical properties. *Density* is a measure of the mass per unit volume. The number of water molecules occupying the space of a unit volume determines the magnitude of the density. As temperature (which measures internal energy) increases or decreases, the molecules vibrate more or less strongly and frequently (which changes the distance between them), expanding or diminishing the volume occupied by the molecules.

As discussed previously, liquid water reaches its maximum density at 4°C, and its minimum at 100°C. In soil science work, the density may be considered to be a unit weight (62.4 lb/ft^3 or 1 g/cm^3).

Viscosity is the measure of the internal flow resistance of a liquid or gas. Stated differently, viscosity is the ease of flow of a liquid, or the capacity of a fluid to convert energy of motion (kinetic energy) into heat energy. Viscosity is the result of the cohesion between fluid particles and the interchange of molecules between layers of different viscosities. High-viscosity fluids flow slowly, while low-viscosity fluids flow freely. Viscosity decreases as temperature rises for liquids.

Have you ever wondered why a needle can float on water? Or why can some insects stand on water? The reason is *surface tension*. Surface tension (or cohesion) is the property that causes the surface of a liquid to behave as if it were covered with a weak elastic skin. It is caused by the exposed surface's tendency to contract to the smallest possible area because of unequal cohesive forces between molecules at the surface.

What does surface tension have to do with soil? The surface tension property of water markedly influences the behavior of water in soils. Consider an example you may be familiar with, one that will help you understand surface tension and the other important physical properties of the water–soil interface. Water commonly rises in clays, fine silts, and other soils, and surface tension plays a major role. The rise of water through clays, silts, and other soils is termed *capillarity* or *capillary action* (the property of the interaction of the water with a solid), and the two primary factors of capillary rise are surface tension (cohesion) and adhesion (the attraction of water for the solid walls of channels through which it moves).

Why does the water rise? The water rises because the water molecules are attracted to the sides of the tube (or soil pores) and starts moving up the tube in response to this attraction. The cohesive force between individual water molecules ensures that water not directly in contact with the side walls is also pulled up the tube (or soil pores). This action continues until the weight of water in the tube counterbalances the cohesive and adhesive forces.

Keep in mind that for water in soil, the rate of movement and the rise in height of soil water are less than one might expect on the basis of soil pore size, because soil pores are not straight like glass tubes, nor are the openings uniform. Also, many soil pores are filled with air, which may prevent or slow down the movement of water by capillarity. A final word on capillarity—keep in mind that capillarity means movement in any direction, not just upward. Since the attractions between soil pores and water are as effective with horizontal pores as with vertical ones, water movement in any direction occurs.

THE BOTTOM LINE

Have you ever wondered what happens to water after it enters the soil? For the average person, probably not, but if you are to work in the soil science field, the answer to this question is one that you definitely need not only to know but must also have a full and complete understanding of. Water that enters the soil has (in simple terms) four ways it may go:

1. It may move on through the soil and percolate out of the root zone, where it may eventually reach the water table.
2. It may be drawn back to the surface and evaporate.
3. It may be taken up (transpired—used) by plants.
4. Finally, it may be held in storage in the water profile.

What determines how much water ends up in each of these categories? It depends. Climate and the properties of the particular soil and the requirements of the plants growing in that soil all have an impact on how much water ends up in each of the categories. But don't forget the influence of anthropogenic actions (what we like to call the heavy hand of humans)—people alter the movement of water, not only by irrigation and stream diversion practices and by building, but also by choosing which crops to plant and the types of tillage practices employed. Note that a basic understanding of water hydraulics is foundational to understanding land subsidence; thus, Chapter 4 covers this important topic.

NOTES

1 Modification from F. Spellman. (2017). *Land Subsidence Mitigation: Aquifer Recharge Using Treated Wastewater Injection*. Boca Raton, FL: CRC Press.
2 Much of the following information is adapted from F.R. Spellman. (2014). *The Science of Water*, 3rd ed. Boca Raton, FL: CRC Press.

REFERENCES

Kehew, A.E. (2021). *Geology for Engineers and Environmental Scientists*, 4th ed. Long Grove, IL: Waveland Press.
Terzaghi, K., (1942). Theoretical Soil Mechanics. New York: Wiley.
Winegardner, D. (1996). *An Introduction to Soils for Environmental Professionals*. Boca Raton, FL: CRC Press.

4 Basic Water Hydraulics

WHAT IS WATER HYDRAULICS?

The practice and study of water hydraulics is not new. Even in medieval times, water hydraulics was not new. "Medieval Europe had inherited a highly developed range of Roman hydraulic components" (Magnusson, 2001). In studying "modern" water hydraulics, it is important to remember that the science of water hydraulics is the direct result of two immediate and enduring problems: the acquisition of freshwater and access to continuous strip of land with a suitable gradient between the source and the destination (Magnusson, 2001).

The word "hydraulic" is derived from the Greek words "hydro" (meaning water) and "aulis" (meaning pipe); originally, the term hydraulics referred only to the study of water at rest and in motion (flow of water in pipes or channels). Today, it is taken to mean the flow of any "liquid" in a system.

What is a liquid? In terms of hydraulics, a liquid can be either oil or water. In fluid power systems used in modern industrial equipment, the hydraulic liquid of choice is oil. Some common examples of hydraulic fluid power systems include automobile braking and power steering systems, hydraulic elevators, and hydraulic jacks or lifts. Probably the most familiar hydraulic fluid power systems in industrial operations are used on dump trucks, front-end loaders, graders, and earth-moving and excavation equipment. In this text, we are concerned with liquid water.

Many find the study of water hydraulics difficult and puzzling, but we know it is not mysterious or difficult. It is the function or output of practical applications of the basic principles of water physics.

FUNDAMENTAL CONCEPTS

$$\text{Air Pressure} (@ \text{ Sea Level}) = 14.7 \text{ pounds per square inch (psi)}$$

This relationship is important because our study of hydraulics begins with air. A blanket of air, many miles thick, surrounds the Earth. The weight of this blanket on a given square inch of the Earth's surface will vary according to the thickness of the atmospheric blanket above that point. As shown above, at sea level, the pressure exerted is 14.7 pounds per square inch (psi). On a mountaintop, air pressure decreases because the blanket is not as thick.

$$1 \text{ft}^3 \text{H}_2\text{O} = 62.4 \text{ lb}$$

This relationship is also important: Both cubic feet and pounds are used to describe volume of water. There is a defined relationship between these two methods of measurement. The specific weight of water is defined relative to a cubic foot. One cubic

DOI: 10.1201/9781003461265-5

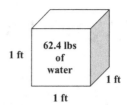

FIGURE 4.1   One cubic foot of water weighs 62.4 lb.

foot of water weighs 62.4 pounds (see Figure 4.1). This relationship is true only at a temperature of 4°C and at a pressure of 1 atmosphere, conditions referred to as *standard temperature and pressure* (STP). Note that 1 atmosphere equals 14.7 lb/in.² at sea level and 1 ft³ of water contains 7.48 gal.

The weight varies so little that, for practical purposes, this weight is used for temperatures ranging from 0°C to 100°C. One cubic inch of water weighs 0.0362 lb. Water 1 feet deep will exert a pressure of 0.43 lb/in.² on the bottom area (12 in. × 0.0362 lb/in³). A column of water 2 feet high exerts 0.86 psi (2 ft × 0.43 psi/ft), and one 55 feet high exerts 23.65 psi (55 ft × 0.43 psi/ft). A column of water 2.31 feet high will exert 1.0 psi (2.31 ft × 0.43 psi/ft). To produce a pressure of 50 psi requires a 115.5-feet water column (50 psi × 2.31 ft/psi).

Remember the following important points:

1. $1 ft^3 H_2O = 62.4$ lb (see Figure 4.1)
2. A column of water 2.31 feet high will exert 1.0 psi

Another relationship is also important:

$$1 \text{ gal } H_2O = 8.34 \text{ lb}$$

At STP, 1 ft³ of water contains 7.48 gal. With these two relationships, we can determine the weight of 1 gal of water. This is accomplished by

$$\text{Weight of gal of water} = 62.4 \text{ lb} / 7.48 \text{ gal} = 8.34 \text{ lb/gal}$$

Thus,

$$1 \text{ gal } H_2O = 8.34 \text{ lb}$$

Note: This information allows cubic feet to be converted to gallons by simply multiplying the number of cubic feet by 7.48 gal/ft³.

Example 4.1

Problem: Find the number of gallons in a reservoir that has a volume of 855.5 ft³.

Solution:

$$855.5 \text{ ft}^3 \times 7.48 \text{ gal/ft}^3 = 6399 \text{ gal (rounded)}$$

Note: As mentioned earlier, the term *head* is used to designate water pressure in terms of the height of a column of water in feet. For example, a 10-foot column of water exerts 4.3 psi. This can be called 4.3-psi pressure or 10 feet of head.

STEVIN'S LAW

Stevin's law deals with water at rest. Specifically, the law states, "The pressure at any point in a fluid at rest depends on the distance measured vertically to the free surface and the density of the fluid." Stated as a formula, this becomes

$$p = w \times h \qquad (4.1)$$

where
  p = pressure in pounds per square foot (psf)
  w = density in pounds per cubic foot (lb/ft³)
  h = vertical distance in feet.

Example 4.2

Problem: What is the pressure at a point 18 feet below the surface of a reservoir?

Solution:

To calculate this, we must know that the density of the water (w) is 62.4 lb/ft³.

$$p = w \times h$$

$$= 62.4 \ \text{lb/ft}^3 \times 18 \ \text{ft}$$

$$= 1123 \ \text{lb/ft}^2 \text{or } 1123 \ \text{psf}$$

Engineers and water/wastewater professionals generally measure pressure in pounds per square inch rather than pounds per square foot; to convert, divide by 144 in.²/ft² (12 in. × 12 in. = 144 in.²):

$$p = \frac{1123 \ \text{psf}}{144 \ \text{in.}^2/\text{ft}^2} = 7.8 \ \text{lb/in.}^2 \text{or psi (rounded)}$$

DENSITY AND SPECIFIC GRAVITY

Table 4.1 shows the relationship between temperature, specific weight, and density of water. When we say that iron is heavier than aluminum, we say that iron has greater density than aluminum. In practice, what we are really saying is that a given volume of iron is heavier than the same volume of aluminum.

Note: What is density? *Density* is the *mass per unit volume* of a substance. Density is typically used to characterize the mass of a fluid system.

Suppose you had a tub of lard and a large box of cold cereal, each having a mass of 600 g. The density of the cereal would be much less than the density of the lard

TABLE 4.1

Water Properties (Temperature, Specific Weight, and Density)

Temperature (°F)	Specific Weight (lb/ft³)	Density (Slugs/ft³)	Temperature (°F)	Specific Weight (lb/ft³)	Density (Slugs/ft³)
32	62.4	1.94	130	61.5	1.91
40	62.4	1.94	140	61.4	1.91
50	62.4	1.94	150	61.2	1.90
60	62.4	1.94	160	61.0	1.90
70	62.3	1.94	170	60.8	1.89
80	62.2	1.93	180	60.6	1.88
90	62.1	1.93	190	60.4	1.88
100	62.0	1.93	200	60.1	1.87
110	61.9	1.92	210	59.8	1.86
120	61.7	1.92			

because the cereal occupies a much larger volume than the lard occupies. The density of an object can be calculated by using the following formula:

$$\text{Density} = \frac{\text{Mass}}{\text{Volume}} \tag{4.2}$$

In water and wastewater treatment, perhaps the most common measures of density are pounds per cubic foot (lb/ft³) and pounds per gallon (lb/gal):

- 1 ft³ of water weights 62.4 lb—Density = 62.4 lb/cu/ft
- One gallon of water weighs 8.34 lb—Density = 8.34 lb/gal

The density of a dry material, such as cereal, lime, soda, and sand, is usually expressed in pounds per cubic foot. The density of a liquid, such as liquid alum, liquid chlorine, or water, can be expressed either as pounds per cubic foot or as pounds per gallon. The density of a gas, such as chlorine gas, methane, carbon dioxide, or air, is usually expressed in pounds per cubic foot.

As shown in Table 4.1, the density of a substance like water changes slightly as the temperature of the substance changes. This occurs because substances usually increase in volume (size—they expand) as they become warmer. Because of this expansion with warming, the same weight is spread over a larger volume, so the density is lower when a substance is warm than when it is cold.

Note: What is specific gravity? Specific gravity is the weight (or density) of a substance compared to the weight (or density) of an equal volume of water. The specific gravity of water is 1.

This relationship is easily seen when a cubic foot of water, which weighs 62.4 lb, is compared to a cubic foot of aluminum, which weights 178 lb. Aluminum is 2.8 times heavier than water.

It is not that difficult to find the specific gravity of a piece of metal.

All you have to do is to weigh the metal in air, then weigh it under water. Its loss of weight is the weight of an equal volume of water. To find the specific gravity, divide the weight of the metal by its loss of weight in water.

$$\text{Specific gravity} = \frac{\text{Weight of substance}}{\text{Weight of equal volume of water}} \qquad (4.3)$$

Example 4.3

Problem: Suppose a piece of metal weighs 150 lb in air and 85 lb under water. What is the specific gravity?

Solution:

Step 1: 150 lb subtract 85 lb = 65 lb loss of weight in water
Step 2:

$$\text{Specific gravity} = \frac{150}{65} = 2.3$$

Note: In a calculation of specific gravity, it is *essential* that the densities be expressed in the same units.

As stated earlier, the specific gravity of water is 1, which is the standard, the reference that all other liquid or solid substances are compared. Specifically, any object that has a specific gravity greater than one will sink in water (rocks, steel, iron, grit, floc, and sludge). Substances with a specific gravity of less than 1 will float (wood, scum, and gasoline). Considering the total weight and volume of a ship, its specific gravity is less than one; therefore, it can float.

The most common use of specific gravity in water/wastewater treatment operations is in gallon-to-pound conversions. In many cases, the liquids being handled have a specific gravity of 1.00 or very nearly 1.00 (between 0.98 and 1.02), so 1.00 may be used in the calculations without introducing significant error. However, in calculations involving a liquid with a specific gravity of less than 0.98 or greater than 1.02, the conversions from gallons to pounds must consider specific gravity. The technique is illustrated in the following example.

Example 4.4

Problem: There is 1455 gal of a certain liquid in a basin. If the specific gravity of the liquid is 0.94, how many pounds of liquid are in the basin?

Solution:

Normally, for a conversion from gallons to pounds, we would use the factor 8.34 lb/gal (the density of water) if the substance's specific gravity were between 0.98 and 1.02. However, in this instance, the substance has a specific gravity outside this range, so the 8.34 factor must be adjusted by multiplying 8.34 lb/gal by the specific gravity to obtain the adjusted factor:
Step 1: (8.34 lb/gal) (0.94) = 7.84 lb/gal (rounded)
Step 2: Then convert 1455 gal to pounds using the corrected factor:
(1455 gal) (7.84 lb/gal) = 11,407 lb (rounded)

FORCE AND PRESSURE

Water exerts force and pressure against the walls of its container, whether it is stored in a tank or flowing in a pipeline. Force and pressure are different, although they are closely related. *Force* is the push or pull influence that causes motion. In the English system, force and weight are often used in the same way. The weight of a cubic foot of water is 62.4 lb. The force exerted on the bottom of a 1-foot cube is 62.4 lb (see Figure 4.1). If we stack two cubes on top of one another, the force on the bottom will be 124.8 pounds. *Pressure* is a force per unit of area. In equation form, this can be expressed as

$$P = \frac{F}{A} \tag{4.4}$$

where
  P = pressure
  F = force
  A = area over which the force is distributed.

Earlier we pointed out that pounds per square inch (lb/in.^{2} or psi) or pounds per square foot (lb/ft^2) are common expressions of pressure. The pressure on the bottom of the cube is 62.4 lb/ft^2 (see Figure 4.1). It is normal to express pressure in pounds per square inch (psi). This is easily accomplished by determining the weight of 1 in.^{2} of 1 feet cube. If we have a cube that is 12 inches on each side, the number of square inches on the bottom surface of the cube is $12 \times 12 = 144$ in.^{2} Dividing the weight by the number of square inches determines the weight on each square inch.

$$psi = \frac{62.4 \text{ lb/ft}}{144 \text{ in.}^2} = 0.433 \text{ psi/ft}$$

This is the weight of a column of water 1 inch square and 1 feet tall. If the column of water were 2 feet tall, the pressure would be 2 ft $\times$ 0.433 psi/ft = 0.866.
  Note: 1 foot of water = 0.433 psi
  With the above information, feet of head can be converted to psi by multiplying the feet of head times 0.433 psi/ft.

Example 4.5

Problem: A tank is mounted at a height of 90 feet. Find the pressure at the bottom of the tank.

Solution:

$$90 \text{ ft} \times 0.433 \text{ psi/ft} = 39 \text{ psi (rounded)}$$

Note: To convert psi to feet, you would divide the psi by 0.433 psi/ft.

Example 4.6

Problem: Find the height of water in a tank if the pressure at the bottom of the tank is 22 psi.

Solution:

$$\text{Height in feet} = \frac{22\text{ psi}}{0.433\text{ psi/ft}} = 51 \text{ ft (rounded)}$$

Important point—One of the problems encountered in a hydraulic system is storing the liquid. Unlike air, which is readily compressible and is capable of being stored in large quantities in relatively small containers, a liquid such as water cannot be compressed. Therefore, it is not possible to store a large amount of water in a small tank, as 62.4 lb of water occupies a volume of 1 ft³, regardless of the pressure applied to it.

HYDROSTATIC PRESSURE

Figure 4.2 shows a number of differently shaped, connected, open containers of water. Note that the water level is the same in each container, regardless of the shape or size of the container. This occurs because pressure is developed, within water (or any other liquid), by the weight of the water above. If the water level in any one container were to be momentarily higher than that in any of the other containers, the higher pressure at the bottom of this container would cause some water to flow into the container having the lower liquid level. In addition, the pressure of the water at any level (such as Line T) is the same in each of the containers. Pressure increases because of the weight of the water. The farther down from the surface, the more pressure is created. This illustrates that the weight, not the volume, of water contained in a vessel determines the pressure at the bottom of the vessel.

Nathanson (1997) pointed out some very important principles that always apply for hydrostatic pressure.

1. The pressure depends only on the depth of water above the point in question (not on the water surface area).
2. The pressure increases in direct proportion to the depth.
3. The pressure in a continuous volume of water is the same at all points that are at the same depth.
4. The pressure at any point in the water acts in all directions at the same depth.

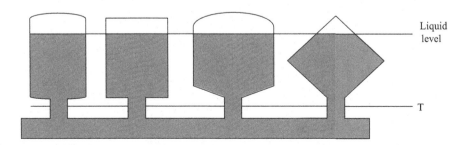

FIGURE 4.2 Hydrostatic pressure.

EFFECTS OF WATER UNDER PRESSURE

Hauser (1996) points out that water under pressure and in motion can exert tremendous forces inside a pipeline. One of these forces, called hydraulic shock or *water hammer*, is the momentary increase in pressure that occurs when there is a sudden change of direction or velocity of the water. When a rapidly closing valve suddenly stops water flowing in a pipeline, pressure energy is transferred to the valve and pipe wall. Shock waves are set up within the system. Waves of pressure move in horizontal yo-yo fashion—back and forth—against any solid obstacles in the system. Neither the water nor the pipe will compress to absorb the shock, which may result in damage to pipes, valves, and shaking of loose fittings.

Another effect of water under pressure is called thrust. *Thrust* is the force which water exerts on a pipeline as it rounds a bend. As shown in Figure 4.3, thrust usually acts perpendicular (at 90°) to the inside surface its pushes against. It affects not only bends but also reducers, dead ends, and tees. Uncontrolled, the thrust can cause movement in the fitting or pipeline, which will lead to separation of the pipe coupling away from both sections of pipeline, or at some other nearby coupling upstream or downstream of the fitting.

Two types of devices commonly used to control thrust in larger pipelines are thrust blocks and thrust anchors. A *thrust block* is a mass of concrete cast in place onto the pipe and around the outside bend of the turn. An example is shown in Figure 4.4.

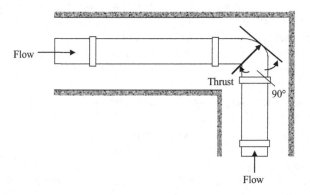

FIGURE 4.3   Direction of thrust in a pipe in a trench (viewed from above).

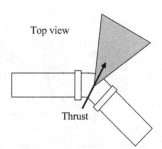

FIGURE 4.4   Thrust block.

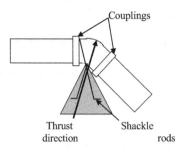

FIGURE 4.5 Thrust anchor.

These are used for pipes with tees or elbows that turn left or right or slant upward. The thrust is transferred to the soil through the larger bearing surface of the block. A *thrust anchor* is a massive block of concrete, often a cube, cast in place below the fitting to be anchored (see Figure 4.5). As shown in Figure 4.5, imbedded steel shackle rods anchor the fitting to the concrete block, effectively resisting upward thrusts. The size and shape of a thrust control device depend on pipe size, type of fitting, water pressure, water hammer, and soil type.

HEAD

Head is defined as the vertical distance water must be lifted from the supply tank to the discharge, or as the height a column of water would rise due to the pressure at its base. A perfect vacuum plus atmospheric pressure of 14.7 psi would lift the water 34 feet. When the top of the sealed tube is opened to the atmosphere and the reservoir is enclosed, the pressure in the reservoir is increased; the water will rise in the tube. Because atmospheric pressure is essentially universal, we usually ignore the first 14.7 psi of actual pressure measurements and measure only the difference between the water pressure and the atmospheric pressure; we call this *gauge pressure*. For example, water in an open reservoir is subjected to the 14.7 psi of atmospheric pressure; subtracting this 14.7 psi leaves a gauge pressure of 0 psi, indicating that the water would rise 0 feet above the reservoir surface. If the gauge pressure in a water main were 120 psi, the water would rise in a tube connected to the main:

$$120 \text{ psi} \times 2.31 \text{ ft/psi} = 277 \text{ ft (rounded)}$$

The *total head* includes the vertical distance the liquid must be lifted (static head), the loss to friction (friction head), and the energy required to maintain the desired velocity (velocity head).

$$\text{Total head} = \text{Static head} + \text{friction head} + \text{velocity head} \qquad (4.5)$$

STATIC HEAD

Static head is the actual **vertical** distance the liquid must be lifted.

$$\text{Static head} = \text{Discharge elevation} - \text{supply elevation} \qquad (4.6)$$

Example 4.7

Problem: The supply tank is located at elevation 118 feet. The discharge point is at elevation 215 feet. What is the static head in feet?

Solution:

Static head, ft $= 215$ ft $- 118$ ft $= 97$ ft

FRICTION HEAD

Friction head is the equivalent distance of the energy that must be supplied to overcome friction. Engineering references include tables showing the equivalent vertical distance for various sizes and types of pipes, fittings, and valves. The total friction head is the sum of the equivalent vertical distances for each component.

$$\text{Friction head, ft} = \text{Energy losses due to friction} \qquad (4.7)$$

VELOCITY HEAD

Velocity head is the equivalent distance of the energy consumed in achieving and maintaining the desired velocity in the system.

$$\text{Velocity head, ft} = \text{Energy losses to maintain velocity} \qquad (4.8)$$

$$\text{Total Dynamic Head} \left(\text{Total System Head} \right)$$

$$\text{Total head} = \text{Static head} + \text{friction head} + \text{velocity head} \qquad (4.9)$$

PRESSURE AND HEAD

The pressure exerted by water and/or wastewater is directly proportional to its depth or head in the pipe, tank, or channel. If the pressure is known, the equivalent head can be calculated.

$$\text{Head, ft} = \text{Pressure, psi} \times 2.31 \text{ ft/psi} \qquad (4.10)$$

Example 4.8

Problem: The pressure gauge on the discharge line from the influent pump reads 72.3 psi. What is the equivalent head in feet?

Solution:

Head, ft = 72.3 × 2.31 ft/psi = 167 ft

HEAD AND PRESSURE

If the head is known, the equivalent pressure can be calculated by

$$\text{Pressure, psi} = \frac{\text{Head, ft}}{2.31 \text{ ft/psi}} \qquad (4.11)$$

Example 4.9

Problem: A tank is 22 feet deep. What is the pressure in psi at the bottom of the tank when it is filled with water?

Solution:

$$\text{Pressure, psi} = \frac{22 \text{ ft}}{2.31 \text{ ft/psi}} = 9.52 \text{ psi (rounded)}$$

FLOW AND DISCHARGE RATES: WATER IN MOTION

The study of fluid flow is much more complicated than that of fluids at rest, but it is important to understand these principles because the water is nearly always in motion.

Discharge (or flow) is the quantity of water passing a given point in a pipe or channel during a given period. Stated another way for open channels, the flow rate through an open channel is directly related to the velocity of the liquid and the cross-sectional area of the liquid in the channel.

$$Q = A \times V \qquad (4.12)$$

where

$$Q = \text{Flow-discharge in cubic feet per second (cfs)}$$

$$A = \text{Cross-sectional area of the pipe or channel } (\text{ft}^2)$$

$$V = \text{Water velocity in feet per second (fps or ft/s)}$$

Example 4.10

Problem: The channel is 6 feet wide and the water depth is 3 feet. The velocity in the channel is 4 fps. What is the discharge or flow rate in cubic feet per second?

Solution:

Flow, cfs = 6 ft × 3 ft × 4 ft/s = 72 cfs

Discharge or flow can be recorded as gallons/day (gpd), gallons/minute (gpm), or cubic feet (cfs). Flows are often referred to in million gallons per day (MGD). The discharge or flow rate can be converted from cfs to other units such as gallons per minute (gpm) or MGD by using appropriate conversion factors.

Example 4.11

Problem: A 12-inch-diameter pipe has water flowing through it at 10 feet per second. What is the discharge in (a) cfs, (b) gpm, and (c) MGD?

Solution

Before we can use the basic formula, we must determine the area (A) of the pipe. The formula for the area of a circle is as follows:

$$\text{Area } (A) = \pi \times \frac{D^2}{4} = \pi \times r^2 \qquad (4.13)$$

Note: π is the constant value 3.14159 or simply 3.14.
where:

$$D = \text{diameter of the circle in feet}$$

$$r = \text{radius of the circle in feet}$$

Therefore, the area of the pipe is as follows:

$$A = \pi \frac{D^2}{4} = 3.14 \times \frac{(1 \text{ ft})^2}{4} = 0.785 \text{ ft}^2$$

(a) Now we can determine the discharge in cfs:

$$Q = V \times A = 10 \text{ ft/s} \times 0.785 \text{ ft}^2 = 7.85 \text{ ft}^3/\text{s or cfs}$$

(b) We need to know that 1 cfs is 449 gpm, so 7.85 cfs $\times$ 449 gpm/cfs = 3525 gpm (rounded)

(c) 1 MGD is 1.55 cfs, so:

$$7.85 \text{ cfs}$$

$$= 5.06 \text{ MGD}$$

$$1.55 = \frac{\text{cfs}}{\text{MGD}}$$

Note: Flow may be *laminar* (streamline, see Figure 4.6) or *turbulent* (see Figure 4.7). Laminar flow occurs at extremely low velocities. The water moves in straight parallel lines, called streamlines or laminae, which slide upon each other as they travel, rather than mixing up. Normal pipe flow is turbulent flow, which occurs because of

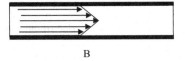

B

FIGURE 4.6   Laminar (streamline) flow.

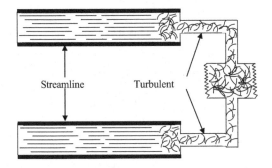

Streamline   Turbulent

FIGURE 4.7   Turbulent flow.

friction encountered on the inside of the pipe. The outside layers of flow are thrown into the inner layers; the result is that all the layers mix and are moving in different directions and at different velocities; However, the direction of flow is forward.

Note: Flow may be steady of unsteady. For our purposes, we consider steady-state flow only, that is, most of the hydraulic calculations in this manual assume steady-state flow.

AREA AND VELOCITY

The *law of continuity* states that the discharge at each point in a pipe or channel is the same as the discharge at any other point (if water does not leave or enter the pipe or channel). That is, under the assumption of steady-state flow, the flow that enters the pipe or channel is the same flow that exits the pipe or channel. In equation form, this becomes

$$Q_1 = Q_2 \text{ or } A_1 V_1 = A_2 V_2 \qquad (4.14)$$

Note: In regard to the area/velocity relationship, Equation 4.14 also makes clear that for a given flow rate, the velocity of the liquid varies indirectly with changes in the cross-sectional area of the channel or pipe. This principle provides the basis for many of the flow measurement devices used in open channels (weirs, flumes, and nozzles).

Example 4.12

Problem: A 12-inch-diameter pipe is connected to a 6-inch-diameter pipe. The velocity of the water in the 12-inch pipe is 3 fps. What is the velocity in the 6-inch pipe?

Solution:

Using the equation $A_1V_1 = A_2V_2$, we need to determine the area of each pipe:

$$12 \text{ in.: } A = \pi \times \frac{D^2}{4}$$

$$= 3.14 \times \frac{(1 \text{ ft})^2}{4}$$

$$= 0.785 \text{ ft}^2$$

$$6 \text{ in.: } A = 3.14 \times \frac{(0.5)^2}{4}$$

$$= 0.196 \text{ ft}^2$$

The continuity equation now becomes

$$\left(0.785 \text{ ft}^2\right) \times \left(\frac{\text{ft}}{\text{sec}}\right) = \left(0.196 \text{ ft}^2\right) \times V_2$$

Solving for V_2:

$$V_2 = \frac{\left(0.785 \text{ ft}^2\right) \times (3 \text{ ft/sec})}{\left(0.196 \text{ ft}^2\right)}$$

$$= 12 \text{ ft/sec or fps}$$

PRESSURE AND VELOCITY

In a closed pipe flowing full (under pressure), the pressure is indirectly related to the velocity of the liquid. This principle, when combined with the principle discussed in the previous section, forms the basis for several flow measurement devices (Venturi meters and rotameters).

$$\text{Velocity}_1 \times \text{Pressure}_1 = \text{Velocity}_2 \times \text{Pressure}_2 \qquad (4.15)$$

or

$$V_1P_1 = V_2P_2$$

PIEZOMETRIC SURFACE AND BERNOULLI'S THEOREM

Most applications of water hydraulics involve water in motion—in pipes under pressure or in open channels under the force of gravity. The volume of water flowing past any given point in the pipe or channel per unit time is called the *flow rate* or *discharge rate*—or just *flow*. The *continuity of flow* and the *continuity equation* have been discussed (i.e., Equation 4.15). Along with the continuity of flow principle and continuity equation, the law of conservation of energy, piezometric surface, and Bernoulli's theorem (or principle) are also important to our study of basic water hydraulics.

CONSERVATION OF ENERGY

Many of the principles of physics are important to the study of hydraulics. When applied to problems involving the flow of water, few of the principles of physical science are more important and useful to us than the *law of conservation of energy*. Simply, the law of conservation of energy states that energy can neither be created nor destroyed, but it can be converted from one form to another. In a given closed system, the total energy is constant.

ENERGY HEAD

In hydraulic systems, two types of energy (kinetic and potential) and three forms of mechanical energy (potential energy due to elevation, potential energy due to pressure, and kinetic energy due to velocity) exist. Energy has the units of foot pounds (ft-lbs). It is convenient to express hydraulic energy in terms of *energy head*, in feet of water. This is equivalent to foot pounds per pound of water (ft-lb/lb = ft).

PIEZOMETRIC SURFACE

As mentioned earlier, we have seen that when a vertical tube, open at the top, is installed onto a vessel of water, the water will rise in the tube to the water level in the tank. The water level to which the water rises in a tube is the *piezometric surface*. That is, the piezometric surface is an imaginary surface that coincides with the level of the water to which water in a system would rise in a *piezometer* (an instrument used to measure pressure). In groundwater, piezometric surface is a synonym of potentiometric surface, which is an imaginary surface that defines the level to which water in a confined aquifer would rise were it completely pieced with wells (Younger, 2007).

The surface of water that is in contact with the atmosphere is known as *free water surface*. Many important hydraulic measurements are based on the difference in height between the free water surface and some point in the water system. The piezometric surface is used to locate this free water surface in a vessel, where it cannot be observed directly.

To understand how a piezometer actually measures pressure, consider the following example. If a clear, see-through pipe is connected to the side of a clear glass or plastic vessel, the water will rise in the pipe to indicate the level of the water in the vessel. Such a see-through pipe, the piezometer, allows you to see the level of the top of the water in the pipe; this is the piezometric surface.

In practice, a piezometer is connected to the side of a tank or pipeline. If the water-containing vessel is not under pressure (as is the case in Figure 4.8), the piezometric surface will be the same as the free water surface in the vessel, just as it would if a drinking straw (the piezometer) were left standing a glass of water.

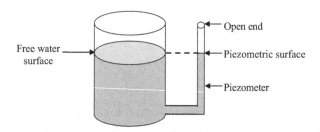

FIGURE 4.8 A container not under pressure where the piezometric surface is the same as the free water surface in the vessel.

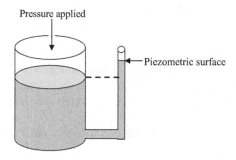

FIGURE 4.9 A container under pressure where the piezometric surface is above the level of the water in the tank.

When a tank and pipeline system is pressurized, as they often are, the pressure will cause the piezometric surface to rise above the level of the water in the tank. The greater the pressure, the higher the piezometric surface (see Figure 4.9). An increased pressure in a water pipeline system is usually obtained by elevating the water tank.

Note: In practice, piezometers are not installed on water towers because water towers are hundreds of feet high, or on pipelines. Instead, pressure gauges are used that record pressure in feet of water or in psi.

Water only rises to the water level of the main body of water when it is at rest (static or standing water). The situation is quite different when water is flowing. Consider, for example, an elevated storage tank feeding a distribution system pipeline. When the system is at rest, all valves closed, all the piezometric surfaces are the same height as the free water surface in storage. On the other hand, when the valves are opened and the water begins to flow, the piezometric surface changes. This is an important point because as water continues to flow down a pipeline, less and less pressure is exerted. This happens because some pressure is lost (used up), keeping the water moving over the interior surface of the pipe (friction). The pressure that is lost is called *head loss*.

HEAD LOSS

Head loss is best explained by the following example. Figure 4.10 shows an elevated storage tank feeding a distribution system pipeline. When the valve is closed (Figure 4.10a), all the piezometric surfaces are the same height as the free water surface in storage. When the valve opens and water begins to flow (Figure 4.10b), the piezometric surfaces *drop*. The farther along the pipeline, the lower the piezometric surface because some of the pressure is used up, keeping the water moving over the rough interior surface of the pipe. Thus, pressure is lost and is no longer available to push water up in a piezometer; this is the head loss.

HYDRAULIC GRADE LINE

When the valve shown in Figure 4.10 is opened, flow begins with a corresponding energy loss due to friction. The pressures along the pipeline can measure this loss. In Figure 4.10b, the difference in pressure heads between sections 1, 2, and 3 can be seen in the piezometer tubes attached to the pipe. A line connecting the water surface in the tank with the water levels at section 1, 2, and 3 shows the pattern of continuous pressure loss along the pipeline. This is called the *hydraulic grade line (HGL)* or *hydraulic gradient* of the system.

Note: It is important to point out that in a static water system, the HGL is always horizontal. The HGL is a very useful graphical aid when analyzing pipe flow problems.

Key point—Changes in the piezometric surface occur when water is flowing.

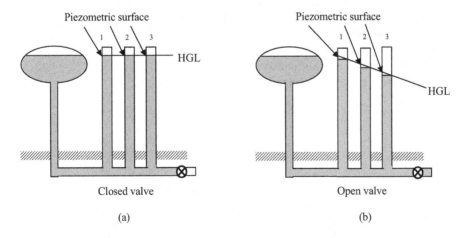

FIGURE 4.10   Head loss and piezometric surface changes when water is flowing.

BERNOULLIS'S THEOREM

Nathanson (1997) noted that Swiss physicist and mathematician Samuel Bernoulli developed the calculation for the total energy relationship from point to point in a steady-state fluid system in the 1700s. Before discussing Bernoulli's energy equation, it is important to understand the basic principle behind Bernoulli's equation. Water (and any other hydraulic fluid) in a hydraulic system possesses two types of energy—kinetic and potential. *Kinetic energy* is present when the water is in motion. The faster the water moves, the more the kinetic energy is used. *Potential energy* is a result of the water pressure. The *total energy* of the water is the sum of the kinetic and potential energy. Bernoulli's principle states that the total energy of the water (fluid) always remains constant. Therefore, when the water flow in a system increases, the pressure must decrease. When water starts to flow in a hydraulic system, the pressure drops. When the flow stops, the pressure rises again. The pressure gauges shown in Figure 4.11 indicate this balance more clearly.

Note: This discussion of Bernoulli's equation ignores friction losses from point to point in a fluid system employing steady-state flow. We assume viscous effects are negligible, and the flow is considered incomprehensible.

BERNOULLI'S EQUATION

In a hydraulic system, the total energy head is equal to the sum of three individual energy heads. This can be expressed as follows:

Total head = Elevation head + pressure head + velocity head

where
    elevation head is the pressure due to the elevation of the water
    pressure head is the height of a column of water that a given hydrostatic pressure in a system could support
    velocity head is the energy present due to the velocity of the water
    Note: the equation is applicable along a streamline.
    This can be expressed mathematically as

$$E = z + \frac{p}{w} + \frac{v^2}{2g} \tag{4.16}$$

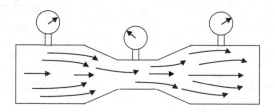

FIGURE 4.11   Bernoulli's principle.

where

E = total energy head
z = height of the water above a reference plane, ft
p = pressure, psi
w = unit weight of water, 62.4 lb/ft³
v = flow velocity, ft/s
g = acceleration due to gravity, 32.2 ft/s²

Consider the constriction in the section of the pipe shown in Figure 4.11. We know, based on the law of energy conservation, that the total energy head at section A, E_1, must equal the total energy head at section B, E_2, and using Equation 4.16, we get Bernoulli's equation.

$$z_A = \frac{P_A}{w} + \frac{v_A^2}{2g} = z_B + \frac{P_B}{w} + \frac{v_B^2}{2g} \tag{4.17}$$

The pipeline system shown in Figure 4.11 is horizontal; therefore, we can simplify Bernoulli's equation because $z_A = z_B$. Because they are equal, the elevation heads cancel out from both sides, leaving

$$\frac{P_A}{w} + \frac{v_A^2}{2g} + \frac{P_B}{w} + \frac{v_B^2}{2g} \tag{4.18}$$

As water passes through the constricted section of the pipe (section B), we know from continuity of flow that the velocity at section B must be greater than the velocity at section A because of the smaller flow area at section B. This means that the velocity head in the system increases as the water flows into the constricted section. However, the total energy must remain constant. For this to occur, the pressure head, and therefore the pressure, must drop. In effect, pressure energy is converted into kinetic energy in the constriction.

The fact that the pressure in the narrower pipe section (constriction) is less than the pressure in the bigger section seems to defy common sense. However, it does follow logically from continuity of flow and conservation of energy. The fact that there is a pressure difference allows measurement of flow rate in the closed pipe.

Example 4.13

Problem: In Figure 4.12, the diameter at section A is 8 inches, and at section B, it is 4 inches. The flow rate through the pipe is 3.0 cfs and the pressure at section A is 100 psi. What is the pressure in the constriction at section B?

Solution:

Compute the flow area at each section as follows:

$$A_A = \frac{p(0.666 \text{ ft})^2}{4} = 0.349 \text{ft}^2 p(0.666 \text{ ft})^2$$

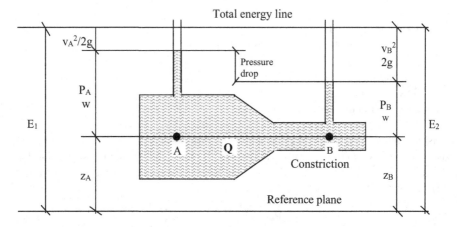

FIGURE 4.12 The law of conservation. Because the velocity and kinetic energy of the water flowing in the constricted section must increase, the potential energy may decrease. This is observed as a pressure drop in the constriction. Source: Adaptation from Nathanson (1997).

$$A_B = \frac{\pi(0.333 \text{ ft})^2}{4} = 0.087 \text{ ft}^2$$

From $Q = A \times V$ or $V = Q/A$, we get

$$V_A = \frac{3.0 \text{ ft}^3/s}{0.349 \text{ ft}^2} = 8.6 \text{ ft/s (rounded)}$$

$$V_B = \frac{3.0 \text{ ft}^3/s}{0.087 \text{ ft}^2} = 34.5 \text{ ft/s (rounded)}$$

And we get

$$\frac{100 \times 144}{62.4} + \frac{8.6^2}{2 \times 32.2} = \frac{p_B \times 144}{62.4} + \frac{34.5^2}{2 \times 32.2}$$

Note: The pressures are multiplied by 144 in.^{2}/ft^2 to convert from psi to lb/ft^2 to be consistent with the units for w—the energy head terms are in feet of head. Continuing, we get

$$231 + 1.15 = 2.3p_B + 18.5$$

and

$$p_B = \frac{232.2 - 18.5}{2.3} = \frac{213.7}{2.3} = 93 \text{ psi (rounded)}$$

REFERENCES

Hauser, B.A. (1996). *Practical Hydraulics Handbook*, 2nd ed. Boca Raton, FL: Lewis
    Publishers.
Magnusson, R.J. (2001). *Water Technology in the Middle Ages*. Baltimore, MD: The John
    Hopkins University Press.
Nathanson, J.A. (1997). *Basic Environmental Technology: Water Supply Waste Management,
    and Pollution Control*, 2nd ed. Upper Saddle River, NJ: Prentice Hall.
Younger, P. (2007). *Groundwater in the Environment*. Oxford, UK: Blackwell Publishing Ltd.

5 Groundwater Hydraulics: Q = kiA

INTRODUCTION

Groundwater is one of nature's most important and indispensable natural resources. Exclusive of machinery and hydropower cooling is used for other purposes. The science of hydrology would be relatively simple if water were unable to penetrate below the earth's surface. Some hydrologists believe that a predevelopment water budget for a groundwater system (i.e., a water budget for the natural conditions before humans used the water) can be used to calculate the amount of water available for consumption (or the safe yield). This concept has been referred to as the "Water-Budget Myth."

Note to reader: Groundwater hydraulics, the movement of water from the surface to underground aquifers is Mother Nature's natural process of stocking and restocking, in many areas, the ocean of fresh water beneath our feet. This chapter about groundwater hydraulics is important to readers in allowing them to gain an understanding of one of the main means by which land subsidence is triggered and maintained.

GROUNDWATER[1]

Unbeknownst to most of us, our Earth possesses an unseen ocean. This ocean, unlike the surface oceans that cover most of the globe, is fresh water: the groundwater that lies contained in aquifers beneath Earth's crust. This gigantic water source forms a reservoir that feeds all the natural fountains and springs of the earth. But how does water travel into the aquifers that lie under the earth's surface?

Groundwater sources are replenished from a percentage of the average of approximately 3 feet of water that falls to earth each year on every square foot of land. Water falling to earth as precipitation follows three courses. Some run off directly to rivers and streams (roughly 6 inches of the 3 feet), eventually working back to the sea. Evaporation and transpiration through vegetation take up about 2 feet. The remaining 6 inches seep into the ground, entering and filling every interstice, each hollow and cavity. Gravity pulls water toward the center of the Earth. That means that water on the surface will try to seep into the ground below it. Although groundwater comprises only one-sixth of the total (1,680,000 miles of water), if we could spread out this water over the land, it would blanket it to a depth of 1000 feet.

The science of groundwater hydraulics is concerned with evaluating the occurrence, availability, and quality of groundwater. In particular, groundwater hydraulics is concerned with the natural or induced movement of water through permeable rock formations. To understand groundwater hydraulics, it is important to understand the operation of the natural plumbing system within it. Moreover, Earth's natural

DOI: 10.1201/9781003461265-6

plumbing system can only be understood if the geologic framework—flow through permeable rock formations—is understood.

To get even closer to understanding basic groundwater hydraulics an understanding of the fundamental properties of unconfined and confined aquifers is necessary.

Part of the precipitation that falls on land infiltrates the land surface, percolates downward through the soil under the force of gravity and becomes groundwater. Groundwater, like surface water, is extremely important to the hydrologic cycle and to our water supplies. Almost half of the people in the U.S. drink public water from groundwater supplies. Overall, more water exists as groundwater than surface water in the U.S., including the water in the Great Lakes. But sometimes, pumping it to the surface is not economical, and in recent years, pollution of groundwater supplies from improper disposal has become a significant problem.

We find groundwater in saturated layers called aquifers under the earth's surface. Hydrologists call groundwater aquifers hydrologic shock absorbers that lessen the impact of rainfall events; they level out the flow in rivers—keep in mind that the base flow of rivers is due to gradual groundwater discharge. Three types of aquifers exist: unconfined, confined, and springs. Aquifers are made up of a combination of solid materials such as rock and gravel and open spaces called pores. Regardless of the type of aquifer, the groundwater in the aquifer is in a constant state of motion. This motion is caused by gravity or by pumping.

The actual amount of water in an aquifer depends upon the amount of space available between the various grains of material that make up the aquifer. The amount of space available is called porosity. The ease of movement through an aquifer is dependent upon how well the pores are connected. For example, clay can hold a lot of water and has high porosity, but the pores are not connected, so water moves through the clay with difficulty. The ability of an aquifer to allow water to infiltrate is called permeability.

UNCONFINED AQUIFERS

The aquifer that lies just under the earth's surface is called the zone of saturation, an unconfined aquifer (water table) (see Figure 5.1). This type of aquifer is composed of granular materials, such as mixtures of clay, silt, sand, and gravel. The top of the zone of saturation is the water table. An unconfined aquifer is only contained on the bottom and is dependent on local precipitation for recharge. Unconfined aquifers are a primary source of shallow well water (see Figure 5.1); these shallow wells are not desirable as a public drinking water source. When a well is sunk a few feet into an unconfined aquifer, the water level remains, for a time, at the same altitude at which it was first reached in drilling, but of course, this level may fluctuate later in response to many factors. They are subject to local contamination from hazardous and toxic materials—fuel and oil, and septic tanks and agricultural runoff providing increased levels of nitrates and microorganisms. These wells may be classified as groundwater under the direct influence of surface water (GUDISW), and therefore require treatment for control of microorganisms. The water level in wells sunk to greater depths in unconfined aquifers may stand at, above, or below the water table, depending upon whether the well is in the discharge or recharge area of the aquifer.

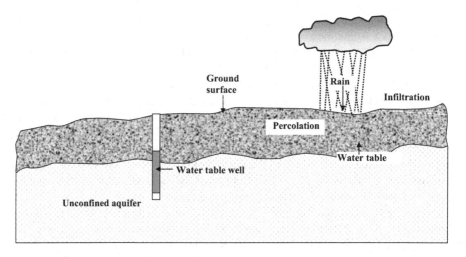

FIGURE 5.1   Unconfined aquifer.

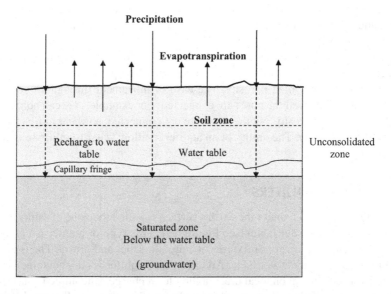

FIGURE 5.2   The unsaturated zone, capillary fringe, water table, and saturated zone.

This type of aquifer is often called a water table aquifer. The unconfined ground-water below the water table is under pressure greater than atmospheric. Unconfined aquifers consist of an unsaturated and a saturated zone (see Figure 5.2). In the unsaturated zone, the spaces between particle grains and the cracks in rocks contain both air and water. Although a considerable amount of water can be present in the unsaturated zone, this water cannot be pumped by wells because capillary forces hold it too tightly. In contrast to the unsaturated zone, the voids in the saturated zone are completely filled with water.

SATURATED ZONE

The approximate upper surface of the saturated zone is referred to as the water table. Water in the saturated zone below the water table is referred to as groundwater. Below the water table, the water pressure is high enough to allow water to enter a well as the water level in the well is lowered by pumping, thus permitting groundwater to be withdrawn for use. Between the unsaturated zone and the water table is a transition zone, capillary fringe. In this zone, the voids are saturated or almost saturated with water that is held in place by capillary forces (USGS, 2016b).

Capillary Fringe

The capillary fringe (see Figure 5.2) acts like a sponge sucking water up from the water table. The capillary fringe ranges in thickness from a small fraction of an inch in coarse gravel to more than 5 feet in silt. The lower part is completely saturated, like the material below the water table, but contains water under less than atmospheric pressure, and hence the water in it normally does not enter a well. The capillary fringe rises and declines with fluctuations of the water table, and may change in thickness as it moves through materials of different grain sizes.

Unsaturated Zone

The *unsaturated zone* contains water in the gas phase under atmospheric pressure, water temporarily or permanently under less than atmospheric pressure, and air or other gases. Fine-grained materials may be temporarily or permanently saturated with waste under less than atmospheric pressure, but coarse-grained materials are unsaturated and generally contain liquid water only in rights surrounding the contacts between grains.

Capillarity

The rise of water in the interstices in rocks or soil may be considered to be caused by (1) the molecular attraction (adhesion) between the solid material and the fluid and (2) the surface tension of the fluid, an expression of the attraction (cohesion) between the molecules of the fluid.

Have you ever noticed that water would wet and adhere to a clean floor whereas it will remain in drops without wetting a dust-covered floor? This, along with the molecular attraction between the solid material and the fluid, is important because it also points to the tendency of the fluid's attraction to the cleanliness of the material. In addition, the height of the capillary rise is governed by the size of the opening.

The surface of water resists considerable tension without losing its continuity. Thus a carefully placed greased needle floats on water, as do certain insects having greasy pads on the feet or water-resistant hair on the underbodies. A good insect example is the Water Strider ("Jesus bugs") (Order: Hemiptera). These ride the top of the water, with only their feet making dimples in the surface film. Like all insects, the water striders have a three-part body (head, thorax, and abdomen), six jointed legs, and two antennae. It has a long, dark, narrow body (see Figure 5.3).

In Figure 5.4, the water has risen to h_c in a tube of radius r immersed in a container of water. The relations shown in Figure 5.4 may be expressed

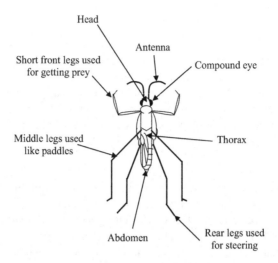

FIGURE 5.3   Water Strider ("Jesus bugs") (Order: Hemiptera).

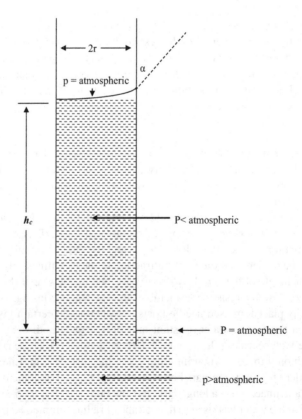

FIGURE 5.4   Capillary rise of water in a tube (exaggerated).

$$\pi r^2 pgh_c = 2\pi r \, T \cos\alpha \qquad (5.1)$$

where
  r = radius of capillary tube
  p = density of fluid
  g = acceleration due to gravity
  h_c = height of capillary rise
  T = surface tension of fluid
  α = angle between meniscus and tube

Note that, according to Equation 5.1, weight equals lift by surface tension. Solving Equation 5.1 for h_c,

$$h_c = \frac{2T}{rpg} \cos\alpha \qquad (5.2)$$

For pure water in a clean glass, α = 0, and cos α = 1. At 20°C, T = 72.8 dyne/cm, may be taken as 1 g/cm³, and g = 980.665 cm/s², whence

$$h_c = \frac{0.15}{r} \qquad (5.3)$$

Surface tension is sometimes given in grams per centimeter and for pure water in contact with air, at 20°C; its value is 0.074 g/cm. In order to express it in grams per centimeter, we must divide 72.8 by g, the standard acceleration of gravity; thus 72.8 dyne/cm/980.665 cm/sect² = 0.074 g/cm.

From Equation 5.3, it is seen that the height of capillary rise in tubes is inversely proportional to the radius of the tube. The rise of water in interstices of various sizes in the capillary fringe may be likened to the rise of water in a bundle of capillary tubes of various diameters, as shown in Figure 5.5. In Table 5.1, note that the capillary rise is nearly inversely proportional to the grain size.

HYDROLOGIC PROPERTIES OF WATER-BEARING MATERIALS

POROSITY

Porosity is defined as the ratio of (1) the volume of the void spaces to (2) the total volume of the rock or soil mass. Stated differently, the porosity of a rock or solid is simply its property of containing interstices. It can be expressed quantitatively as the ratio of the volume of the interstices to the total volume and may be expressed as a decimal fraction or as a percentage. Thus,

$$\theta = \frac{v_i}{V} = \frac{v_w}{V} = \frac{V - V_m}{V} = 1 - \frac{V_m}{V} \, [\text{dimensionless}] \qquad (5.4)$$

where
  θ = porosity, as a decimal fraction
  v_i = volume of interstices
  V = total volume
  v_w = volume of water (in a saturated sample)
  v_m = volume of mineral particles

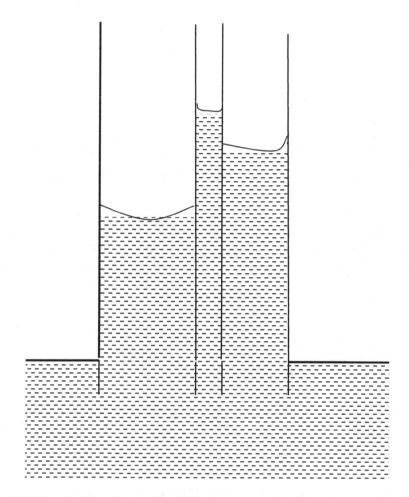

FIGURE 5.5 Rise in capillary tubes of different diameters (exaggerated).

TABLE 5.1
Capillary Rise in Samples Having Virtually the Same Porosity, 41%, after 72 Days

Material	Grain Size (mm)	Capillary Rise (cm)
Fine gravel	5–2	2.5
Very coarse sand	2–1	6.5
Coarse sand	1–0.5	13.5
Medium sand	0.5–0.2	24.6
Fine sand	0.2–0.1	42.8
Silt	0.1–0.05	105.5
Silt	0.05–0.02	200

Source: Based on Atterberg, cited in Terzaghi (1942).

Porosity may be expressed also as

$$\theta = \frac{p_m - p_d}{p_m} = 1 - \frac{p_d}{p_m} [\text{dimensionless}] \tag{5.5}$$

where
  p_m = mean density of mineral particles (grain density)
  p_d = density of dry sample (bulk density)

Multiplying the right-hand aides of Equations 5.4 and 5.5 by 100 gives the porosity as a percentage.

PRIMARY POROSITY

Primary porosity is porosity associated with the original depositional texture of the sediment. That is, in soil and sedimentary rocks the primary interstices are the spaces between grains or pebbles. In intrusive igneous rocks, the few primary interstices result from cooling and crystallization. Extrusive igneous rocks may have large openings and high porosity resulting from the expansion of gas, but the openings may or may not be connected. With time, the metamorphism of igneous or sedimentary rocks generally reduces the primary porosity and may virtually obliterate it.

SECONDARY POROSITY

Secondary porosity is porosity that develops after the deposition and burial of the sediment in the sedimentary basin. Fractures such as joints, faults, and openings along planes of bedding or schistosity in consolidated rocks having low primary porosity and permeability may afford appreciable secondary porosity.

CONTROLLING POROSITY OF GRANULAR MATERIALS

In describing the conditions that control the porosity of granular materials, it is convenient to use the time-proven approach originally used by Slichter in 1899. He explained that if a hypothetical granular material were composed of spherical particles of equal size, the porosity would be independent of particle size (whether the particles were the size of silt or the size of the earth) but would vary with the packing arrangement of the particles. Slichter explained that the lowest porosity of 25.95% (about 26%) would result from the most compact rhombohedral arrangement (Figure 5.6a) and the highest porosity of 47.64% (about 48%) would result from the least compact cubical arrangement (Figure 5.6c). The porosity of the other arrangements, such as that shown in Figure 5.6b, would be between these limits.

In addition to the arrangement of grains (as shown in Figure 5.7) having an impact on controlling the porosity of granular materials, the shape of the grains (i.e., their angularity) and their degree of assortment (i.e., range in particle size) also have an impact on porosity. The angularity of particles causes wide variations in porosity and may increase or decrease it, according to whether the particles tend to bridge openings

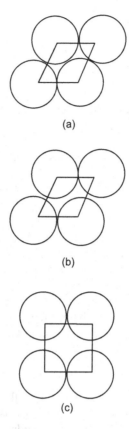

(a)

(b)

(c)

FIGURE 5.6 Sections of four contiguous spheres of equal size. (a) most compact arrangement, lowest porosity; (b) less compact arrangement, higher porosity; (c) least compact arrangement, highest porosity. (Modified from Slichter, 1899, p. 1.)

or pack together like pieces of mosaic. The greater the range in particle size the lower the porosity, as the small particles occupy the voids between the larger ones.

VOID RATIO

The void ratio of a rock or soil is the ratio of the volume of its interstices to the volume of its mineral particles. It may be expressed:

$$\text{Void ratio} = \frac{v_i}{v_m} = \frac{v_w}{v_m} = \frac{\theta}{1-\theta} [\text{dimensionless}] \tag{5.6}$$

where
    θ = Porosity, as a decimal fraction
    v_i = volume of interstices
    v_w = volume of water (in a saturated sample)
    v_m = volume of mineral particles

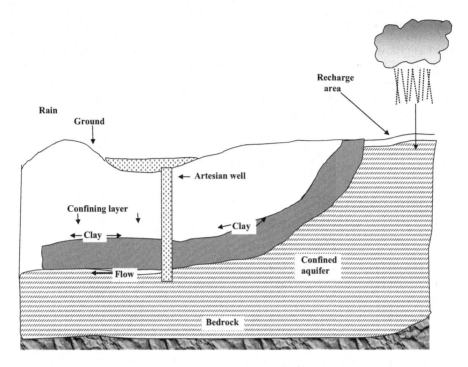

FIGURE 5.7 Confined aquifer.

PERMEABILITY

The permeability of a rock or soil is a measure of how easily water can travel (i.e., how easily it transmits fluid) such as water, under a hydropotential gradient. Soil and loose sediments, such as sand and gravel, are porous and permeable. They can hold a lot of water, and it flows easily through them. The permeability is approximately proportional to the square of the mean grain diameter,

$$K \approx Cd^2 \qquad (5.7)$$

where
   k = intrinsic permeability
   C = a dimensionless constant depending upon porosity, range and distribution of particle size, shape of grains, and other factors
   d = the mean grain diameter of some workers and the effective grain diameter of others

---

DID YOU KNOW?

Although clay and shale are porous and can hold a lot of water, the pores in these fine-grained materials are so small that water flows very slowly through them. Clay has a low permeability.

INTRINSIC PERMEABILITY

The term intrinsic permeability, adopted by the U.S. Geological Survey, states that the permeability in question is an intensive property (not a spatial average of a heterogeneous block of material), that it is a function of the material structure only (and not of the fluid), and explicitly distinguishes the value from that of relative permeability. Intrinsic permeability may be expressed

$$k = -\frac{qv}{g(dh/dl)} = -\frac{qv}{(dp/dl)} \qquad (5.8)$$

where
   k = intrinsic permeability
   q = rate of flow per unit area = Q/A
   v = kinematic viscosity
   g = acceleration of gravity
   dh/dl = gradient, or unit change in head per unit length of flow
   dp/dl = potential gradient, or unit change in potential per unit length of flow

From Equation 5.8, it may be stated that a porous medium has an intrinsic permeability of one unit of length squared if it will transmit in unit time a unit volume of fluid of unit kinematic viscosity through a cross section of the unit area measured at right angles to the flow direction under a unit potential gradient.

HYDRAULIC CONDUCTIVITY

Hydraulic conductivity (HC), represented as K, is a property of soils and rocks that describes the ease with which water can move through pore spaces or fractures. It is said that a medium has a HC of unit length per unit time if it will transmit in unit time a unit volume of groundwater at the prevailing viscosity through a cross section of unit area, measured at right angles to the direction of flow, under a hydraulic gradient of a unit change in the head through a unit length of flow. The suggested units are:

$$K = \frac{q\left(\text{rate of flow per unit area} = Q/A\right)}{dh/dl\left(\text{gradient}\right)} \qquad (5.9)$$

$$= -\frac{ft^3}{ft^2 day\left(-ft\ ft^{-1}\right)} = ft/day \qquad (5.10)$$

or

$$K = \frac{m^3}{m^2 day\left(-m\ m^{-1}\right)} = m/day \qquad (5.11)$$

Note that the minus signs in Equations 5.10 and 5.11 result from the fact that the water moves in the direction of decreasing head.

TRANSMISSIVITY

The *Transmissivity* (*T*) is the rate at which water of the prevailing kinematic viscosity flows horizontally through an aquifer, such as to a pumping well. Transmissivity is a preferred parameter because aquifers are not uniform; they have variable HC, gradients, and cross-sectional area. The standard equation for Transmissivity is shown below:

$$T = Q/W*I \tag{5.12}$$

where
   T = transmissivity
   W = the width of the aquifer
   Q = flow
   I = hydraulic gradient

WATER YIELDING AND RETAINING CAPACITY

SPECIFIC YIELD

Specific yield (S_y) is defined as the ratio of (1) the volume of water that a saturated rock or soil will yield by gravity to (2) the total volume of the rock or soil. This may be expressed

$$S_y = \frac{v_g}{V}[\text{dimensionless}] \tag{5.13}$$

where
   S_y = specific yield, as a decimal fraction
   v_g = volume of water drained by gravity
   V = total volume

DID YOU KNOW?

Specific yield is usually expressed as a percentage. The value is not definitive, because the quantity of water that will drain by gravity depends on variables such as duration of drainage, temperature, mineral composition of the water, and various physical characteristics of the rock or soil under consideration. Values of specific yield, nevertheless, offer a convenient means by which hydrologists can estimate the water-yielding capacities of earth materials and, as such, are very useful in hydrologic studies.

SPECIFIC RETENTION

The *specific retention* of a rock or soil with respect to water has been defined by Meinzer (1923) as the ratio that will be retained against gravity drainage from a saturated rock to the volume of the rock. It may be expressed

$$S_r = \frac{v_r}{V} = \theta - S_y \tag{5.14}$$

where

S_r = specific retention, as a decimal fraction

v_r = volume of water retained against gravity, mostly by molecular attraction

CONFINED AQUIFERS

A confined aquifer is sandwiched between two impermeable layers that block the flow of water.

The water in a confined aquifer is under hydrostatic pressure. It does not have a free water table (see Figure 5.7). Confined aquifers are called artesian aquifers. Wells drilled into artesian aquifers are called artesian wells and commonly yield large quantities of high-quality water. An artesian well is any well where the water in the well casing would rise above the saturated strata. Wells in confined aquifers are normally referred to as deep wells and are not generally affected by local hydrological events.

A confined aquifer is recharged by rain or snow in the mountains where the aquifer lies close to the surface of the earth. Because the recharge area is some distance from areas of possible contamination, the possibility of contamination is usually very low. However, once contaminated, confined aquifers may take centuries to recover.

Groundwater naturally exits the earth's crust in areas called springs. The water in a spring can originate from a water table aquifer or from a confined aquifer. Only water from a confined spring is considered desirable for a public water system.

STEADY FLOW OF GROUNDWATER

In a steady flow of groundwater through permeable material, there is no change in the head with time. Mathematically, this statement is symbolized by $dh/dt = 0$, which says that the change in head, dh, with respect to the change in time, dt, equals zero. Note that steady flow generally does not occur in nature, but it is a very useful concept in that steady flow can be closely approached in nature and in aquifer tests, and this condition may be symbolized dh/dt 0.

Figure 5.8 shows a hypothetical example of true steady radial flow. Here steady radial flow will be reached and maintained when all the recoverable groundwater in the cone of depression has been drained by gravity into the well discharging at a constant rate Q.

DARCY'S LAW

Hagen (1839) and Poiseuille (1846) found that the rate of flow through capillary tubes is proportional to the hydraulic gradient; however, it was Darcy, a French engineer in 1856 who experimented with the flow of water through sand and determined that the rate of laminar (viscous) flow of water through sand is proportional to the hydraulic gradient. This is known as *Darcy's law* and it is generally expressed as shown in Equation 5.15,

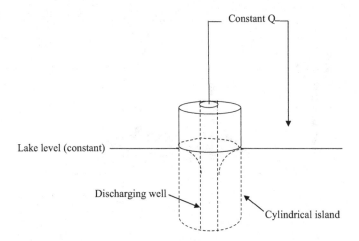

Constant Q

Lake level (constant)

Discharging well

Cylindrical island

FIGURE 5.8 Hypothetical example of steady flow (well discharging at constant rate Q from a cylindrical island in a lake of constant level). Source: *Ground Water Hydraulics.* Department of Interior Geological Survey Paper 708 (1975) Washington, DC.

$$q = \frac{Q}{A} = \frac{Kdh}{dl} \qquad (5.15)$$

It will be noted that K, the constant of proportionality in Darcy's law, is the HC. Darcy's law is also often shown as in Equation 5.16.

$$Q = kiA \qquad (5.16)$$

where
   Q = rate of flow
   k = hydraulic conductivity
   i = slope
   A = cross-sectional area

Units for Darcy's law (Equation 5.16): If Q = cubic meters per second, then K must be in meters per second, A in square meters (I is a unitless quantity: a 1% slope would be 0.01).

To illustrate the use of Equation 5.15 (Darcy's law), assume that we wish to compute the total rate of groundwater movement in a valley where A, the cross-sectional area, in 100 feet deep times 1 mile wide, where K = 500 ft/day, and dh/dl = 5 ft per mile. Then

$$Q = -(100 \text{ ft})(5280 \text{ ft})(500 \text{ ft/day})\left(\frac{-5 \text{ ft}}{5280 \text{ ft}}\right)$$

$$= 250,000 \text{ ft}^3/\text{day}$$

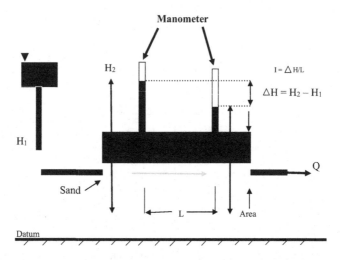

FIGURE 5.9   Rough diagram of the experiment conducted by Darcy. A column is filled.

Sand manometers, which are hollow tubes used to measure water pressure (height), are inserted into the upper end of the column and after some time water will begin to flow out of the lower end. Water will push up into the manometers to heights that will indicate the gradient. Source: Adaptation from Idaho Department of Environmental Quality (Boise, Idaho): accessed 8/31/23 @ www.deq.idaho.gov.

EXPERIMENT 5.1—DETERMINING HYDRAULIC CONDUCTIVITY

In determining the HC of samples collected by my students in water/wastewater short courses that I taught at Virginia Tech University, I had the students assemble and use the Darcy tube (permeameter) (Figure 5.10) shown below. This particular assembly is modeled and adapted after a training lesson from Idaho Department of Environmental Quality, Boise, Idaho (2023).

PROCEDURE

1. Add water to funnel until equal return to drain and the water level is holding.
2. Cycle water through the permeameter until all the air has been displaced and you are receiving a steady flow.
3. Measure the lengths of h, L, and D (see Figure 5.10) and record values.
4. Measure the volume of water (Q) that flows from the permeameter over a measure length of time (t) and record and continue recording for several measurements.
5. Convert all the length measurements to feet and time to days.

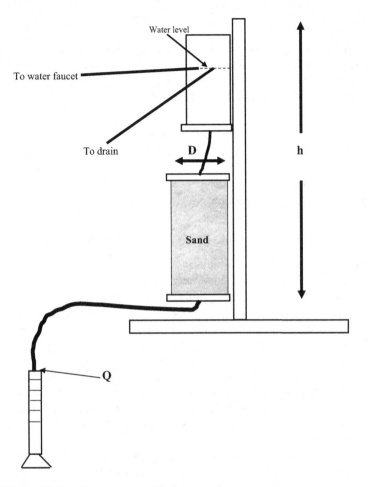

FIGURE 5.10 Makeshift permeameter for lab experiments.

6. Calculate the HC (ft/day) for each Q and t measurement using the rear-ranged Darcy's law equation;

$$K = QL/At$$

7. Calculate the average hydraulic value.
8. Compare to your recorded values.
9. Using the K value from above and an effective porosity value from your recorded values, calculate the velocity of the water moving through the Darcy tube.

$$V = Ki/ne$$

where
I = gradient (ft/ft) = h(ft)/L (ft)
He = effective [porosity = 0.25]

VELOCITY

Because the HC, K, has the dimensions of velocity some might mistake this for the particle velocity of the water, whereas K is actually a measure of the volume rate of flow through a unit cross-sectional area. For the average particle velocity, v, we must also know the porosity of the material. Thus,

$$Q = vA\theta = - KA \frac{dh}{dl}$$

where
    v = average velocity, in feet per day
    θ = porosity, as a decimal fraction

GROUNDWATER FLOW AND EFFECTS OF PUMPING

Water pumped from the groundwater system not only causes the water table to lower, alters the direction of groundwater movement and leads to land subsidence but also can have a debilitating impact on the quality of the water within an aquifer. Generally, groundwater possesses high physical quality. When pumped from an aquifer composed of a mixture of sand and gravel, if not directly influenced by surface water, groundwater is often used without filtration. It can also be used without disinfection if it has a low coliform count (depending on location). However, groundwater can become contaminated. When septic systems fail, saltwater intrudes, improper disposal of wastes occurs, improperly stockpiled chemicals leach, underground storage tanks lead, hazardous materials spill, fertilizers and pesticides are misplaced, and when mines are improperly abandoned, groundwater can become contaminated.

Note—Word to the wise: if you are planning on digging a well, you need to tap productive aquifers; locate near users; locate near power; drill as shallow as possible.

Figure 5.11 shows a hypothetical flow of water from a lake. The point made in Figure 5.8 is that in some circumstances the withdrawal of water is relatively simple and easy—this is not the case with groundwater.

To understand how an underground aquifer becomes contaminated, you must understand what occurs when pumping is taking place within the well. When groundwater is removed from its underground source (i.e., from the water-bearing stratum) via a well, water flows toward the center of the well. In a water table aquifer, this movement causes the water table to sag toward the well. This sag is called the cone of depression. The shape and size of the cone depend on the relationship between the pumping rate and the rate at which water can move toward the well. If the rate is high, the cone is shallow, and its growth stabilizes. The area that is included in the cone of depression is called the cone of influence and any contamination in this zone will be drawn into the well.

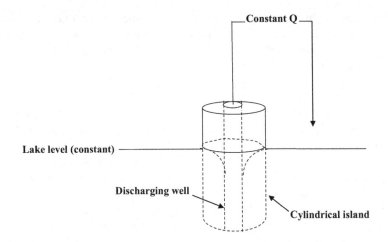

FIGURE 5.11 Hypothetical example of steady flow (well discharging at constant rate Q from a cylindrical island in a lake of constant level). Source: *Ground Water Hydraulics*. Department of Interior Geological Survey Paper 708 (1975) Washington, DC.

THE 411 ON WELLS

Water supply wells may be characterized as shallow or deep. In addition, wells are classified as follows:

1. Class I—cased and grouted to 100 feet
2. Class II A—cased to a minimum of 100 feet and grouted to 20 feet
3. Class II B—cased and grouted to 50 feet

Note: During the well development process, mud/silt forced into the aquifer during the drilling process is removed, allowing the well to produce the best-quality water at the highest rate from the aquifer.

Issues for Any Well

- Type of well
- Location
- Design
- Installation
- Operation
- Abandonment

Shallow Wells

Shallow wells are those that are less than 100 feet deep. Such wells are not particularly desirable for municipal supplies since the aquifers are likely to fluctuate considerably in depth, making the yield somewhat uncertain. Municipal wells in such aquifers cause a reduction in the water table (or phreatic surface) that affects nearby private wells, which are more likely to utilize shallow strata. Such interference with private wells may result in damage suits against the community. Shallow wells may be dug, bored, or driven.

1. *Dug wells*—Dug wells are the oldest type of well and date back many centuries; they are dug by hand or by a variety of unspecialized equipment. They range in size from approximately 4 to 15 feet in diameter and are usually about 20–40 feet deep. Such wells are usually lined or cased with concrete or brick. Dug wells are prone to failure from drought or heavy pumpage. They are vulnerable to contamination and are not acceptable as a public water supply in many locations.
2. *Driven wells*—Driven wells consist of a pipe casing terminating in a point slightly greater in diameter than the casing. The pointed well screen and the lengths of pipe attached to it are pounded down or driven in the same manner as a pile, usually with a drop hammer, to the water-bearing strata. Driven wells are usually 2–3 inches in diameter and are used only in unconsolidated materials. This type of shallow well is not acceptable as a public water supply.
3. *Bored wells*—Bored wells range from 1 to 36 inches in diameter and are constructed in unconsolidated materials. The boring is accomplished with augers (either hand- or machine-driven) that fill with soil and then are drawn to the surface to be emptied. The casing may be placed after the well is completed (in relatively cohesive materials) but must advance with the well in non-cohesive strata. Bored wells are not acceptable as a public water supply.

Deep Wells

Deep wells are the usual source of groundwater for municipalities. Deep wells tap thick and extensive aquifers that are not subject to rapid fluctuations in water (piezometric surface—the height to which water will rise in a tube penetrating a confined aquifer) level and that provide a large and uniform yield. Deep wells typically yield water of more constant quality than shallow wells, although the quality is not necessarily better. Deep wells are constructed by a variety of techniques; we discuss two of these techniques (jetting and drilling) below.

1. *Jetted wells*—Jetted well construction commonly employs a jetting pipe with a cutting tool. This type of well cannot be constructed in clay, hardpan, or where boulders are present. Jetted wells are not acceptable as a public water supply.
2. *Drilled wells*—Drilled wells are usually the only type of well allowed for use in most public water supply systems. Several different methods of drilling are available; all of which are capable of drilling wells of extreme depth and diameter. Drilled wells are constructed using a drilling rig that creates a hole into which the casing is placed. Screens are installed at one or more levels when water-bearing formations are encountered.

Well Casing

A well is a hole in the ground called the borehole. The hole is protected from collapse by placing a casing inside the borehole. The well casing prevents the walls of the hole from collapsing and prevents contaminants (either surface or subsurface) from entering the water source. The casing also provides a column of stored water and housing for the pump mechanisms and pipes. Well casings constructed of steel or plastic material are acceptable. The well casing must extend a minimum of 12 inches above grade.

Grout

To protect the aquifer from contamination, the casing is sealed to the borehole near the surface and near the bottom where it passes into the impermeable layer with grout. This sealing process keeps the well from being polluted by surface water and seals out water from water-bearing strata that have undesirable water quality. Sealing also protects the casing from external corrosion and restrains unstable soil and rock formations. Grout consists of near cement that is pumped into the annular space (it is completed within 48 hours of well construction); it is pumped under continuous pressure starting at the bottom and progressing upward in one continuous operation.

Well Pad

The well pad provides a ground seal around the casing. The pad is constructed of reinforced concrete 6 ft × 6 ft (6-inch thick) with the well head located in the middle. The well pad prevents contaminants from collecting around the well and seeping down into the ground along the casing.

Sanitary Seal

To prevent contamination of the well, a sanitary seal is placed at the top of the casing. The type of seal varies depending upon the type of pump used. The sanitary seal contains openings for power and control wires, pump support cables, a drawdown gauge, discharge piping, a pump shaft, and an air vent, while providing a tight seal around them.

Well Screen

The well screen is the most important single factor affecting the efficiency of a well, and because of its importance, it is often called the heart of the well. "Screens can be installed at the intake point(s) on the end of a well casing or on the end of the inner casing on gravel packed well. These screens perform two functions: (1) supporting the borehole, and (2) reducing the amount of sand that enters the casing and the pump. They are sized to allow the maximum amount of water while preventing the passage of sand/sediment/gravel."

CASING VENT

The well casing must have a vent to allow air into the casing as the water level drops. The vent terminates 18 inches above the floor with a return bend pointing downward. The opening of the vent must be screened with #24 mesh stainless steel to prevent the entry of vermin and dust.

DROP PIPE

The drop pipe or riser is the line leading from the pump to the well head. It assures adequate support so that an aboveground pump does not move and so that a submersible pump is not lost down the well. This pipe is either steel or PVC. Steel is the most desirable.

WELL HYDRAULICS

When the source of water for a water distribution system is from a groundwater supply, knowledge of well hydraulics is important to the operator. Basic well hydraulics terms are presented and defined, and they are related pictorially (see Figure 5.12).

BASIC WELL HYDRAULICS

- *Static water level*—The water level in a well when no water is being taken from the groundwater source (i.e., the water level when the pump is off; see Figure 5.12). Static water level is normally measured as the distance from

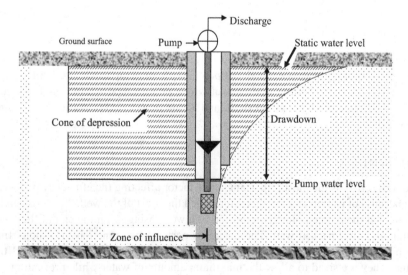

FIGURE 5.12   Hydraulic characteristics of a well.

the ground surface to the water surface. This is an important parameter because it is used to measure changes in the water table.

- *Pumping water level*—The water level when the pump is off. When water is pumped out of a well, the water level usually drops below the level in the surrounding aquifer and eventually stabilizes at a lower level; this is the pumping level (see Figure 5.12).
- *Drawdown*—The difference, or the drop, between the Static Water Level and the Pumping Water Level, measured in feet. Simply, it is the distance the water level drops once pumping begins (see Figure 5.12).
- *Cone of depression*—In unconfined aquifers, there is a flow of water in the aquifer from all directions toward the well during pumping. The free water surface in the aquifer then takes the shape of an inverted cone or curved funnel line. The curve of the line extends from the Pumping Water Level to the Static Water Level at the outside edge of the Zone (or Radius) of Influence (see Figure 5.12).
- *Note*—The shape and size of the cone of depression are dependent on the relationship between the pumping rate and the rate at which water can move toward the well. If the rate is high, the cone will be shallow and its growth will stabilize. If the rate is low, the cone will be sharp and continue to grow in size.
- *Zone (or radius) of influence*—The distance between the pump shaft and the outermost area affected by drawdown (see Figure 5.12). The distance depends on the porosity of the soil and other factors. This parameter becomes important in well fields with many pumps. If wells are set too close together, the zones of influence will overlap, increasing the drawdown in all wells. Obviously, pumps should be spaced apart to prevent this from happening.

Two important parameters not shown in Figure 5.12 are well yield and specific capacity.

1. *Well yield* is the rate of water withdrawal that a well can supply over a long period, or, alternatively, the maximum pumping rate that can be achieved without increasing the drawdown. The yield of small wells is usually measured in gallons per minute (liters per minute) or gallons per hour (liters per hour). For large wells, it may be measured in cubic feet per second (cubic meters per second).
2. *Specific capacity* is the pumping rate per foot of Drawdown (gpm/ft), or

$$\text{Specific capacity} = \text{Well yield} \div \text{drawdown} \qquad (5.17)$$

Example 5.1

Problem: If the well yield is 300 gpm and the drawdown is measured to be 20 feet, what is the specific capacity?

Solution:

Specific capacity = 300 ÷ 20
Specific capacity = 15 gpm per ft of drawdown
Specific capacity is one of the most important concepts in well operation and testing. The calculation should be made frequently in the monitoring of well operation. A sudden drop in specific capacity indicates problems such as pump malfunction, screen plugging, or other problems that can be serious. Such problems should be identified and corrected as soon as possible.

Well Drawdown Calculations

As mentioned and shown in Figure 5.12 *drawdown* is the drop in the level of water in a well when water is being pumped. Drawdown is usually measured in feet or meters. One of the most important reasons for measuring drawdown is to make sure that the source water is adequate and not being depleted. The data that is collected to calculate drawdown can indicate if the water supply is slowly declining. Early detection can give the system time to explore alternative sources, establish conservation measures or obtain any special funding that may be needed to get a new water source. Well drawdown is the difference between the pumping water level and the static water level.

$$\text{Drawdown, ft} = \text{Pumping water level, ft} - \text{static water level, ft} \qquad (5.18)$$

Example 5.2

Problem: The static water level for a well is 70 feet. If the pumping water level is 90 feet, what is the drawdown?

Solution:

$$\text{Drawdown, ft} = \text{Pumping water level, ft} - \text{static water level, ft}$$

$$= 90 \text{ ft} - 70 \text{ ft}$$

$$= 20 \text{ ft}$$

Example 5.3

Problem: The static water level of a well is 122 feet. The pumping water level is determined using the sounding line. The air pressure applied to the sounding line is 4.0 psi and the length of the sounding line is 180 feet. What is the drawdown?

Solution:

First calculate the water depth in the sounding line and the pumping water level:
(1) Water depth in sounding line = (4.0 psi) (2.31 ft/psi) = 9.2 ft
(2) Pumping water level = 180 ft − 9.2 ft = 170.8 ft
Then calculate drawdown as usual:
Drawdown, ft = Pumping water level, ft − static water level, ft
    = 170.8 ft − 122 ft
    = 48.8 ft

Well Yield Calculations

Well yield is the volume of water per unit of time that is produced from the well pumping. Usually, well yield is measured in terms of gallons per minute (gpm) or gallons per hour (gph). Sometimes, large flows are measured in cubic feet per second (cfs). Well yield is determined by using the following equation.

$$\text{Well yield, gpm} = \frac{\text{Gallons produced}}{\text{Duration of test, min}} \qquad (5.19)$$

Example 5.4

Problem: Once the drawdown level of a well stabilized, it was determined that the well produced 400 gallons during a 5-minute test.

Solution:

$$\text{Well yield, gpm} = \frac{\text{Gallons produced}}{\text{Duration of test, min}}$$

$$= \frac{400 \text{ gallons}}{5 \text{ minutes}}$$

$$= 80 \text{ gpm}$$

Example 5.5

Problem: During a 5-minute test for well yield, a total of 780 gallons are removed from the well. What is the well yield in gpm? In gph?

Solution:

$$\text{Well yield, gpm} = \frac{\text{Gallons removed}}{\text{Duration of test, min}}$$

$$= \frac{780 \text{ gallons}}{5\text{-minutes}}$$

$$= 156 \text{ gpm}$$

Then convert gpm flow to gph flow:

$$(156 \text{ gal/min}) \ (60/\text{hr}) = 9360 \text{ gph}$$

Specific Yield Calculations

Specific yield is the discharge capacity of the well per foot of drawdown. The specific yield may range from 1 gpm/ft drawdown to more than 100-gpm/ft drawdown for a properly developed well. Specific yield is calculated using the Equation 5.20.

$$\text{Specific yield, gpm/ft} = \frac{\text{Well yield, gpm}}{\text{Drawdown, ft}} \qquad (5.20)$$

Example 5.6

Problem: A well produces 260 gpm. If the drawdown for the well is 22 feet, what is the specific yield in gpm/ft, what is the specific yield in gpm/ft of drawdown?

Solution:

$$\text{Specific yield, gpm/ft} = \frac{\text{Well yield, gpm}}{\text{Drawdown, ft}}$$

$$= \frac{260 \text{ gpm}}{22 \text{ ft}}$$

$$= 11.8 \text{ gpm/ft}$$

Example 5.7

Problem: The yield for a particular well is 310 gpm. If the drawdown for this well is 30 feet, what is the specific yield in gpm/ft of drawdown?

Solution:

$$\text{Specific yield, gpm/ft} = \frac{\text{Well yield, gpm}}{\text{Drawdown, ft}}$$

$$= \frac{310 \text{ gpm}}{30 \text{ ft}}$$

$$= 10.3 \text{ gpm/ft}$$

DEPLETING THE GROUNDWATER BANK ACCOUNT

Normally, as shown in Figure 5.13, water is recharged to the groundwater system by percolation of water from precipitation and then flows to a stream or other water body through the groundwater system. Where surface water, such as lakes and rivers, is scarce or inaccessible, groundwater supplies many of the hydrologic needs of people everywhere. In the United States, it is the source of drinking water for about half the total population and nearly all of the rural population, and it provides over 50 billion gallons per day for agricultural needs. Water stored in the ground for use can be compared to money in a bank account. However, water pumped from the groundwater system causes the water table to lower and alters the direction of groundwater movement. Some water that flowed to a stream or other water body no longer does so and some water may be drawn from the steam into the groundwater system, thereby reducing the amount of streamflow. Groundwater depletion, the term used to define long-term water level, is a key issue associated with groundwater use. Many areas of the United States are experiencing groundwater depletion (USGS, 2016a).

Overdrawing the Groundwater Bank Account

If you withdraw money at a faster rate than you deposit new money in your bank account you will eventually start having account-supply problems. Likewise, if you

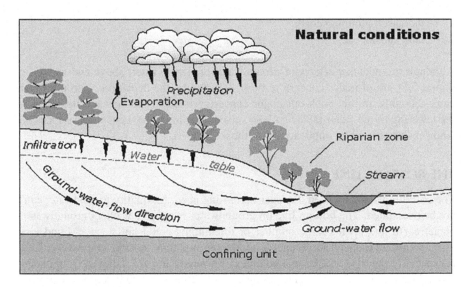

FIGURE 5.13   Normal groundwater conditions. Source: USGS (2016a).

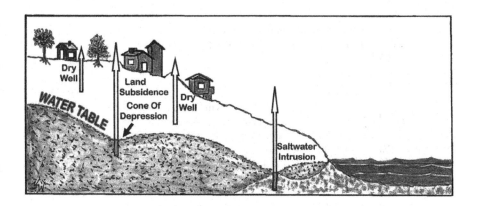

FIGURE 5.14   Impacts of over-pumping of groundwater and groundwater depletion. Source: Adapted from USGS (2016b). Illustration by F.R. Spellman and K. Welsh.

pump water out of the ground faster than it is replenished over the long-term, similar problems will develop. The volume of groundwater in storage is decreasing in many areas of the United States in response to pumping.

Groundwater depletion is primarily caused by sustained groundwater pumping. Some of the negative effects of groundwater depletion and the focus of this book are listed below and depicted in Figure 5.14:

- drying up of wells
- reduction of water in streams and lakes
- deterioration of water quality

- increased pumping costs
- land subsidence

Although the negative effects of groundwater depletion listed above and shown in Figure 5.14 are of major concern to those affected by the depletion of groundwater bank accounts, in this book our major concern related to groundwater depletion is the lowering of the water table and land subsidence. In the next chapter, we focus on land subsidence and its implications for those affected.

THE BOTTOM LINE

The absolute value and need, the ever-increasing need, for quality groundwater cannot be overstated. The bottom line on groundwater is simple: Without groundwater, planet earth and all it's large number of inhabitants could not survive. Land loss associated with induced subsidence is more common than most people realize. This is especially the case where large volumes of groundwater are removed from underground formations. Human-induced subsidence is also responsible for the creation of sinkholes; this topic is covered in the following chapter.

NOTE

1 The following information is adapted from F. Spellman (2019). *Sustainable Water Initiative for the Future (SWIFT)*. Lanham, MD: Bernan Press.

REFERENCES

Hagen, GHL (1839). Uber die Bewgung des Wassers in Engen cylindrische Rohren, *Pegendorfs Annalen der Physick unde Chemie* (2)46, pp. 432–442.

Meinzer, O.E. (1923). Outline of groundwater hydrology, with definitions: U.S. Geological Survey Water-Supply Paper 494, p. 71.

Poiseuille, H. (1846). Physiques—Recherché experimeles sur de mouvement des liquids dn les tubes de tres petits diametres. *Académie des Sciences, Comptes Rendus*, 11, pp. 961–967.

Slichter, C.S. (1899). Theoretical investigations of the motion of ground waters: U.S. Geological Survey 19th Annual Report pt. II-C, pp. 295–384.

Terzaghi, K., (1942). *Theoretical Soil Mechanics*. New York: Wiley.

USGS. (2016a). Groundwater Flow and Effects of Pumping. Accessed from https://water.usgs.gov/edu/earthgwdeclien.html.

USGS. (2016b). Groundwater Depletion. Accessed from https://water.usgs.gov/edu/gwdepletion.html.

6 Sinkholes

INDUCED SINKHOLES

Groundwater pumping, construction, and development practices produce sinkholes. New sinkholes have been connected to land-use practices (Newton, 1986). There are two types of sinkholes: those, already described, resulting from groundwater pumping and those resulting from construction and development practices (Sinclair, 1982). The problem with construction and development is they tend to modify drainage and divert surface water which can lead to focused infiltration of surface runoff, flooding, and erosion of sinkhole-prone earth materials.

There are two types of sinkholes, natural and induced (caused or accelerated by humans). Humanmade impoundments used to treat or store industrial-process water, wastewater (sewage) effluent, or runoff can also create a significant increase in the load bearing on the supporting geologic materials, causing sinkholes to form. Other construction activities that can induce sinkholes include the erection of structures, well drilling, dewatering foundations, and mining. The sudden development of both types results from the collapse of the roof of a cavity or cavern in rock, or from the downward migration of unconsolidated deposits into solutionally enlarged openings in deep bedrock. Note that it is rare when the occurrence of sinkholes results from bedrock roof collapses, in comparison with the occurrence of sinkholes resulting from downward migration of unconsolidated deposits.

DID YOU KNOW?

The difference between a sinkhole and a pothole is that a sinkhole is a closed natural depression in the ground surface caused by removal of material below the ground and either collapse or gradual subsidence of the surface into the resulting void (USGS, 1999).

Structural influence that causes sinkholes is triggered by overburden sediments that cover buried cavities in aquifers which are delicately balanced by groundwater fluid pressure. This is especially the case in sinkhole areas where the lowering of groundwater levels plus increasing the load at land surface, the combination of both work to exasperate the problem.

Note that aggressive pumping induces sinkholes by abruptly changing groundwater levels and disturbing the equilibrium between a buried cavity and the overlying earth materials (Newton, 1986). Quick declines in water levels can cause a loss of fluid-pressure support, bringing more weight to bear on the soils and rocks bridging buried voids. When the stresses on these supporting materials increase, partially filling with the overburdened material.

It is interesting to note that prior to water-level declines, budding sinkholes are in a slightly stable stress equilibrium with the aquifer system. What the presence of water portends is that it provides support and also increases the cohesion of sediments. When the water table is lowered, unconsolidated sediments may dry out and coarser-grained sediments, in particular, may move without difficulty into openings (USGS, 1999).

DID YOU KNOW?

Induced sinkholes are generally cover-collapse type sinkholes and tend to occur abruptly (see Figure 6.1). They have been forming at increasing rates during the past several decades and pose potential hazards in developed and developing areas of west-central Florida. The increasing incidence of induced sinkholes is expected to continue as our demand for groundwater levels and land resources increases (USGS, 2013).

GROUNDWATER PUMPING = SINKHOLES

To this point, groundwater pumping for urban water supply induces new sinkholes. By the early 1930s, groundwater pumping along the west coast of Florida had lowered hydraulic heads—the higher the water level = hydraulic head—in the fresh-water aquifers and caused the upcoming of saline water. Coastal municipalities began to abandon coastal groundwater sources and develop inland sources (USGS, 1999).

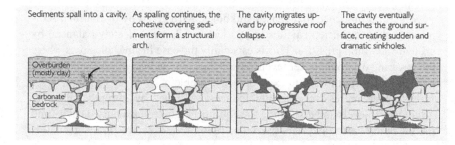

FIGURE 6.1 Cover-collapse type sinkhole. Public Domain. USGS Sinkhole. (2018). Accessed 9/3/23 @ https://www.usgs.gov/special-topics/water-science-school/science/sinkholes.

Case Study 6.1—Tampa Bay Well Fields

United States Geologic Survey (1999) reported that the City of St Petersburg pumps groundwater from well fields in a rural area north of Tampa. By 1978, four well fields had been established in parts of surrounding counties and were pumping an average of almost 70,000 acre-feet per year. The result: sinkholes occurred in conjunction with the development of each of the well fields (Sinclair, 1982).

The area north of Tampa in the municipal well fields is densely pitted with natural sinkholes and sinkhole lakes that have resulted from collapses of surficial sand and clay into solution cavities in the underlying carbonate rocks of the Floridan Aquifer. The Floridan Aquifer system is a principal aquifer of the United States and is one of the most productive aquifers in the world. It covers approximately 10,000 square miles of the southeastern United States (see Figure 6.2) including all of Florida and parts of Georgia, Alabama, Mississippi, and South Carolina.

While it is true that the solution of the underlying carbonate rocks is the basic cause of sinkholes, some have been produced by sudden changes in groundwater levels caused by pumping. Declines in water levels cause loss of support to the bedrock roofs over cavities and surficial material overlying openings in the bedrock. Collapse is promoted by alternate swelling-shrinking and support-withdrawal caused by seasonal fluctuations in water level which tends to disrupt the cohesiveness of unconsolidated surficial material.

In 1978, the constituents of limestone and dolomite (calcium, magnesium, and carbonate) in solution in the water withdrawn from four well fields near Tampa totaled about 240,000 feet^3. It is the surface, however, where the area of induced solution takes place at the limestone surface, and the area of induced recharge is so extensive that the effect of induced limestone solution on sinkhole development is negligible.

The discovery of an alinement of established sinkholes along joint patterns in the bedrock suggests that wells along these lineations (i.e., outlines, contours) might have a direct hydraulic connection with a zone of incipient (emerging) sinkholes. So, what this means is that activities involved with pumping of large amounts of well water are likely to increase the probability of sinkhole development.

Note that although sinkholes generally form abruptly in this particular region, there are indicators that sinkholes may be developing. For example, local changes such as vegetative stress—the lowered water table may result in visible stress to a small area of vegetation; ponding of rainfall may serve as a first indication of actual land subsidence; misalignment of structures, that is, cracks along mortar join in walls and pavements; slumping or sagging, that is, canting of fence posts or other objects from the vertical and doors and windows that fail to open or close properly; and turbidity in well water are all indications that are likely the precursors of subsidence taking place (Sinclair, 1982).

Aggressive pumping, groundwater depletion, and sinkhole development are nowhere as apparent and seemingly common in the Tampa region in what is known as the section 21 well field. The records show that within 1 month of increasing the

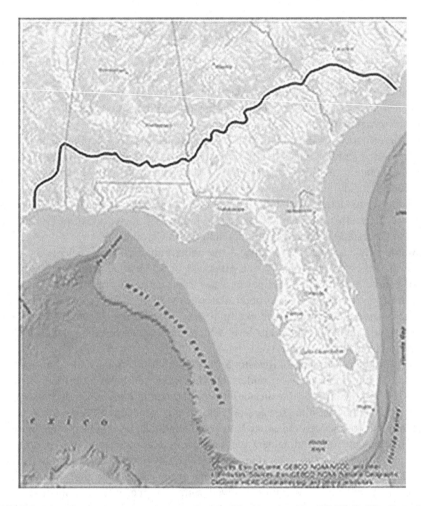

FIGURE 6.2 Shows the area encompassing the Floridan Aquifer. USGS. Public Domain. *Source:* USGS and Sinclair (1982).

pumping rate, 64 new sinkholes formed within a 1-mile radius of the well field. The record also shows that most of the sinkholes were formed in the vicinity of well 21–10, which was pumping at nearly twice the rate of the other wells. Neighboring areas also noticed dramatic declines in lake levels and dewatering of wetland areas.

Research directed at studying the effects of groundwater pumping on lake levels and wetlands is ongoing in section 21 well field, while the field is still in operation.

WINTER FREEZING AND SINKHOLE PRODUCTION

Aurit et al. (2013) point out that Florida is riddled with sinkholes due to its karst topography. It has been implied that agricultural practices, such as sprinkler irrigation methods used to protect crops, can increase the development of sinkholes, for the most part when temperatures drop below freezing, causing groundwater levels to drop abruptly during groundwater pumping. In Dover/Plant City area of Hillsborough County, east of Tampa, Florida, in the strawberry growing region 5800 acres are harvested annually (USDA, 2007), the effects have caused water shortages resulting in dry wells and ground subsidence through the development of sinkholes that can be costly to maintain and repair.

The mild winters are an important growing season for west-central Florida strawberry and nursery. However, occasional freezing temperatures can result not only in property damage (see Figure 6.3) but also can result in sizeable crop losses. To prevent freeze damage, growers pump warm (about 73°F) groundwater from the upper Floridan aquifer and spray it on crops to form an insulating coat of ice. When extended freezes occur intense and prolonged groundwater pumping is required, causing large drawdowns in the upper Floridan aquifer and the rapid appearance

FIGURE 6.3 More than 110 sinkholes formed in the Dover area of Florida during a freeze event in January 2010. Groundwater levels dropped to record-setting levels low as farmers pumped water to irrigate their plants for protection from cold temperatures. *Source:* USGS photo by Ann Tihansky. Freeze Event. Accessed 0/6/23 @ https://www.usgs.gov/special-optics/water-science-school/science/sinkholes.

of sinkholes. The truth be told the relation between freezing weather, prolonged groundwater withdrawals, and sinkhole manifestation has been well documented in the Dover area about 10 miles east of Tampa (Bengtson, 1987).

In January 1977, extended freezes and associated groundwater withdrawals led to the sudden formation of 22 new sinkholes (Metcalf and Hall, 1984). During a 6-day period of record-breaking cold weather, groundwater was pumped at night when temperatures fell below 39°F. The new sinkholes were attributed to the movement of sandy overburden material through a breached clay confining unit into cavities in the limestone below.

Note that one of the problems with the relation between pumping wells and the distribution of induced sinkholes is the nonuniform hydraulic connection between the well and various buried cavities. Water flows like electricity; they both chose the path of least resistance. And this is certainly the case when water seeks to move from place to place, so to speak, in ground and in wells. Dissolution cavities such as carbonate rocks such as limestone, composed mostly of the mineral calcite ($CaCO_3$) are very susceptible to dissolution by groundwater during the process of chemical weathering. Such dissolution can result in sinkholes.

This dissolution process is generated whenever water in the atmosphere dissolves small amounts of carbon dioxide (CO_2). This results in precipitation having a small amount of carbonic acid (H_2CO_3) when it falls on the Earth's surface. During the infiltration of the water into the groundwater system, it encounters carbonate rocks like limestone where it begins to dissolve the calcite in the limestone as calcite reacts with carbonic acid to produce dissolved Calcium ion plus dissolved Bicarbonate ion as shown by the following chemical reaction:

$$CaCO_3 + H_2CO_3 = Ca^{+2} + 2HCO_3^{-2}$$

This reaction takes place as the water moves along structural weaknesses in the limestone, such as bedding planes, joints, or factures—places where water can more easily infiltrate the rock—the path of least resistance. The result of this is a dissolution of the limestone if the reaction continues to take place over a long period of time. The distribution of cavities can be controlled by the presence of these features and thus may be preferentially oriented (USGS, 1999). It is interesting to note that it is common for a pumping well to have more impact on distant cavities that are well-connected hydraulically—even though further away from the pumping well—than on nearby cavities that are not as well-connected hydraulically. The point is that proximity to pumping wells is not always a reliable indicator for predicting induced sinkholes.

One of the important engineering practices learned over time has been to construct buildings and highways in such a way whereby surface-water drainage is diverted away from the foundations to avoid their structural integrity. Related activities may include grading slopes and removal or addition of vegetative cover, installing foundation piles and drainage systems, and ditching for storm drainages and conduits for service utilities. During construction activities and the accompanying altering of landscapes, local changes to established pathways of surface-water runoff, infiltration, and groundwater recharge typically occur. Storm-drainage systems, pavements, roofs, and cleared lands can drastically increase the rate of groundwater recharge to a local area, thus increasing flow velocity in the bedrock and potentially inducing sinkholes. An established cause of induced sinkholes in urban areas is broken water or sewer pipes. Experience has also demonstrated that when pipelines are strung through karst terrane they are subject to uneven settling as soils compact or are piped into dissolution cavities. The result can be cracked water pipes or the separation of raw sewer line sections, further intensifying erosion and carrying out the process.

Experience has also shown that loading by heavy equipment during construction or, later, by the weight of the structures themselves may induce sinkholes. Thus it is necessary to check out the construction site to ensure that hazards to building foundations are not present including (Sowers, 1986):

- pinnacles
- slots
- chimney-like openings in rock
- alternate strata of hard rock
- weak and compressible soils in cone-shaped depressions over claps domes and cavities
- open sinkholes from soil or rock collapse and soil erosion.

A number of engineering methods are commonly used to ward off this type of sinkhole damage (Sowers, 1986), including drilling and driving pilings into competent limestone for support, injecting cement into subsurface, cavities, and construction of reinforced and spread foundations that can span cavities and support the weight of the construction. Compaction by hammering, vibratory rollers, and heavy clock drops may be used to induce collapse so that areas of weakness can be reinforced prior to construction (USGS, 1999).

INDUCING SINKHOLES BY SURFACE LOADING

In this book, the term "construction" applies not only to the erection of a structure but to any modification of the land surface and the diversion and impoundment of surface water. Also, any diversion of drainage includes any activity that results in changes in the downward percolation or movement of recharge to the water table. Note that these activities include removal of timber and drilling, coring, and auguring where pumpage is not involved.

Having stated the preamble for this section with basic definitions it is important to point out that in April 1988, several cover-collapse sinkholes developed in an area when effluent from a wastewater treatment plant was sprayed for irrigation in northwestern Pinellas County. The likely cause was an increased load on the sediments at the land surface due to waste-disposal activities, including construction practices. In the interior terrane, grading removes the upper tough or cohesive weathered residual soils and exposes the underlying less cohesive clay. Loading with addition on thinned roofs of existing cavities in residual clay or on roofs of shallow bedrock openings can cause collapse during or after construction. The weight of a structure on unconsolidated deposits overlying the irregular surface of the top of bedrock gives rise to differential compaction which results in subsidence and foundation problems. Roof failure due to rainfall and saturation of roofs of cavities in residual clay after grading can also occur.

Cavity roofs can also fail when shocks or vibrations resulting from blasting or other causes can result in the failure of cavity roofs in bedrock and unconsolidated deposits. Experience has shown that sinkholes resulting from grading and loading are rare in coastal terrane in comparison to the interior terrane.

Case Study 6.2—Phosphate Mining and Sinkholes

In west-central Florida, the sands and clays of the overburdened sediments support a large phosphate mining and processing industry. A gaping sinkhole formed abruptly on June 27, 1994, within a 400-acre, 220-foot-high gypsum stack at a phosphate mine. The gypsum stack is a flat-topped pile of accumulated phosphogypsum— a byproduct of phosphate-ore chemical processing. Acidic mineralized water with a pH in the 1.5 range causes phosphogypsum to precipitate in the processing of the ore and is circulated and evaporated from the top of the continually growing stack of waste gypsum. The waste slurry of slightly radioactive phosphogypsum stems from the manufacture of phosphoric acid, a main ingredient in many forms of fertilizer (USGS, 1999).

It is likely that the sinkhole formed from the collapse of a preexisting dissolution cavity that had developed in limestone deposits beneath the stack. The acidic waste slurry may have accelerated its development. Note that infiltration of the applied waste slurry into the underlying earth materials was unrestrained because there was no natural or engineered physical barrier directly beneath the stack. Dissolution and erosion worked to enlarge the cavities and the increasing weight of the stack would have facilitated the sinkhole collapse. This effect probably was exacerbated by the reduction of fluid-pressure support for the overburden weight due to localized groundwater-level declines; the phosphate industry withdraws groundwater from the upper Floridan aquifer to supply water to the ore-refining plant (USGS, 1999).

Case Study 6.3—The Forest Is Not Immune

In an upland forest region that straddles parts of Pasco and Hernando counties hundreds of sinkholes ranging in diameter from 1 foot to more than 150 feet formed within a 6-hour period on February 25, 1998, during the development of a newly drilled irrigation well; the well was flushed in order to obtain maximum production efficiency. A 2-foot-thick sediment cover composed primarily of sand with little clay is underlain by caverned limestone bedrock. The well was drilled through 140 feet of limestone, and a cavity was reported in the interval from 148 to 160 feet depth, where drilling ceased. Very shortly after development began, two small sinkholes developed near the drill rig. As well progressed, additional new sinkholes of varying sizes began to appear throughout the area. This uprooted and toppled trees as sediment collapsed and slumping took place, and concentric extensional cracks and crevices formed throughout the landscape. Along the margins of the larger sinkholes, the unconsolidated sandy material slumped and caved as the sinkholes continued to expand. The largest of the sinkholes expanded during a 6-hour development period. More than 20 acres of forest and numerous tall trees were swallowed up and left the well standing on a small bride of land.

Test borings and hydrogeologic data indicated susceptibility to sinkholes. The affected land contains several ancient ponds formed by sinkholes long ago. Stability was tested along the margins of these ponds to determine if the site had a higher-than-normal risk of sinkhole development. Several test borings were made to measure the structural integrity of the bedrock, uncovering a highly irregular limestone surface. Two of the borings, approximately 100 feet apart, were made within a few hundred feet of the well site. One boring indicated that there was firm limestone at depth, whereas the other never came upon a firm foundation.

Irregularity in the limestone surface is typical of much of west-central Florida. During drilling in this area, it is common for cavities, sudden bit drops, and lost circulation to be frequently reported. These drilling characteristics reveal the presence of significant cavernous porosity in the underlying limestone and, while commonly noted in drilling logs, only occasionally cause trouble during well construction.

THE BOTTOM LINE

Although sinkholes have very localized structural impacts, they may have far-reaching effects on groundwater resources. Recognition of specific conditions that can affect the type and frequency including a general lowering of groundwater levels, reduced runoff, increased recharge, or significant surface loading is the first step in minimizing the impact of sinkholes.

REFERENCES

Aurit, M.D., Peterson, R.O., and Blanford, J.I. (2013). A GIS analysis of the relationship between sinkholes, dry-well complaints and groundwater pumping for frost-freeze protection of winter strawberry production in Florida. *PLoS ONE*, v. 8, no. 1, pp. 32–34. doi:10.1371/journal.pone.0053832.

Bengtson, T.O. (1987). The hydrologic effects from intense groundwater pumpage in east-central Hillsborough County, Florida: Proceedings of a Conference Sponsored by the Florida Sinkhole Research Institute, February 9–11, 1987, College of Engineering, University of Central Florida, Orland, [Beck, B.F. and Wilson, W.L., eds., Karst hydrogeology engineering and environmental applications: Boston, MA, A.A. Balkema], pp. 109–114.

Metcalf, S.L. and Hall, L.E. (1984). Sinkhole collapse due to groundwater pumpage for freeze protection irrigation near Dover, Florida, January 1977: Multidisciplinary Conference on Sinkholes, 1st, Orlando, Florida, October 15–17, [Proceedings, Beck, B.F., ed., Sinkholes-Their geology, engineering and environmental impact: Boston, MA, A.A. Balkema], pp. 29–33.

Newton, J.G. (1986). Development of sinkholes resulting from man's activities in the eastern United States: U.S. Geological Survey Circular 968, 54 p.

Sinclair, W.C. (1982). Sinkhole development resulting from groundwater development in the Tampa Aare, Florida: U.S. Geological Survey Water-Resources Investigations Report 81-50, 19 p.

Sowers, T.G. (1986), *Building on Sinkholes Design and Construction of foundation in Karst Terrain*. NY: ASCE Press.

USDA. (2007). *Census of Agriculture-County Data*. Washington, DC: USDA, National Agricultural Statistics Service.

USGS. (1999). *Land Subsidence in the United* States. Washington, DC: U.S. Department of the Inter, U.S. Geological Survey.

USGS. (2013). Circular 1182-Land Subsidence in the United States. Accessed 8/29/23 @ https://pubs.usgs.gov/publication/cir1182.

7 Organic Soil Subsidence

DRAINAGE OF ORGANIC SOILS

The foundation was set for this chapter on organic soils subsidence from the soil(s) information that is provided in Chapters 2 and 3 of this book. Organic soils subsidence occurs principally with drainage and development of peat for agriculture. Subsidence occurs either from loss of buoyancy, shrinkage, and compaction (densification), or from actual shortfall of mass (biological oxidation, burning, hydrolysis and leaching, erosion, and mining). The loss of buoyancy, shrinkage, and compaction usually occurs soon after drainage is established. Biological oxidation causes a slow, continuous loss of mass. Erosion is minor except in specific sites. The removal of materials by mining varies greatly.

In the United States system of soil taxonomy, organic soils or histosols are 1 of 11 soil orders (Chapter 2, Table 2.6). They are defined as having more than 50% organic matter in the upper 30 inches but may be of less thickness if they overlie fragmental rock permeated by organic remains. Organic soil and accumulations by other names are commonly termed "peat," "bog," "fen," "moor", and "muskeg" if fibrous plant remains are still visible or "muck" where plant remains are more fully decomposed.

It is in wetlands where organically formed soils are found. These are locations where roots, stems, and leaves (plant litter) accumulate faster than they can decompose. The remains of sedges and reeds in shallow water typically make up the fibrous peats. Swamp forests are where "woody" peats and mucks form; these commonly consist of granular, well-disintegrated residues of woody plants mixed with some wood fragments. In northerly latitudes with cool, moist climates, many peats are composed mainly of sphagnum moss and associated species. The total area of organic soils in the United States is about 80,000 square miles, about half of which is "moss peat" located in Alaska (Lucas, 1982). About 70% of the organic soil area in the contiguous 48 States occurs in northerly, formerly glaciated areas, where moss peats are also common (Stephens et al., 1984).

DID YOU KNOW?

Increasing fire frequency in Alaska is likely to favor feather moss proliferation and decrease sphagnum abundance, which will reduce soil moisture retention and decrease peat accumulation, likely leading to deeper burning during wildfire and accelerated permafrost thaw (Turetsky et al., 2010).

Land subsidence regularly occurs when organic soil is drained for agriculture or other purposes—shrinkage is inevitable with drainage. As the water table is lowered,

DOI: 10.1201/9781003461265-8

soil pores allow more air into the pores. Organic soil is oxidized by the action of aerobic bacteria converting such matter to carbon dioxide which escapes into the atmosphere and water. Drainage of water causes the weight of upper soil layers to compact lower layers. The operation of farming equipment in preparing and compacting seedbeds consolidates surface layers by pulverizing the soil and eliminating larger soil voids (USGS, 2013). So, we can sum up the causes of land subsidence in drained organic soil as follows:

- Compaction
- Desiccation
- Erosion by wind or water
- Prescribed or accidental burning.

Because organic soils have extremely low density and high porosity or saturated water content (in the neighborhood of about 85%), the effects of compaction and desiccation after initial draining can be dramatic.

The Disappearing Act

The most important cause of organic soil subsidence is the oxidation process. Note that the balance between accumulation and decomposition of organic material shifts dramatically when peat wetlands are drained. Under undrained conditions, anaerobic microbial decomposition of pant litter—that is, decomposition occurs in the absence of oxygen—can't keep pace with the rate of accumulation. Lignin, an important cell wall component of all vascular plants, is the reason because it is much more vulnerable to decomposition under aerobic conditions. The organic carbon is converted to carbon dioxide gas and water during oxidation. It is about efficiency. Aerobic decomposition under drained conditions is very efficient.

In 1930, laboratory experiments with Florida peat that balanced the loss of dry soil weight with rates of carbon dioxide production indicated that the biochemical origin of much organic soil was proven (Waksman and Stevens, 1929; Waksman and Purvis, 1932). This early laboratory work also suggested the optimal temperature ranges and moisture contents for microbial decomposition. As time progressed and field studies and observations continued, it was found and confirmed that "oxidation" is the dominant subsidence process in many instances.

Note that whereas natural rates of accumulation of organic soil are on the order of a few inches per 100 years, the loss of drained organic soil can be 100 times greater, up to a few inches per year in extreme cases. Thus, deposits that have accumulated over many millennia can disappear—the disappearing act—over time scales that are very relevant to human activity (USGS, 2013).

Case Study 7.1—Disappearing Act in Florida Everglades

Note that in the Florida Everglades, sod fields and residential areas—where causal mechanisms, such as erosion, burning, and compaction, are minimized or absent— have sunk as rapidly as the cultivated land (Stephens et al., 1984). Wosten et al.

(1997) point out that many experts feel that oxidation-related soil loss can be halted only by complete restoration of the soil or complete consumption of its organic carbon content.

So, let's take a closer look at the Florida Everglades. And in taking a closer look at the Everglades and subsidence, it is best to begin with a dated but classic description of the Everglades provided by Samuel Sanford in Matson and Sandford (1913):

> It is difficult for one who has not seen the Everglades to form even an approximate idea of that far extending expanse of sedge, with its stretches and its many islands. Photographs fail to convey the impressions of distance, or remoteness, and of virgin wildness which strikes the visitor who for the first time looks out across that vast expanse.

In 2023, only about 50% of the unique and treasured original Everglades (wetlands) remain. This is because since around 1900, much of the Everglades has been drained for agriculture and urban development. The Everglades ecosystem is important; it includes Lake Okeechobee and its tributary areas, as well as the roughly 40- to 50-mile-wide, 130-mile-long wetland mosaic that once extended continuously from Lake Okeechobee to the southern tip of the Florida peninsula at Florida Bay—from a subtropical wetland region that once encompassed three million acres—from Orlando to the Florida Keys, it is now halved and only a memory; it's only a memory now because via Mother Nature and the actions of humans, there is no such thing as a static environment on Earth. The Everglades has been described as a long-pointed spoon (Douglas, 1947).

Today, water levels and water flows are largely controlled by an extensive system of man-made levees and canals. The purpose of the control system was to achieve multiple objectives of flood control, land drainage, and water supply. The planning for all these effects of human control was directed at preserving the essential lives of fish and wildlife, but the system also provides huge benefits to people, as it:

• Protects communities from floods and hurricanes.
• Provides drinking water for millions of Floridians.
• Supports the fishing industry.
• Supports tourism.
• Supports a large and rich environment for plants and animals (see Figure 7.1).

As the old saying goes, "the times they are a changing", and now the focus is on ecosystem restoration. Sounds good, and the newly found movement of ecosystem restoration in the Everglades is complicated and threatens the future of agriculture in the Everglades. The Monkey Wrench in the movement to restore, to a degree, the Everglades to what it had been is complicated by extensive land subsidence caused by drainage and oxidation of soils. The truth be told the Everglades ecosystem has been badly degraded. This is the case despite establishment of Everglades National Park in the southern Everglades in 1947. The symptoms of the ecosystem decline are an 80% reduction in wading bird populations since the 1930s (Ogden, 1994);

FIGURE 7.1 Great Egret, *Ardea alba*. Public Domain Photo by R. Commauf U.S. National Park. Service. Source: NPS accessed 9/10/23 @ https://www.nps.gov/ever/learn/newsevergladeswildifeimages.

the near-extinction of the Florida panther (Smith and Bass, 1994); invasion of exotic species (Bodle et al., 1994), and declining water quality in Florida Bay, which is likely due to but not entirely responsible for the actions but has played a role in it, to decreased freshwater inflow (McIvor et al., 1994).

Note that a thin rim of bedrock in South Florida protects the region from the ocean. Keep in mind we are talking about a thin rim a few inches above sea level in some places; moreover, the limestone bedrock ridge that separates the Everglades from the Atlantic coast extends 20 feet or less above sea level. What this means is that under natural conditions, all of southeast Florida, except for a 5- to 15-mile-wide strip along this bedrock ridge, is subject to annual floods. Much of the area is perennially inundated with freshwater. Water levels in Lake Okeechobee and local rainfall drive slow-moving sheets flowing through the Everglades under the topographic and hydraulic gradient of only 2 inches/mile (USGS, 2013). Note that Lake Okeechobee, which once overflowed its southern bank at water levels in the range of 20–21 feet above sea level, today is artificially maintained at about 13–16 feet above sea level by a dike system and canals to the Atlantic and Gulf coasts.

Not all the problems with the Everglades region are brought about by flooding—the drying-out process is also an issue. It was around 1913 when the first successful farming ventures were begun on a slightly elevated natural levee south of Lake Okeechobee (Snyder and Davidson, 1994). Earlier efforts to farm the sawgrass area had little success, being plagued by flooding, winter freezes, and trace-nutrient deficiencies. Later the early farmers found out that the soil beneath the sawgrass was too low in vital nutrients, such as copper. Today a network of dikes and canals controls water movement, providing optimum irrigation and drainage for sugar cane.

As the area around or within the Everglades was developed, land subsidence accompanied the development activities. At the present time, in the Everglades agricultural area, evidence of substantial land subsidence can readily be detected from the relative elevations of the land surface, the drainage-canal system, and the lake, and from the elevation of older buildings that were built on piles extending to bedrock. Note that it is only at certain locations within the Everglades is it possible to obtain precise measurements. Looking at the long term, average rate of subsidence is generally considered to have been between 1 and 1.2 inches/year (Stephens and Johnson, 1951; Shih et al., 1979; Stephens et al., 1984). Trends of average subsidence in the agricultural area from 1968 to 2009 were 0.55 inches/year. From 2010 to 2019, the average subsidence rate was 0.25 inches/year (Bhadha et al., 2021).

DID YOU KNOW?

Conventional surveying has always been extremely difficult in the Everglades. Stable bedrock benchmarks are nonexistent or very distant; moreover, the surficial material is soft and yielding, and access is difficult.

Synder and Davidson (1994) pointed out that the Everglades agricultural area is now mainly devoted to sugarcane, with considerably smaller areas used for vegetables, sodgrass, and rice. The thinking, theorizing about the eventual demise of agriculture predates many of the present observations of the effect of Everglade subsidence on agriculture (Douglas, 1947; Stephens and Johnson, 1951). Note that agriculture depends upon a relatively thin, continually shrinking layer of peat soil that directly overlies limestone bedrock. These histosols, which form in undrained or poorly drained areas, will subside when drained and cultivated. The causes include mechanical compaction; burning; shrinkage due to dehydration; and most importantly, oxidation of organic matter. Oxidation is a microbially mediated process that converts organic carbon in the soil to (mainly) carbon dioxide gas and water (USGS, 2013).

DID YOU KNOW?

Organic soils or histosols are defined by the Food and Drug Administration as having 40 cm (15.7 inches) or more of organic materials with an organic content of 12%–18% or more.

With enough time and sunlight (photosynthesis), vegetation converts carbon dioxide and water into carbohydrates. During periods of moderate drainage and under natural conditions, aerobic microorganisms convert dead sawgrass root (mostly) to peat (see Figure 7.2). A gradual increase in peat thickness is caused when vegetative debris is deposited faster than it can fully decompose. Over time (about 5000 years), in the agricultural area of the Everglades, a delicate balance of 9–12-month flood and 0–3-month slight (0–12 inches) drainage, with sawgrass the dominant species, all led

FIGURE 7.2   Sea of sawgrass in Everglades. Public Domain photo from National Park Service (2015). Accessed 9/14/23 @ https://www.nps.gov/ever/learn/news/evrgladeshabita-timages.htm.

to a peat accretion rate of about 0.03 inches/year. Drainage disrupted this balance so that instead of accretion, there has been subsidence at a rate of about 1 inch/year.

Note that although there has been an ongoing progression of subsidence in the Everglades, in comparison to other subsidence areas, such as the Sacramento-San Joaquin Delta of California (in the Delta, average subsidence rates have been up to 3 inches/year) and the Hampton Roads Lower Chesapeake Bay region, and other major areas where peat-oxidation subsidence is in progress in the United States. The pre-agricultural peat thickness, however, was much greater in the Delta (up to 60 feet) than in the Everglades, where initial thicknesses were less than 12 feet (USGS, 2013). The subsidence rate observed in the Everglades is similar to those observed in the deep peat soils of the English fends during the past 100 years (Lucas, 1982; Stephens et al., 1984).

It is interesting to note that in the Everglades agricultural area, the initial peat thickness tapered southward from approximately 12 feet near Lake Okeechobee to about 5 feet near the southern boundary.

THE BOTTOM LINE

Organic land subsidence makes true restoration of the Everglades agricultural area itself technically impossible. This is the case even in the event that if it were politically and economically feasible.

REFERENCES

Bhadha, J.H., Wright, A.L., and Synder, G.H. (2021). *Everglades Agricultural Area Soil Subsidence and Sustainability*. Gainesville, FL: University of Florida Institute of Food and Agricultural Science.

Bodle, M.J., Ferriter, A.P., and Thayer, D.D. (1994). The biology, distribution, and ecological consequences of Melaleuca quinquenervia in the Everglades. In *The Everglades-The Ecosystem and Its Restoration*. Eds. Davis, S.M. and Ogen, J.C. Delray Beach, FL: St. Lucie Press, pp. 341–355.

Douglas, M.S. (1947). *The Everglades-River of Grass*. St. Simons Island, FL: Mockingbird Press, 308p.

Lucas, R.E. (1982). Organic soils (histosols)-Formation, distribution, physical and chemical properties and management for crop production: Michigan State University Farm Science Research Report 435, 77 p.

Matson, G.C. and Sandford, S. (1913). Geology and groundwater of Florida: U.S. Geological Survey Water Supply Paper 319, 445 p.

McIvor, C.C., Ley, J.A., and Bjork, R.D. (1994). Changes in freshwater inflow from the Everglades to Florida Bay including effects on biota and biotic processes: A review. In *The Everglades-The Ecosystem and Its Restoration*. Eds. Davis, S.M. and Ogden, J.C. Delray Beach, FL: St. Lucie, pp. 117–146.

Ogden, J.C. (1994). A comparison of wading bird nesting colony dynamics (1931-1946 and 1974-1989) as an indication of ecosystem conditions in the southern Everglades. In *Everglades-The Ecosystem and Its Restoration*. Eds. Davis, J.C. Delray Bech, FL: St. Lucie Press, pp. 533–570.

Shih, S.F., Stewart, G.H., and Hilliard, J.E. (1979). Variability in depth to bedrock in Everglades organic soil. Soil and Crop Society of Florida Proceedings, v. 38, pp. 66–71.

Smith, T.S. and Bass, O.L., Jr. (1994). Landscape, white-tailed deer, and the distribution of Florida panthers in the Everglades. In *The Everglades-The Ecosystem and Its Restoration*. Eds. Davis, S.M. and Ogden, J.C. Delray Beach, FL: St. Lucie Press, pp. 693–708.

Stephens, J.C., Allen, I.H., Jr., and Chen, E. (1984). Organic soil subsidence, in Holzer, T. L. ed., Man-induced land subsidence. *Geological Society of America Review in Engineering Geology*, v. 6, pp. 107–123.

Stephens, J.C. and Johnson, L. (1951). Subsidence in organic soils in the upper Everglades region of Florida. *Soil Science Society of Florida Proceedings*, v. XIX, pp. 191–237.

Synder, G.H. and Davidson, J.M. (1994). Everglades agriculture: Past, present, and future. In *The Everglades-The Ecosystem and Its Restoration*. Eds. Davis, S.M. and Ogden, J.C. Delray Beach, FL: St. Lucie Press, pp. 85–115.

Turetsky, M.R., Mack, M.C., Hollingsworth, T.N., and Harden, J.W. (2010). The role of mosses in ecosystem succession and function of Alaska's boreal forest. *Canadian Journal of Forest Research*, v. 40, pp. 1237–1264.

USGS. (2013). Circular 1182-Land Subsidence in the United States. Accessed 8/29/23 @ https://pubs.usgs.gov/publication/cir1182.

Waksman, S.A. and Purvis, E.R. (1932). The influence of moisture upon the rapidity of decomposition of Low Moor peat. *Soil Science*, v. 34, pp. 323–336.

Waksman, S.A. and Stevens, K.R. (1929). Contribution to the chemical composition of peat, part 5. *The Role of Microorganisms in Peat Formation and Decomposition*, v. 28, p. 315–340.

Wosten, J.H.M., Ismail, A.B., and van Wijk, A.L.M. (1997). Peat subsidence and its practical implications: A case study in Malaysia. *Geoderma*, v. 78, pp. 25–36.

Section II

That Sinking Feeling

8 Arrested Subsidence

INTRODUCTION

In Part 1 of this book, the basics and fundamentals of land subsidence were presented to provide an understanding of land subsidence phenomena that is currently occurring in various regions of the United States. Moreover, in this part of this book, seven U.S. locations are discussed along with their particular land subsidence issues. In Part 3 of this book, a discussion of actual ongoing land subsidence being arrested (i.e., it is hoped; it is theorized; the jury is still out, so to speak) in the Hampton Roads Region of Virginia in the Lower Region of the Chesapeake Bay by using a human-engineered water cycle (aka SWIFT).

In the most simple of ways and to this point, it has been stated that land subsidence is a general term for downward movement of the Earth's surface. While it is true that land subsidence has been characterized by some as a cave-in, a mass wasting, a structural settlement, a sinkhole, and soil liquefication and it is true that these may be correctly associated, land subsidence in general is the key term expressed herein. The point has been made that subsidence is caused by both natural processes and human activities. Keep in mind that little to no horizontal movement occurs in land subsidence; instead, downward, vertical movement is what it is all about. And it is this feature that distinguishes land subsidence from slope movement (Jackson, 1997; Allaby, 2013; Fleming and Varnes, 1991).

There are several causes of land subsidence including dissolution of underlying carbonate rock by groundwater; gradual compaction of sediments; withdrawal of fluid lava from beneath a solidified crust of rock; mining; pumping of subsurface fluids, such as groundwater and petroleum; or warping of the Earth's crust by tectonic forces.

In Part 2 of this book, actual examples of land subsidence occurring in various regions of the United States are presented.

SANTA CLARA VALLEY, CALIFORNIA

Ten miles southeast of downtown, San Jose is the densely populated area known as the "Silicon Valley," the birthplace of the worldwide electronics industry (basically known as the Tech Hub of the United States) is part of a geological trough extending 90 miles from San Francisco. San Francisco Bay takes up the northern reach of the trough, the central third by the Santa Clara valley, and the southern third by the San Beito valley. In the first half of the last century, the Santa Clara Valley was intensely farmed, mainly for fruit and vegetables. The Valley region was dubbed as a Garden of Eden or the Valley of the Hearts Delight because of its extensive orchards, dominated by plums, cherries, apricots, and pears. After World War II, rapid population growth was associated with the transition from an agriculturally based economy to an industrial and urban economy. It was this changing of land and water use that was the prime mover of subsidence in the Valley.

According to Tolman and Poland (1940), the Santa Clara Valley was the first area in the United States where land subsidence due to groundwater withdrawal was recognized. More importantly, it was also the first area where organized (organized is the key term here) remedial action was undertaken, and by 1969 subsidence was effectively halted. One practice has not changed, groundwater is still extensively used, but importation of surface water has reduced groundwater pumping and allowed an effective program of groundwater recharge that prevents groundwater levels from approaching the historic lows of the 1960s.

Groundwater pumping supplied orchards and, eventually, cities. It is all about climate' the moderate climate of the Santa Clara Valley has distinct wet and dry periods. During November to April (the wet season), average rainfall ranges from a high of about 40 inches in the low, steep mountain ranges to the southwest to a low of about 14 inches on the valley floor. Here's the point: 14 inches is generally insufficient to support specialty crops. In the early days, farmers diverted water to local surface water; the problem was this was only enough water to irrigate crops in a very limited manner. So, by the 1860s, wells were in common use.

As time advanced to the late 1800s, technology also advanced. Construction of railroads, refrigerator cars, and improved canning techniques gave farmers access to the growing California and eastern markers for spoilable crops. The pumping of groundwater for orchards increased rapidly into the 1900s.

In the later 1880s, most wells in the area between downtown Alviso and San Jose along the Bay northwest and northeast of Alviso were artesian. The advantage is that the water flowed freely without needing to be pumped. The bad news is that a substantial amount of groundwater was wasted and flowed here, there, and everywhere because it was not tapped. It was the natural hydrogeology of the Santa Clara Valley that caused the widespread use of artesian water.

It was in the 1920s when two-thirds of the Santa Clara Valley was irrigated, including about 90% of the orchards, and new wells were being drilled at the rate of 1700 per year (California History Center, 1981). By the late 1920s, about 130,000 acre-feet of groundwater was pumped annually to irrigate crops and support a total population of about 100,000 (USGS, 2013).

DID YOU KNOW?

The term acre-feet is used to describe the volume of water. One acre-foot is the volume of water that will cover an area of 1 acre to a depth of 1 foot. The term is especially useful where large volumes of water are being described. One acre-foot is equivalent to about 325,851.4 gallons.

Well, the problem was nothing new in the history of draining a water source or aquifer; groundwater was being used faster than it could be replenished—the water level dropped, big time, and artesian wells were becoming harder to find. By 1930, the water level in a formerly artesian United States Geological Survey (USGS monitoring well in downtown San Jose had fallen 80 feet below the land surface. One of my former students called this "water depletion on steroids."

An average of approximately 10,000 acre-feet per year of groundwater was used to irrigate crops during the 1920–1960 timeframe (USGS, 2013). Then in the 1940s, the nonagricultural use of groundwater began to increase substantially—by 1960 total groundwater withdrawals approached 200,000 acre-feet per year. USGS (2013) points out that in 1964 the water lever in the USGS monitoring well in downtown San Jose had fallen to a historic low of 235 feet below the land surface.

Massive groundwater withdrawal caused the ground to subside. For example, substantial land subsidence occurred in the northern Santa Clara Valley as a result of the massive groundwater overdrafts. Detectable subsidence of the land surface (>0.1 feet) took place over much of the area. It was the downtown area of San Jose where maximum subsidence occurred, where land-surface elevations decreased from about 98 feet above sea level in 1910 to about 84 feet above sea level in 1995.

Downtown San Jose is not the only region where groundwater levels decreased. Land adjacent to the southern end of San Francisco Bay sank from 2 to 8 feet by 1969, putting 17 square miles of dry land below the high-tide level. To prevent the landward movement of seawater, a flood-control system of dikes now rings the bayward ends of the stream channels. Now the stream channels are maintained above the surrounding land to provide a gradient for flow to the Bay. In the land section that has sunk below the high-tide level, local storm discharge must be captured and pumped over levees in order to prevent widespread flooding (USGS, 2013).

In 1993, it became abundantly clear that Santa Clara County was subsiding. Even more clarity (via science, of course) was shed upon the ongoing subsidence occurring in Santa Clara County when the benchmarks that were established in 1912 were resurveyed and found to have subsided 4 feet. Because of this finding, the U.S. Coast Guard and Geodetic Survey established a network of benchmarks tied to stable rock on the margins of the Valley. The mapping of subsidence continues thanks to the benchmark network.

PROBLEM RECOGNIZED

Recognition of the Santa Clara County subsidence problem brought about the realization that subsidence had to be stopped, fixed, and mitigated. So, during the 1935–1936 timeframe, the Santa Clara Valley Water District built five storage dams on local streams to capture stream flows. This allowed for the controlled releases to recharge groundwater through streambeds. Artificial recharge was enhanced by natural and artificial recharge during the early 1940s by periods of heavy precipitation. These actions briefly arrested subsidence during World War II, but these measures proved inadequate to halt water level declines over the long term, subsidence resumed at an accelerated rate between 1950 and 1965. In 1965, the Santa Clara Valley Water District recognized the beneficial impact of importing surface water which allowed a significant expansion of the program of groundwater recharge. When substantial recovery of groundwater levels resulted, there has been little additional subsidence since about 1969 (USGS, 2013).

The Santa Clara Valley Water District's actions have lowered the subsidence problem in the area.

The bottom line: The total subsidence has been significant and largely permanent—the good news is that future subsidence can be controlled if groundwater levels are maintained safely above their subsidence limits.

SURFACE WATER TO THE RESCUE

Surface water has been imported from northern and eastern California via aqueducts to balance Santa Clara Valley's water-use deficit. The aqueduct inputs are provided by Hetch Hetchy (San Francisco Water Department) a valley, a reservoir, and a water system; the California State Water Project (a multi-purpose water storage and delivery system that extends more than 700 miles); and the San Felipe Project (plays an important role providing water for agriculture, industry, and municipal use in our country). Much of the imported water also feeds into various local distribution systems. However, about one-fourth of the water imported by the Santa Clara Valley Water District (about 40,000 of the 150,000 acre-feet total) is used for groundwater recharge (USGS, 2013).

In order to save time, money, construction, and other issues, the aquifer systems are used for natural storage and conveyance systems. Groundwater levels are maintained well above their historic lows to prevent the recurrence of the land subsidence that plagued the area prior to 1969. This is the procedure even during drought periods.

THE BOTTOM LINE

It is important to note that subsidence issues in many parts of the State of California are complexly interrelated.

REFERENCES

Allaby, M. (2013). *Subsidence. A Dictionary of Geology and Earth Sciences*, 4th ed. Oxford: Oxford University Press.

California History Center. (1981). *Water in the Santa Clara Valley-A History*. Cupertino, CA: De Anza College California History Center Local History Studies, v. 27, 155 p.

Fleming, R.W. and Varnes, D.J. (1991). *Slope Movements. The Heritage of Engineering Geology: The First Hundred Years*. McLean, VA: Geoscience World.

Jackson, J.A., ed. (1997). Subsidence. *Glossary of Geology*, 4th ed. Alexandria, VA: American Geological Institute.

Tolman, C.F. and Poland, J.F. (1940). Groundwater infiltration, and ground-surface recession in Santa Clara Valley, Santa Clara County, California. *Transactions American Geophysical Union*, pp. 23–34.

USGS. (2013). Circular 1182-Land Subsidence in the United States. Accessed 8/29/23 @ https://pubs.usgs.gov/publication/cir1182.

9 Human Alteration of the Earth's Surface

INTRODUCTION

Land subsidence in the San Joaquin Valley was first noted in 1935. Mining groundwater for agriculture has enabled the San Joaquin Valley of California to become one of the world's most productive agricultural regions, while at the same time contributing to one of the single largest alterations of the land surface attributed to humankind. The San Joaquin Valley is the backbone of California's modern and highly technological agricultural industry. California ranks as the largest agricultural-producing state in the nation, producing a significant portion of the total U.S. agricultural value. Cone (1997) points out that the Central Valley of California, which includes the San Joaquin Valley, the Sacramento Valley, and the Sacramento-San Joaquin Delta, produces about 25% of the nation's table food on only 1% of the country's farmland.

In 1975, a comprehensive survey of land subsidence was made, showing that subsidence in excess of 1 foot had affected more than 5200 square miles of irrigable land—one-half the entire San Joaquin Valley (Poland et al., 1975). Note that the maximum subsidence was more than 28 feet near the Mendota area.

Note that land subsidence continues today in San Joaquin Valley, but at a lesser rate than before when reductions in groundwater pumpage and the accompanying recovery of groundwater levels were made possible by supplemental use of surface water for irrigation. The source of surface water comes from diversions from the Sacramento-San Joaquin Delta and the San Joaquin, Kings, Kern, and Feather Rivers. The flow input rate from these sources is occasionally disrupted during periods of drought, and subsidence continues when groundwater pumpage is increased.

Experience in the San Joaquin via observation, study, and practice sums up the land subsidence in the simplest form as:

Agriculture + groundwater pumpage = Land subsidence

Further agricultural development without accompanying subsidence is dependent on the continued availability of surface water, which is subject to ambiguities due to regulatory decisions and the indeterminate toing and froing of climate.

THE GREAT TROUGH

When (while) the tectonic forces caused (cause) the collision of the Pacific and North American Plates, a great sediment-filled trough was (is continuing to be) formed, which includes the San Joaquin Valley of California. The San Joaquin Valley includes the southern two-thirds of the Central Valley of California. The valley is

DOI: 10.1201/9781003461265-11

situated between the Sierra Nevada on the east, the Diablo and Temblor Ranges to the west, and the Tehachapi Mountains to the south. The trough is filled with marine sediments overlain by continental sediments that are, in some places, thousands of feet deep. The sediments were mainly contributed by streams draining the mountain ranges.

With regard to the composition of the sediments, it has been found that over half the thickness of the continental sediments is composed of clay, sandy clay, sandy silt, and silt (basically fine-grained clay), fluvial (stream), and lacustrine (lake) deposits susceptible to compaction (USGS, 2013).

The 10,000 square miles of the arid to semiarid valley floor receive an average of 5–16 inches of rainfall annually. It is from the east side of the valley where most of the streamflow drains from the western Sierra Nevada snowbanks. The San Joaquin River starts off high in the Sierra Nevada and descends onto the valley floor, where it takes a northerly path toward the Sacramento-San Joaquin Delta. Along its northward course to the Delta, it collects streamflow from the central and northern portions of the valley. At the southern valley, streamflow is received from the Kings, Kaweah, and Kern Rivers, which flow from steep canyons onto broad, extensive alluvial fans. Over time, the natural flow of these rivers issued networks of streams and washes on the slopes of the alluvial fans and came to an end in topographically closed sinks, such as Tulare Lake, Kern Lake, and Buena Vista Lake. Note that precipitation and streamflow in the valley vary greatly from year to year. Consider, for example, the streams draining the drier western slopes and Coast Ranges adjacent to the valley that are intermittent (ephemeral), flowing only episodically.

PUMPING THE AQUIFERS DRY

Groundwater occurs in shallow, unconfined, or partially confined aquifers throughout the San Joaquin Valley. A laterally extensive lacustrine clay known as the Corcoran Clay is a significant hydrogeologic feature that varies in thickness from a feather edge to about 160 feet beneath the present bed of Tulare Lake and is distributed throughout the central and western valley. The clay layer divides the aquifer system into two distinct zones: an upper unconfined to semi-confined aquifer and a lower confined aquifer system. Overlying these formations are flood-plain deposits. Note that most of the subsidence measured in the valley has been correlated with the distribution of groundwater pumpage and the reduction of water levels in the deep-confined aquifer system.

Before development, under natural conditions, groundwater in the alluvial sediments was replenished primarily by infiltration via stream channels near the valley fringes. Sierra Nevada streams carry runoff that is the primary source of recharge for valley aquifers. Precipitation falling on the valley floor and some stream and lake seepage add to the additional recharge. After a long time, natural replenishment was dynamically balanced by natural depletion through groundwater discharge, which occurred primarily through evapotranspiration and contributions to streams flowing into the Delta. It is interesting to note that the areas of natural discharge in the valley generally corresponded with areas of flowing, artesian wells mapped in an early USGS investigation. It is thought that direct outflow to the Delta has been negligible.

Today, it is more than 170 years since water was first diverted at Peoples Weir on the Kings River and more than 140 years after the first irrigation colonies were established in the valley. It is important to note that almost all of the Kings River's water is consumed for agriculture. The river irrigates about 1.1 million acres (4500 km²) of very productive farmland; in 2009, the Kings delta produced crops valued at more than $3 billion (Central Valley Water Awareness Committee, 2003). The main crops grown in the Kings River service area are grapes, citrus, grain, and carious fruits and nuts. Other crops include alfalfa, berries, rice, and miscellaneous nursery and field crops (Kings Basin Water Authority, 2018).

The intensive development of groundwater resources for agricultural uses has drastically altered the valley's water budget. Natural recharge of the aquifers has remained about the same, but more water has discharged than recharged the aquifer system; in the late 1960s, the deficit may have amounted to as much as 800,000 acre-feet per year (Williamson et al., 1989). Most of the surface water presently being imported is transpired by the crops of evaporation from the soil. The quantity of surface water outflow from the valley has essentially been reduced compared to pre-development conditions. Note that groundwater in the San Joaquin Valley has generally been depleted and redistributed from the deeper aquifer system to the shallow aquifer system. This creates an additional problem to the quantity (Q) of freshwater availability by adding the quality (Q) issue with freshwater—what this author calls the Q and Q problem. The quality issue has to do with the problems created with groundwater quality and drainage into the shallow aquifer. The problem with Q is that the water infiltrates with passengers: excess irrigation includes water that has been exposed to agricultural chemicals and natural salts.

DID YOU KNOW?

Water budgets provide a means for evaluating the availability and sustainability of a water supply. A water budget simply states that the rate of change in water stored in an area, such as a watershed, is balanced by the rate at which water flows into and out of the area. Keep in mind that water flows into aquifers via infiltration slowly. If you ever visit and hike the Narrows in Zion National Park in Utah (highly recommended), there is a waterspout at the lodge that, when opened, pours water that took an estimated 3000 years to flow from the top of the canyon to the tap. Having personally consumed gallons of this water, I can confirm that it is the tastiest water I have ever consumed.

Example 9.1 Water Budget Model for a Lake

The water budget model for lake evaporation is used to make estimations of lake evaporation in some areas. It depends on an accurate measurement of the inflow and outflow of the lake. It is expressed as

$$\Delta S = P + R + GI - GO - E - T - O \tag{9.1}$$

where
  ΔS = change in lake storage, mm
  P = precipitation, mm
  R = surface runoff or inflow, mm
  GI = groundwater inflow, mm
  GO = groundwater outflow, mm
  E = evaporation, mm
  T = transpiration, mm
  O = surface water release, mm

If a lake has little vegetation and negligible groundwater inflow and outflow, lake evaporation can be estimated by:

$$E = P + R - O \pm \Delta S \qquad (9.2)$$

A STABLE WATER SUPPLY

After the 1849 Gold Rush and again in 1857, irrigated water surged in the San Joaquin Valley. This later surge is attributed to the California Legislature passing an act in 1857 that promoted the drainage and reclamation of river bottom lands (Manning, 1967). By 1900, much of the flow of the Kern River and the entire flow of the Kings River had been diverted through canals and ditches to irrigate lands throughout the southern part of the valley (Nady and Laragueta, 1983). There is more to diverting from a flowing source designed to aid these diversions. However, when constructed, the diversions had no significant storage facilities; thus, without the significant storage facilities accompanying these earliest diversions, the agricultural water supply, and hence crop demand, was largely limited by the summer low flow. Because of the need for constant surface water flows, coupled with a drought in 1880 and by 1910, nearly all the available surface water supply in the San Joaquin Valley had been diverted, which promoted the development of groundwater resources—they had to have water from whatever source they could tap.

It was in the central valley locations near old lake basins where shallow groundwater was plentiful and easily accessible that the first development of the groundwater resource occurred; this was especially the case where flowing wells were commonplace. In due course, the yields of flowing wells shrank as water levels were reduced. When this occurred, it was necessary to install pumps in wells to sustain flow rates. In the 1930s, the development of an improved deep-well turbine pump and rural electrification enabled additional groundwater development for irrigation. It has been said that history repeats itself, and this was certainly the case after the groundwater resource was established as a stable, reliable water supply for irrigation. Many other basins in California and throughout the Southwest followed suit, where surface water was limited and groundwater was readily available.

DELTA-MENDOTA CANAL

Extensive groundwater withdrawal from the unconsolidated sediments in the San Joaquin Valley and aquifer-system compaction resulted in land subsidence from 1926 to 1970—locally exceeding 8.5 m (approximately 28 feet). The importation of surface water beginning in the early 1950s through the Delta-Mendota canal and in the early 1970s through the California Aqueduct resulted in decreased pumping, the initiation of water-level recovery, and a reduced rate of compaction in some areas of the San Joaquin Valley.

Note that after the construction of the Delta-Mendota canal, California land and water officials, along with Bureau of Reclamation personnel, observed and became concerned about the land subsidence problem caused by groundwater withdrawal in the northern San Joaquin Valley.

This was the case because it wasn't long after the 117-mile-long Delta-Mendota canal was completed in 1951 that subsidence was caused by groundwater withdrawal in the northern San Joaquin Valley; the point is, land subsidence throughout the San Joaquin Valley was unrecognized prior to the construction of the canal—it can be said that the canal was the canary in the coal mine. Officials were concerned largely because of the looming threat to the canal and the possibility of remedial repairs. Because of the threat to the canal and in order to help plan other major canals and engineering projects proposed for construction in the subsiding areas, the USGS, in cooperation with the California Department of Water Resources, began a rigorous investigation into land subsidence in the San Joaquin Valley. The objectives were to determine the causes, rates, and extent of land subsidence and to create scientific criteria for the estimation and reining in of subsidence. At the same time that the California Department of Water Resources was studying the subsidence problem, the USGS began a federally funded research project with the goal of determining the physical principles and mechanisms governing the expansion and compaction of aquifer systems resulting from changes in aquifer hydraulic heads. Note that it is these studies that provide much of the information presented here.

In 1955, almost 8 million acre-feet (about one-fourth) of the total groundwater extracted in the U.S. was pumped in the San Joaquin Valley. It was in the deep, confined aquifer system in the western and southern portions of the valley where the maximum changes in water levels occurred. More than 400 feet of water-level decline occurred in some west-side areas in the deep aquifer system. After the canal was constructed, the discrepancies in elevation were first blamed on earthquakes. By 1966, Prokopovic and Hebert (1968) pointed out that about 35 miles of the canal showed a drop of 1 foot, a 3-foot elevation drop in the 15 mile stretch, a 5-foot decrease in the 5-mile portion, and 2 feet of the canal showed a drop of 6 feet. Keep in mind that, until 1968, irrigation water in these areas was supplied almost entirely by groundwater. As of 1960, water levels in the deep aquifer system were declining at a rate of roughly 10 feet/year. This drop today seems significant and a head-scratcher (and it really is) to many, but in the western and southern portions of the valley, more than 100 feet of water decline occurred; however, a large portion of the valley did experience declines of more than 40 feet. Note that in some areas on the northwest side, the water-table aquifer rose up to 40 feet due to infiltration of excess irrigation water (USGS, 2013).

It is now known that accelerated groundwater pumpage and water-level declines, mainly in the deep aquifer system during the 1950s and 1960s, caused about 75% of the total volume of land subsidence in the San Joaquin Valley. It was in the late 1960s when surface water was diverted to agricultural interests from the Sacramento-San

Joaquin Delta and the San Joaquin River through federal reclamation projects and from the Data through the newly completed, massive California State Water Project. A significant amount of less expensive water for crop irrigation came from the Delta-Mendota canal, the Friant-Kern canal, and the California Aqueduct. Groundwater levels began a striking period of recovery, and subsidence slowed or was arrested over a large part of the affected area. Ireland et al. (1984) point out that water levels in the deep aquifer system recovered as much as 200 feet in the 6 years from 1967 to 1974.

As soon as water levels began to recover in the aquifer system, aquifer-system compaction and land subsidence began to decline, although many areas continued to subside, although at a lower rate. Although the rate of compaction declined and, in a few locations, measured water levels increased by as much as 200 feet, subsidence continued to increase. The problem is related to the time delay in compaction of the aquitards in the aquifer system. The time delay is all about pore-fluid pressure in the aquitards lagging behind the pressure changes in the aquifers, which are more responsive to the current volume of groundwater being pumped (or not pumped) from the aquifer system (USGS, 2013). Typically, in the San Joaquin Valley, centuries will be required for most of the balance in pressure to occur and, therefore, for the ultimate compaction to be realized.

Since 1974, land subsidence has been greatly slowed or largely arrested but remains poised to resume. For instance, during the severe droughts in 1976–1977 and 1987–1991, old pumping plants were refurbished, new wells were drilled, and the pumping of groundwater was increased to make up for cutbacks in the imported water supply. The decision to pump groundwater was encouraged because the groundwater levels had recovered nearly to predevelopment levels. Groundwater levels during the 1976–1978 drought rapidly nosedived more than 150 feet in a wide area, and subsidence resumed. In 1977, nearly 0.5 feet of subsidence was measured near Cantua Creek. Again, during the 1987–1991 drought, this same situation was repeated. What this demonstrates is that there is a sensitive dependence between subsidence and the dynamic state of imported water availability and use.

The renewed pumpage of groundwater caused a rapid decline in water levels as a result of compaction, which reduced water storage capacity because of lost pore space. This demonstrates the nonrenewable nature of the resource embodied in "water of compaction"—sometimes referred to as hydrocompaction (see Example 9.2). What all this points out is that extraction of this resource, available only on the first cycle of large-scale drawdown, must be viewed, like more traditional forms of mining, in terms not only of its obvious economic return but also of its less readily identifiable costs.

Example 9.2 Hydrocompaction

Hydrocompaction (aka hydroconsolidation; soil collapse)—compaction due to artificial wetting—is a near-surface phenomenon that produces land-surface subsidence through a mechanism entirely different from the compaction of deep, overpumped aquifer systems. Both of these processes accompanied the expansion of irrigated agriculture onto the arid, gentle slopes of the alluvial fans along the west side and south end of the San Joaquin Valley. At first, the difference between them and their relative contributions to the overall subsidence problem were not fully recognized.

During the 1940s and 1950s, farmers bringing virgin soils under cultivation found that standard techniques of flood irrigation (aka surface irrigation; wild flooding) caused an irregular settling of their carefully graded fields, producing a rolling surface of depressions and hillocks with local relief, typically of 3–5 feet. Keep in mind that 'wild flooding' is basically a catchall, an all-encompassing category for situations where water is simply allowed to flow onto an area without any attempt to regulate the application or its uniformity. Where water flowed or ponded continuously for months, very localized settlements of 10 feet or more might occur on susceptible soils. These consequences of artificial wetting seriously disrupted the distribution of irrigation water and damaged pipelines, power lines, roadways, airfields, and buildings. No instruments of any kind were needed to discern the rapid differential subsidence due to hydrocompaction.

The bottom line: recognition of its obvious impact on the design and construction of the proposed California Aqueduct played a major role in the initiation in 1956 of intensive studies to identify, characterize, and quantify the subsidence processes at work beneath the surface of the San Joaquin Valley.

DID YOU KNOW?

One of the risks of surface irrigation is that scalding could occur. Scalding occurs when thin, sandy, or loamy topsoil has been eroded, exposing sub-surface or subsoil material that is physically and/or chemically hostile to plant establishment and growth.

THE BOTTOM LINE

While subsidence in the San Joaquin Valley is recognized for what it is and many of its impacts are obvious, it is also true that the economic impacts are not that well known. Not only have some of the damages related to land subsidence been identified, but some have also been quantified. Increased awareness and acknowledgment of the long-term environmental effects need more focus and study. Some of the obvious, known, and direct damages have included decreased storage in aquifers, partial or complete submergence of canals and associated bridges and pipe crossings, collapse of well casings, and disruption of collector drains and irrigation ditches. Costs associated with these damages have been conservatively estimated at $25,000,000 (1978 dollars; EDAW-ESA, 1978)—keep in mind that many of the cost projections are underreported, especially in the case of well rehabilitation and replacement.

REFERENCES

Central Vally Water Awareness Committee. (2003). The Kings River Handbook. Accessed 9/22/23 @ https://www.centralvalleywaer.org/_pdf/KingsRiverHandbook-03final.pdf.

Cone, T. (1997). The vanishing valley. *San Jose Mercury News West Magazine*, June 29, pp. 9–15.

EDAW-ESA. (1978). Environmental and economic effects of subsidence: Lawrence Berkely Laboratory Geothermal Subsidence Research Program Final Report-Category IV, Project I, [variously paged].

Ireland, R.I., Poland, J.F., and Riley, E.S. (1984). Land subsidence in the San Joaquin Vally, California as of 1980: U.S. Geological Survey Professional Paper 437-1, 93 p.

Kings Basin Water Authority. (2018). Kings Basin Land Use. Accessed 9/22/23 @ https://www.kingsbasinauthority.org/landuse/resoruces/P2_CWF%201a.

Manning, J.C. (1967). Report on the groundwater hydrology in the southern San Joaquin Valley. *American Water Works Association Journal*, v. 59, pp. 1513–1526.

Nady, P. and Larragueta, L.L. (1983). Development of irrigation in the Central Valley of California: U.S. Geological Survey Hydrologic Investigations Atas HA-649, 2 sheets, scale 1:500,000.

Poland, J.F., Lofgren, B.E., Ireland, R.L., and Pugh, R.G. (1975). Land subsidence in the San Joaquin Vally, California, as of 1972: U.S. Geological Survey Professional Paper 437-h, 78 p.

Prokopovic, N.P. and Hebert, D.J. (1968). Land subsidence along the Delta-Mendota Canal, California. *American Water Works Association*, v. 60, no. 8, pp. 915–920.

USGS. (2013). Circular 1182-Land Subsidence in the United States. Accessed 8/29/23 @ https://pubs.usgs.gov/publication/cir1182.

Williamson, A.K., Prudic, D.E., and Swain, L.A. (1989). Groundwater flow in the Central Valley California: U.S. Geological Survey Professional Paper 1401-D, 127 p.

10 Coastal Subsidence

HOUSTON-GALVESTON, TEXAS

We can say that subsidence and the greater Houston area are synonymous. This is the case because the greater Houston area, possibly more than any other metropolitan area in the United States, has been adversely affected by land subsidence. Not only has groundwater pumping caused this problem but also the mining of oil and gas has contributed to the frequency of flooding; caused extensive damage to industrial and transpiration infrastructure; prompted major investments in levees, reservoirs, and surface-water distribution facilities; and caused considerable loss of wetland habitat.

We also can say that subsidence is insidious like types of cancer where tumor cells hide deep inside the body, betraying no symptoms until after the cancer has spread to other organs. With regard to land subsidence, the mechanics of underground action(s) is not normally/typically (key word is normally or typically) discovered until ground settling, ground cracks, and/or cave-ins are apparent.

Although regional land subsidence is often insidious, very subtle, and difficult to detect, there are localities in and near Houston where the effects are quite evident. In this low-lying coastal environment, as much as 10 feet of subsidence has shifted the position of the coastline and changed the distribution of wetlands and aquatic vegetation. What's more, the San Jacinto Battleground State Historical Park, the site of the battle that won Texas independence, is now partly submerged—about 100 acres of park is now under water due to subsidence. Another problem is ground faulting induced by groundwater pumping and subsidence, resulting in visible fracturing, surface offsets, and associated property damage (USGS, 1999).

Because of growing awareness and concern of the subsidence-related problems in a section of the community and business leaders prompted the 1975 Texas legislature to create the Harris-Galveston Coastal Subsidence District (HGCS), "... for the purpose of ending subsidence which contributes to, or precipitates, flooding, inundation, and overflow of any area within the District ...." This unique District was authorized to issue or refuse well permits, promote water conservation and education, and promote conversion from groundwater to surface-water supplies. It has largely succeeded in its primary objective of arresting subsidence in the coastal plain east of Houston. Note, however, that subsidence accelerated in fast-growing inland areas north and west of Houston, which still rely on groundwater.

THE PATH OF LEAST RESISTANCE

The flow of water is often compared to the flow of electricity; both flow via the path of least resistance. The flat, humid Gulf Coast is prone to flooding; its flat terrain provides the least path of resistance for water flow. The flat terrain includes the

DOI: 10.1201/9781003461265-12

Houston-Galveston Bay area including a large bay-estuary-lagoon system consisting of the Trinity, Galveston, and East and West Bays, which are separated from the Gulf of Mexico by Pelican Island, Galveston Island, and the Bolivar Peninsula. Tidal exchange occurs between the Gulf and bay system through the barrier-island and peninsula complex (USGS, 1999).

The Houston-Galveston Bay climate is subtropical; temperatures range from 45°F to 93°F, and on average, roughly, about 47 inches of rain falls each year. The humid coastal plain slopes gently toward the Gulf at a rate of close to 1 foot per mile. Two major rivers, the Trinity and San Jacinto, and many smaller ones traverse the plain before discharging into estuarine areas of the bay system. The Brazos River is another large river that crosses the Fort Bend Subsidence District and discharges directly into Galveston Bay. The warm waters of the Gulf of Mexico are a double-edged sword for the area because these same warm waters attract tourists and commercial fishermen on one end of the sword and also, on the other edge, tropical storms and hurricanes. McGowen et al. (1977) point out that the Texas coast is subject to a tropical storm or hurricane about once every 2 years—some of these storms have resulted in the Galveston area experiencing tidal flooding of more than 15 feet. Note that it is the flat-lying topography of the Gulf region that makes the region particularly prone to flooding from both riverine and coastal sources, and the rivers, their reservoirs, and an extensive system of bayous and human-made canals are managed as part of an extensive flood-control system.

The flooding in the Houston-Galveston area has another problem, adding salt to the wound, so to speak. This insidious salt-in-the-wound-bearer is land subsidence. The truth be known it is the land subsidence that makes the frequency and severity of flooding even worse. Near the coast, the net result of land subsidence is an apparent increase in sea level or a relative sea-level rise: the net effect of global sea-level rise and regional land subsidence in the coastal zone.

Truth be told sea level is in fact rising due to regional and global processes, both natural and human-induced. Global warming is a cyclical phenomenon that we have no control over, but we exacerbate the current warming cycle via human-induced pollution (Spellman, 2020).

Paine (1993) pointed out that the combined effects of the actual sea-level rise and natural consolidation of the sediments along the Texas Gulf Coast yield a relative sea-level rise from natural causes that locally may exceed 0.08 inches per year. Global warming is contributing to the present-day sea-level rise and is expected to result in a sea-level increase of approximately 4 inches by the year 2050 (Titus and Narayanan, 1995). As a point of interest, consider that since the early 1900s, most of the groundwater withdrawals in the Texas Gulf area have been from three hydrogeologic units—the Chicot, Evangeline, and Jasper aquifers (declines in these aquifers of more than 300 feet) and, more recently, from the Catahoula confining unit. Withdrawals from these hydrogeologic units are used for municipal supply, commercial and industrial use, and irrigation purposes (Ellis et al., 2023).

During the 20th century, however, human-induced subsidence has been by far the dominant cause of relative sea-level rise along the Texas Gulf Coast, exceeding 1 inch per year throughout most of the affected area. This subsidence has resulted principally from extraction of groundwater and, to a lesser extent, oil and gas from subsurface reservoirs.

DID YOU KNOW?

Note that subsidence caused by oil and gas withdrawal from subsurface reservoirs is largely restricted to the field of production, as contrasted to the regional-scale subsidence typically caused by groundwater pumpage.

SUBSURFACE RESERVOIRS

In 1897, the population of Houston was about 25,000; since then, Houston has experienced rapid growth and established the Port of Houston mainly due to subsurface reservoirs of gas and oil. In 1907, the first successful oil well was drilled, marking the beginning of the petrochemical industry that provided the economic base on which Houston area was built and still stands. The patterns of subsidence in the Houston area closely follow the temporal and spatial patterns of subsurface fluid extraction from subsurface reservoirs.

In the oilfields of Goose Creek, prior to the 1940s, there was localized subsidence caused chiefly by the extraction of oil and gas along with attendant brine, groundwater, and sand. Between 1906 and 1943, near Texas City, the withdrawal of groundwater for public supply and industry caused more than 1.6 feet of subsidence. During this time, the beginning of a slow but steady development of groundwater resources constituted the sole water supply for communities and industries around Ship Channel, including Houston. Right around the 1937 timeframe, groundwater levels were falling in a growing set of gradually coalescing cones of depression centered on the areas of heavy use. Until 1942, almost all water demand in Houston was supplied by local groundwater. It was in 1943 that subsidence had begun to affect part of the Houston area in amounts generally less than 1 foot.

DID YOU KNOW?

Most subsidence in the Houston area has been caused by groundwater withdrawal, but the earliest subsidence was caused by oil production.

During the early 1940s through the late 1970s, the drilling for oil and gas in the Houston area caused land subsidence problems due to the need of these same petrochemical companies to mine their oil and gas that drove the need for and increased use of groundwater. By the mid-1970s, 6 or more feet of subsidence had occurred throughout an area along the Ship Channel between Bayport and Houston as a result of declining groundwater levels associated with the rapid industrial expansion. At the same time, subsidence problems took on crisis proportions, prompting the creation of the Harris-Galveston Coastal Subsidence District. Approximately 10 feet of subsidence had occurred in 1979, and almost 3200 square miles had subsided more than 1 foot (USGS, 1999).

During the 1940s, upstream reservoirs and canals allowed the first deliveries of surface water to Galveston, Pasadena, and Texas City, but groundwater remained the primary source until the 1970s. In 1973, Lake Houston and Lake Livingston (a reservoir on the Trinity River) supplied surface water to the cities of Pasadena and Texas City.

Note that since the late 1970s, subsidence has largely been arrested along the Ship Channel and in the Baytown-LaPorte and Pasadena areas due to a reduction in groundwater pumpage made possible by the conversion from groundwater to surface-water supplies. It is also interesting to note that by 1995, total annual groundwater pumpage in the Houston area had declined to only 60% of peak amounts pumped during the later 1980s; annual groundwater pumpage in the Houston area had declined to only 60% of peak amounts pumped during the late 1960s. It is important to point out, however, that as subsidence in the coastal area was stabilizing, inland subsidence—north and west of Houston—was accelerating. It was accelerating to the point where groundwater levels declined more than 100 feet in the Evangeline aquifer between 1977 and 1997. And more than 2.5 feet of subsidence was measured in the Addicks area between 1973 and 1996.

THE FLOOD THREAT

Land subsidence by any source is a serious concern for many who live in the Houston area and for those in the entire Gulf Coastal region—for good reason, flooding. The point is subsidence increases the frequency and intensity of flooding. This is especially the case along the low-lying Gulf Coast that is subject to tropical storms—the Houston area is naturally subject to flooding. In coastal areas, subsidence has increased the amount of land subject to the threat of tidal inundation. Inhabitants of islands and coastal communities are endangered by flooding produced by tidal surges, and heavy rains accompanying hurricanes may block emergency evacuation routes—no route of escape.

Case Study 10.1—The Brownwood Memory

Hurricane Alicia in 1983 was like a giant sweeping eraser of the Brownwood subdivision of Baytown—a particularly dramatic example of the dangers of coastal subsidence. Brownwood was constructed in 1938 as an upper-income subdivision on wooded lots along Galveston Bay (Holzschuh, 1991). The problem was that at that time, the area was generally 10 feet or less above sea level. By 1978, more than 8 feet of subsidence had occurred.

An interesting article by Gabrysch in the *Impact of Science on Society* (1983) pretty much sums up what today is only a memory.

> The subdivision is on a small peninsula bordered by three bays. [It] is a community of about 500 single-unit family houses. Because of subsidence, a perimeter road was elevated in 1974 to allow ingress and egress during periods of normal high tide [about 16 inches], and to provide some protection during unusual high tide. Pumps were installed

to remove excess rainfall from inside the leveed area. Because of subsidence after the roadway was elevated, tides of about [4 feet] will cause flow over the road. [In 1979], 12 inches of rain fell on Brownwood causing the flooding of 187 homes.

(Gabrysch, 1983)

The year after this article was published, Hurricane Alicia struck a final blow to Brownwood. The area had sunk approximately 10 feet. As a result, the hurricane completely destroyed it. The site was abandoned and the new swampy area became an area well suited for waterfowl.

Engineered and natural drainageways are altered by subsidence, increasing their risk of flooding and erosion by altering the open channels and pipelines (drainageways). That depend on the gravity-driven flow of stormwater and wastewater (sewerage). Differential subsidence (i.e., difference in subsidence between two points on a structure) depending on where it occurs with respect to the location of drainageways may either reduce or enhance preexisting gradients. The rate of drainage is affected by gradient reductions which increase the rate of drainage and thereby increase the chance of flooding by stormwater runoff. Ponding and backflow of wastewater and stormwater can be the result of gradient reversals. Note that in some areas, the drainage gradients may be enhanced, and the rate of drainage may be increased. What is great locally does not mean it is great downstream. This can be seen in terms of the flooding risk that may have a beneficial effect locally but an adverse effect downstream. In the case of open channels, the changing gradients alter streamflow characteristics leading to potentially damaging consequences for channel erosion and sediment deposition.

Well, we all know there is nothing pleasant or enlightening about unwanted flooding caused by land subsidence, but there are other concerns related to subsidence and flooding—loss of wetlands, for example. White et al. (1995) pointed out that 61% of Galveston Bay, one of the most significant bay ecosystems in the Nation, shoreline is composed of highly productive fringing wetlands, but mainly due to subsidence, more than 26,000 acres of emergent wetlands have been converted to open water and barren flats. There's more, subsidence has also contributed to a significant loss of submerged aquatic vegetation (mostly seagrass) since the 1950s. The loss of fringing wetlands has caused some bay shorelines to be more susceptible to erosion by wave action. At the same time, the reduction of sediment inflows to the bay system resulting from construction of reservoirs along tributary rivers slows the natural rebuilding of shoreline. With the loss of wetlands, the interrelated effects of relative sea-level rise, and reduced sediment supply, the shoreline is eroding at an average rate of 3.4 feet/year (Paine and Morton, 1986). The marsh along the shoreline is drowned as the sea-level rises—a reoccurring problem in many locations in the United States. The potential for the landward migration of marshes is eliminated when residential, commercial, or industrial development is located near the shoreline. The result is a reduction in wetland habitats, which provides the foundation for commercial and recreational fisheries (USGS, 1999).

It is in the lower reaches of the San Jacinto River near is to its confluence with Buffalo Bayou where the most extensive changes in wetland have occurred. This area had subsided by 3 feet or more by 1976, resulting in submergence and changes in wetland environments that progressed inland along the axis of the stream valley. Riverine woodlands and swamps are displaced by open water. Trends along the lower reaches of other rivers, bayous, and creeks have been similar, resulting in an increase in the extent of open water; loss of inland marshes and woodlands; and, in some areas, the development of new marshes inland from the encroaching waters.

The magic ingredients in a healthy and productive bay and salt marsh ecosystem depend on a correct mix of marsh, river, and bay water. Several species of fish, wildlife, aquatic plants, and shellfish in Galveston Bay depend on adequate freshwater inflows for survival. The estuary is used to highly variable inflows of freshwater. For instance, oysters prefer somewhat salty water, but need occasional surges of freshwater. The volume, timing, and quality of freshwater inflows to the estuary are key factors (USGS, 1999).

It is important to point out that the increasing demand for surface-water supplies, driven in past years by efforts to mitigate land subsidence, has led to construction of reservoirs and diversions that have reduced the sediments and nutrients transported to the bay system (Galveston Bay Nation Estuary Program, 1995). The amount of mineral sediment has been reduced due to controlled releases from surface impoundments, such as Lake Houston and Lake Livingston; these flows have changed the natural freshwater inputs to the bay system.

Both development along the coast and anthropogenic climate change have made coast wetlands particularly vulnerable. Note that under normal conditions, wetlands can naturally keep pace with changing sea levels. In the process of *wetland accretion*, sediment is trapped by plants, which increases the elevation of the wetland's surface. This process of wetland accretion is self-regulated through negative feedback between the elevation of the wetland and the relative sea level. Note that when wetland elevations are in balance relative to main sea level, periodic and frequent tidal inundations mobilize sediments and nutrients in the wetland in a way that favors vegetative growth and a balance between sediment deposition and erosion. The problem is that subsidence may upset this balance by submerging the wetland. A submerged or drowned wetland can't support the same floral community, loses its ability to trap sediment as before, and is virtually unregulated by relative sea-level changes. It's the natural processes in the bay and related ecosystems that are affected; that is, the normal rhythm of unregulated streams and rivers is upset.

DID YOU KNOW?

In the Houston, Texas, metropolitan area, there are more than 150 historically active faults (Verbeek et al., 1979). Many of these faults are moving at rates of 0.2–0.8 inch/year. Fault movement has caused substantial, costly damage to private and public infrastructure including buildings, roadways, and utilities.

THE BOTTOM LINE

Rapid growth of the Houston area means that subsidence must continue to be vigilantly monitored and managed. The good news is the Houston area region is better positioned to deal with future problems than many other subsidence-affected areas for several reasons: "a raised public consciousness, the existence of well-established subsidence districts with appropriate regulatory authority, and the knowledge base provided by abundant historical data and ongoing monitoring" (USGS, 1999).

REFERENCES

Ellis, J., Knight, J.E., White, J.T., Sneed, M., Hughes, J.D., Ramage, J.K., Braun, C.L., Teeple, A., Foster, L.K., Rendon, S.H., and Brandt, J.T. (2023). *Hydrogeology, Land-Surface Subsidence, and Documentation of the Gulf Coast Land Subsidence and Groundwater-Flow (GULF) Model, Southeast Texas, 1897-2018. Professional Paper 1877*. Washington, DC: United States Geological Survey.

Gabrysch, R.K. (1983). The impact of land-surface subsidence, in Impact of Science on Society, Managing our freshwater resources: United National Educational, Scientific and Cultural Organization No. 1, p. 17–123.

Galveston Bay National Estuary Program. (1995). The Galveston Bay plan-the comprehensive land management plan for Galveston Bay ecosystem, October 18, 1994: Galveston Bay National Estuary Program Publication GBNEP-49, 457 p.

Holzschuh, J.C. (1991). Land subsidence in Houston, Texas USA: Field-Trip Guidebook for the Fourth International Symposium on Land Subsidence, Houston, Texas, May 12–17, 1991, 22 p.

McGowen, J.M., Garner, L.E., and Wilkinson, B.M. (1977). The Gulf shoreline of Texas-processes. Characteristics, and factors in use: University of Texas, Bureau of Economic Geology, Geological Circular 75-6, 43 p.

Paine, J.G. (1993). Subsidence of the Texas coast-Inferences from historical and late Pleistocene sea levels. *Tectonophysics*, v. 222, pp. 445–458.

Paine, J.G. and Morton, R.A. (1986). Historical shoreline changes in Trinity, Galveston, West and East Bays: Texas Golf University of Texas Bureau of Economic Geology Circular 86-3, 58 p.

Spellman, F.R. (2020). *The Science of Environmental Pollution*, 2nd ed. Boca Raton, FL: CRC Press.

Titus, J.G. and Narayanan, V.K. (1995). *The Probability of Sea Level Rise*. Washington, DC: U.S. Environmental Protection Agency, EPA-230-95-008.

USGS (1999) San Jacinto Battleground State Park. Houston, Texas: Managing Coastal Subsidence. Accessed 10/10/23 @ https://pubs.usgs.gov/circ/1999/circ/182/pdf/part1/pdf.

Verbeek, E.R., Ratzlaff, K.W., and Clanton, U.S. (1979). Faults in parts of north-central and western Houston metropolitan area, Texas: U.S. Geological Survey Miscellaneous Field Studies Map. 1136, 2 plates, Washington, DC, United States Geological Survey.

White, W.A., Tremblay, E.G., Jr., and Handley, L.R. (1995). Trends and status of wetland and aquatic habitats in the Galveston Bay system, Texas: Galveston Bay National Estuary Program Publication GBNEP-31, 225 p.

11 Sinking in "The Meadows"

INTRODUCTION

When in "The Meadows" (aka Las Vegas, Las Vegas Valley, Vegas, or Sin City) that sinking feeling one may get is related almost exclusively to gaming losses. For non-residents, it is a good bet that very little, if any, perception of land subsidence (that sinking feeling) is occurring in the Las Vegas Valley. Exacerbating the problem with ongoing subsidence is that Las Vegas is one of the fastest-growing metropolitan areas in the United States. The underlying issue with population growth and subsidence is the resident's need for water (don't we all?). With population growth comes increasing demand for water and other services. The accelerating demand for water to support the rapid growth of the municipal-industrial section in this desert region is being met with imported Colorado River System supplies and local groundwater. The once plentiful groundwater supplies are now contributing to land subsidence and that "sinking feeling" in Las Vegas Valley. USGS (2013) points out that since 1935, compaction of the aquifer system has caused nearly 6 feet of subsidence and led to the formation of numerous earth fissures and the reactivation of several surface faults, creating hazards and potentially harmful impacts to the environment.

WHEN SUPPLY DOES NOT MEET DEMAND

Visitors who visit Las Vegas are more interested in testing their luck in gaming, visiting world-class entertainment venues, and/or entering a world where they can let their hair down, so to speak, to indulge in whatever Las Vegas—Sin City—makes openly available to them. All of the possible activities in Las Vegas are a powerful attraction for many people.

Visitors come and go as they want to or as bad luck in gaming activities lets them know it is time to depart and maybe tell themselves to come back at another time to try their luck at another time. Fulltime residents of the Las Vegas Valley region are mostly interested in employment opportunities or want to experience desert living. The drawback of living in the Valley is that more and more people are moving into the region. Because of the influx of new residents, it is expected that in the near future, the current water supplies are expected not to satisfy the anticipated water demand. Imported water supplied by the region's Colorado River reservoir—Lake Mead—is federally limited and far-thinkers involved with water supply planning and operations know that a continued need for groundwater will be required to supplement limited reservoir-supplied water. Water supply-and-demand forces at work in

 DOI: 10.1201/9781003461265-13

this growing desert community are likely to perpetuate problems of land subsidence and related ground failures in the Las Vegas Valley unless balanced use of the groundwater resource is achieved.

THE BROWNING

Browning of "The Meadows" is occurring due to the demand for water that is depleting the aquifer system. Prior to development in the Las Vegas Valley, there was a natural, although dynamic, balance between aquifer-system recharge and discharge. Over a relatively short period of time, yearly and decadal climatic variations like drought and El Nino caused large variations in the amount of water available to replenish the aquifer system. But in the longer run, the average amount of water recharging the aquifer system was in balance with the amount discharging, chiefly from springs and by evapotranspiration.

In 1964, Domenico et al. pointed out that in 1907 the first flowing well was drilled by settlers to support the settlement of Las Vegas, and there began to be more groundwater discharge than recharge. At first, untapped artesian wells were permitted to flow freely onto the desert floor, wasting huge quantities of water. In 1911, State Engineer, W.M. Kearney warned that the careless use of groundwater warned that water should be used "... with economy instead of the lavish wasteful manner, which has prevailed in the past" (Maxey and Jameson, 1941).

Needless to say, the intensive use of groundwater use led to a steady decline in spring flows and groundwater levels throughout Las Vegas Valley (Maxey and Jameson, 1941). By 1912, more than 100 wells in Las Vegas Valley (60% of which were free-flowing artesian wells) were discharging nearly 15,000 acre-feet per year. By 1968, the annual groundwater pumpage in the valley reached nearly 90,000 acre-feet (Harrill, 1976). Fast-forwarding to more recently in 2019, the committed groundwater resources total 97,898 acre-feet, with estimated pumpage of approximately 58,908 acre-feet. This figure includes an estimated 4787 acre-feet pumped from domestic wells (State of Nevada, 2019).

DID YOU KNOW?

In 2019, in the Las Vegas Valley, municipal and quasi-municipal were the largest manners of use within the valley, with pumpage totaling 40,418 acre-feet (including 322 acre-feet of recharge recovery), and appropriations of 61,910 acre-feet. The second largest manner of use was recreation, with pumpage of 5255 acre-feet and appropriations totaling 5083 acre-feet. The third largest manner of use was domestic wells (State of Nevada, 2019).

Fast-backward to the 1912 through 1944 timeframe, in the Las Vegas Valley, when groundwater levels declined at an average rate of about 1 foot per year (Domenico et al, 1964). Bell et al. (1981a) point out that between 1944 and 1963, some areas of the valley experienced water-level declines of more than 90 feet. The first area in the Valley to experience large water-level decline was the City of North Las Vegas but, as Las Vegas expanded, new wells were drilled, pumping patterns changed, and groundwater-level declines spread to areas south and west of the City of North Las Vegas. Note that between 1946 and 1960, the area of the valley that could sustain flowing artesian wells shrank from more than 80 square miles (Maxey and Jameson, 1948) to less than 25 square miles (Domenico et al., 1964). By 1962, the springs that had supported the Native Americans, and those who followed, were completely dry (Bell, 1981a).

As of the 1970s, annual groundwater pumpage in the valley has remained between 60,000 and 90,000 acre-feet; most of that has been pumped from the northwestern part of the valley. By the 1990s, areas in the northwest experienced more than 300 feet of decline, and areas in the central (including the Strip) and southeastern (Henderson) sections experienced declines between 100 and 200 feet (Burbey, 1995). In 1996, imports from Lake Mead provided the Las Vegas Valley with approximately 356,000 acre-feet of water (Coache, 1996) and represented the valley's principal source of water. This amount included 56,000 acre-feet of return flow credits for annual streamflow discharging into Lake Mead and from the Las Vegas Wash (see Figure 11.1).

DID YOU KNOW?

Note that the Las Vegas Wash (Figure 11.1) is a 12-mile-long arroyo or wash which feeds most of the Las Vegas Valley's excess water into Lake Mead. The Wash drains 1600 square miles within the Las Vegas Valley

FIGURE 11.1   The Las Vegas Wash. Public Domain Photo from USGS accessed at https://www.usgs.gov/media/images/lake-mead-las-vegas-wash. USGS (2005).

DRAINING THE AQUIFER

Based on comparisons of repeat leveling surveys by the USGS and the U.S. Coast and Geodetic Survey between 1915 and 1948 land subsidence and related ground failures were first recognized by Maxey and Jameson (1948). From then on repeat surveys of various regional networks have shown continuous land subsidence throughout large regions within the valley.

The ongoing surveys have revealed that subsidence continued at a steady rate into the mid-1960s, after which rates began increasing through 1987 (Bell, 1981a; Bell and Price, 1991). Note that surveys made in the 1080s delineate three distinct, localized subsidence basins, bowls, or zones, superimposed on a larger, valley-wide basin. One of these smaller subsidence basins or bowls, located in the northwestern part of the valley, subsided more than 5 feet between 1963 and 1987. Two other subsidence bowls, in the central (downtown) and southern (Las Vegas Strip) parts of the valley, subsided more than 2.5 feet between 1963 and 1987. It is important to point out that the areas of maximum subsidence do not necessarily coincide with areas of maximum water-level declines. Why? Well, one likely explanation is that those areas with maximum subsidence are underlain by a large aggregate thickness of fine-grained, compressible sediments (Bell and Price, 1991).

Aquifer-system compaction creates earth fissures and reduces storage. However, keep in mind that all of the impacts of subsidence in the Las Vegas Valley have not been fully realized. To date, two important impacts that have been documented are (1) ground failures—localized ruptures of the land surface, and (2) the permanent reduction of the storage capacity of the aquifer system. Other potential impacts that are at the earliest stages of study include:

- Creation of flood-prone areas by altering natural and engineered drainage ways
- Creation of earth fissures connecting non-potable or contaminated surface and near-surface water to the principal aquifers
- Replacement costs associated with protruding wells and collapsed well casings and well screens.

Note: Keep in mind that all of these potential damages create legal issues related to mitigation, restoration, compensation, and accountability.

The most dominant and spectacular type of ground failures associated with groundwater withdrawal in the Las Vegas Valley are the tensile failures in subsurface materials commonly known as *earth fissures*. The earth fissures result when differential compaction of sediments pulls apart the earth's materials. These buried, incipient earth fissures become obvious only when they reach the surface and begin to erode, often following extreme rains or surface flooding conditions. Earth fissures have been observed in the Las Vegas Valley as early as 1925 (Bell and Price, 1991), but were not linked directly to subsidence until the later 1950s (Bell, 1981a). Most of the earth fissures are partially and temporarily correlated with groundwater-level declines. The bottom line: Subsidence still continues.

Movement of preexisting surface faults has also been correlated to groundwater-level changes and differential land subsidence in numerous alluvial basins (Holzer, 1979, 1984; Bell, 1981a). In the Las Vegas Valley, earth fissures often occur preferentially along preexisting surface faults in the unconsolidated alluvium. They tend to form as a result of the warping of the land surface that occurs when the land subsides more on one side of the surface fault than the other. This differential land subsidence creates tensional stresses that ultimately result in fissuring near zones of maximum warping. The association of most earth fissures with surface faults suggests a causal relationship. The surface faults may act as partial barriers to groundwater flow, creating a contrast in groundwater levels across the fault, or many offset sediments of differing compressibility (USGS, 1999).

Rigid and precisely leveled structures are capable of being damaged due to land-surface displacements and tilt. Other damage related to fissuring includes cracking and displacement of roads, curbs, sidewalks, playgrounds, and swimming pools; warped sewage lines; ruptured water and gas lines; well failures resulting from shifted, sheared, and/or protruded well casings; differential settlement of railroad tracks; and a buckled drainage canal (Bell, 1981b; USGS, 1999). Earth fissures are also susceptible to erosion and can form wide, steep-walled gullies capable of redirecting surface drainage and creating floods and other hazards. Adverse impacts of ground failures may worsen as the valley continues to urbanize and more developed areas become affected.

As mentioned, reduction of storage capacity in the Las Vegas Valley aquifer system is another important consequence of aquifer-system compaction. The volume of groundwater derived from the irreversible compaction of the aquifer system—"water of compaction"—is approximately equal to the reduced storage capacity of the aquifer system and represents a one-time quantity of water "mined" from the aquifer system.

Loss of aquifer-system storage capacity is cause for concern, especially for a fast-growing desert metropolis that must rely in part on local groundwater resources. A study conducted by the Desert Research Institute (Mindling, 1971) estimated that, at times, up to 10% of the groundwater pumped from the Las Vegas Valley aquitard system has been derived from water of compaction. Assuming conservatively that only 5% of the total groundwater pumped between 1907 and 1996 was derived from water of compaction, the storage capacity of the aquifer system has been reduced by about 187,000 acre-feet. This may or may not be considered "lost" storage capacity: arguably if this water is derived from an irreversible process, this storage capacity has been used in the only way that it could have been. In any case, producing water of compaction represents mining groundwater from the aquifer system. Further, the reduced storage implies that, even if water levels recover completely, any future drawdowns will progress more rapidly.

The bottom line: Las Vegas is dealing with a limited water supply. Moreover, managing land subsided in the Las Vegas Valley is linked directly to the effective use of this limited water supply. With regard to the groundwater supply, at present more is appropriated by law and is being pumped in the Las Vegas Valley than is available to be safely withdrawn from the groundwater basin (Nevada Department of Conservation and Natural Resources, Division of Water Resources and Water

Planning, 1992; Coache, 1996). Historic and recent rates of aquifer-system depletion caused by overuse of the groundwater supply cannot be sustained without contributing further to land subsidence, earth fissures, and the reactivation of surface faults.

In order to arrest subsidence in the valley, groundwater levels must be stabilized or maintained above historic low levels. Stabilization or recovery of groundwater levels throughout the valley will require that the amount of groundwater pumped from the aquifers be less than or equal to the amount of water recharging the system. Eliminating any further decline will reduce the stresses contributing to the compaction of the aquifer system. Even so, a significant amount of land subsidence (residual compaction) will continue to occur until the aquifer system equilibrates fully with the stresses imposed by lowered groundwater levels in the aquifers (Riley, 1969).

THE BOTTOM LINE

Despite the ambitious efforts to artificially recharge the aquifer system in the Las Vegas Valley, valley-wide net groundwater pumpage still exceeds the estimated natural recharge—simply outstripping Nature's bounce-back ability. To minimize any future subsidence some combination of increased recharge and reduced pumpage is needed, especially in areas prone to subsidence.

REFERENCES

Bell, J.W. (1981a). Subsidence in Las Vegas Valley: Nevada Bureau of Mines and Geology Bulletin 95, 83 p, 1 plate, scale 1:62,500.

Bell, J.W. (1981b). Results of leveling across fault scarps in Las Vegas Valley, Nevada, April 1978–June 1981: Nevada Bureau of Mines and Geology Open-File Report 81-5, 7 p.

Bell, J.W. and Price, J.G. (1991). Subsidence in Las Vegas Valley, 1980-91-Final project report: Nevada Bureau of Mines and Geology, Open-File Report 93-4, 10 sect., 9 plates, scale 1:62,500.

Burbey, T.J. (1995). Pumpage and water-level change in the principal aquifer of Las Vegas Valley, 1980-90: Nevad Division of Water Resources Information Report 34, 224 p.

Coache, R. (1996). Las Vegas Valley water usage report, Clark County, Nevada, 1996: Nevada Division of Water Resources Report, [50+] p.

Domenico, P.A., Stephenson, D.A., and Maxey, G.G. (1964). Ground water in Las Vegas Valley: Nevada Department of Conservation and National Resources Division of Water Resources Bulletin 29, 53 p.

Harrill, J.R. (1976). Pumping and ground-water storage depletion in Las Vegas Valley, Nevada, 1955-74: Nevada Department of Conservation and Natural Resources, Division of Water Resources Bulletin No. 44, 70 p.

Holzer, T.L. (1979). Leveling data-Eglington fault scarp, Las Vegas Valley, Nevada: U.S. Geological Survey Open-File Report 79-950, 7 p.

Holzer, T.L. (1984). Ground failure induced by ground-water withdrawal from unconsolidated sediment. In *Man-Induced Land Subsidence*, vol. 6. Ed. Holzer, T.L. Geological Society of American Reviews in Engineering Geology, Boulder, Colorado, 221 p.

Maxey, J. and Jameson, M., (1941). Las Vegas Valley: Land Subsidence and Fissuring due to groundwater withdrawal. Washington, DC: United States Geological Survey.

Maxey, G.B. and Jameson, C.H. (1948). Geology and water resources of Las Vegas, Pahrump, and Indian Springs Valleys, Clark and Nye Counties, Nevada: Nevada State Engineer Water Resources Bulletin 5, 121 p.

Mindling, A.L. (1971). A summary of data relating to land subsidence in Las Vegas Valley: Center for Water Resources Research, Desert Research Institute, University of Nevada, Reno, 55 p.

Nevada Department of Conservation and Natural Resources, Division of Water Resources and Water Planning. (1992). Hydrographic basin summaries, Carson City, Nevada, 1990–1992.

Riley, F.S. (1969). Analysis of borehole extensometer data from central California: International Association of Scientific Hydrology Publication 89, pp. 423–431.

State of Nevada (2019). Las Vegas Valley: Land Subsidence and Fissuring due to groundwater withdrawal. Washington, DC: United States Geological Survey.

USGS (1999). Las Vegas Land Subsidence. Accessed 10/10/23 @ https://pubs.usgs.gov/circ/1999/circ/182/pdf.

USGS. (2013). *Land Subsidence in the United States. Circular 1182*. Washington, DC: U.S. Department of the Interior, U.S. Geological Survey.

12 Subsidence & Fissures in South-Central Arizona[1]

COMPLICATING DESERT WATER RESOURCES

There is a direct relationship between groundwater overdrafts in the deep alluvial basin of southern Arizona and widespread land subsidence within the region. Since 1900, groundwater has been pumped for irrigation, mining, and municipal use, and in some areas, more than 500 times the amount of water that naturally replenishes the aquifer systems has been withdrawn (Schumann and Cripe, 1986). The resulting groundwater-level declines—more than 600 feet in some places—have led to increased pumping costs, degraded the quality of groundwater in many locations, and led to the extensive and uneven permanent compaction of compressible fine-grained silt- and clay-rich aquitards. A total area of more than 3000 square miles has been affected by subsidence, including the expanding metropolitan areas of Phoenix and Tucson and some important agricultural regions nearby.

In southern Arizona, buildings, roads and highways, railroads, flood-control structures, and water and wastewater lines have been damaged by ground fissures from ground failure in areas or uneven or differential compaction. The presence and ongoing threat of subsidence and fissures forced a change in the planned route of the massive, federally financed Central Arizona Project (CAP) aqueduct that has delivered imported surface water from the Colorado River to central Arizona since 1985. In the CAP, Arizona now has a supplemental water supply that has lessened the demand and overdraft of groundwater supplies. Some CAP deliveries have been used in pilot projects to artificially recharge depleted aquifer systems. When fully implemented, recharge of this imported water will help to maintain water levels and forestall further subsidence and fissure hazards in some areas (USGS, 2013).

In Arizona, agriculture is synonymous with and dependent on irrigation. Irrigation is needed to grow corps because of the low annual rainfall and the high rate of potential evapotranspiration—more than 60 inches/year. Precipitation in south-central Arizona ranges from as low as 3 inches/year over some of the broad flat alluvial basins to more than 20 inches/year in the rugged mountain ranges. Large volumes of water can be stored in the intermontane basins, which contain up to 12,000 feet or more of sediments eroded from the various metamorphic, plutonic, volcanic, and consolidated sedimentary rocks that form the adjacent mountains. Groundwater is generally produced from the upper 1000 to 2000 feet of the basin deposits, which constitute the aquifer systems. Groundwater pumped from the aquifer systems became a reliable and heavily drawn on source of irrigation water that fueled the development of agriculture during the early and mid-20th century. In many areas, the aquifer systems include a large fraction of fine-grained

DOI: 10.1201/9781003461265-14

deposits containing silt and clay that are susceptible to compaction when the supporting fluid pressures are reduced by pumping.

Pumping for irrigation has been increasingly used since the 1900s. By the mid-1960s, the expected growth in the metropolitan Phoenix and Tucson areas, coupled with the already large groundwater-level declines and worsening subsidence problems, prompted Arizona water officials to push for and receive congressional approval for the CAP. Since then, growth in the metropolitan areas has exceeded expectations, and municipal–industrial and domestic water use presently accounts for nearly 20% of Arizona's water demand.

Population growth increased agricultural production and industrial expansion combined to exact a toll on groundwater supplies and also increased subsidence. Subsidence first became apparent during the 1940s in several alluvial basins in southern Arizona where large quantities of groundwater were being pumped to irrigate crops. By 1950, earth fissures began forming around the margins of some of the subsiding basins. The areas affected then and subsequently included metropolitan Phoenix in Maricopa Count and Tucson in Pima Count, as well as important agricultural regions in Pinal and Maricopa Counties near Apache Junction, Chandler Heights, Stanfield, and into the Picacho Basin; in Cochise County near Willcox and Bowie; and in La Paz County in the Harquahala Plain. By 1980, groundwater levels had declined at least 100 feet in each of these areas and between 300 and 500 feet in most of the areas.

Land subsidence was first verified in south-central Arizona in 1948 using repeat surveys of benchmarks near Eloy (Robinson and Peterson, 1962). By the late 1960s, installation and monitoring of borehole extensometers at Eloy, Higher Road south of Mesa, and at Lake Air Force Base, as well as analysis of additional repeat surveys, indicated that land subsidence was occurring in several areas. The areas of greatest subsidence corresponded with the areas of greatest water-level decline (Schuman and Poland, 1970; USGS, 2013).

By 1977, nearly 625 square miles had subsided around Eloy, where as much as 12.5 feet of subsidence was measured; another 425-square-mile shad subsided around Stanfield, with a maximum subsidence of 11.8 feet (Laney et al., 1978) Near Queen Creek, and area of almost 230 square miles had subsided more than 3 feet. In northeast Phoenix, as much as 5 feet of subsidence was measured between 1962 and 1982. By contrast, in the Harquahala Plain, only about 0.6 feet of subsidence occurred in response to about 300 feet of water-level decline, whereas near Willcox, more than 5 feet of subsidence occurred in response to 200 feet of water-level decline (Holzer, 1980; Strange, 1983; Schumann and Cripe, 1986). The relation between water-level decline and subsidence varies between and within basins because of differences in the aggregate thickness and compressibility of susceptible sediments (USGS, 2013).

Groundwater decline continued, and by 1992, declines of more than 300 feet had caused aquifer-system compaction and land subsidence of as much as 18 feet near the Luke Air Force base, about 20 miles west of Phoenix. Associated earth fissures occur in three zones of differential subsidence at Luke, which led to a flow reveal in a portion of the Dysart Drain, and engineered flood conveyance. In September 1992, surface runoff from a rainstorm of 4 inches closed the base for 3 days.

The sluggish Dysart Drain spilled over, flooding the base runways along with more than 100 houses and resulted in about $3 million in damage (Schumann, 1995).

Subsidence-related earth fissures, cracks, seams, or separations in the ground are common in many basins; some of the most spectacular examples occur in south-central Arizona where they have been part of the landscape for at least 70 years. Earth fissures are the dominant mode of ground failure related to subsidence in alluvial-valley sediments in Arizona and are typically long linear cracks at the land surface with little or no vertical offset. The temporal and spatial correlation of earth fissures with groundwater-level declines indicates that many of the earth fissures are induced and are related to groundwater pumpage. More than 50 fissure areas had been mapped in Arizona prior to 1980 (Laney et al., 1978).

Most fissures occur near the margins of alluvial basins or near exposed or shallow buried bedrock in regions where differential land subsidence has occurred. They tend to be concentrated where the thickness of the alluvium changes markedly. In a very early stage, fissures can appear as hairline cracks less than 0.02 inch wide interspersed with lines of sink-like depressions resembling rodent holes. When they first open, fissures are usually narrow vertical cracks less than about 1 inch wide and up to several hundred feet long. They can progressively lengthen to thousands of feet. Apparent depths of fissures range from a few feet to more than 30 feet; the greatest recorded depth is 82 feet for a fissure on the northwest flank of Picacho Peak (Johnson, 1980). Fissure depths of more than 300 feet have been speculated based on various indirect measurements, including horizontal movement, volume-balance calculations based on the volume of air space at the surface, and the amount of sediment transported into the fissures. Widening of fissures by collapse and erosion results in fissure gullies (Laney et al., 1978) that may be 30 feet wide and 20 feet deep. No horizontal shear (strike-slip movement) has been detected at earth fissures, and very few fissures show any obvious vertical offset. However, fissures monitored by repeated leaving surveys commonly exhibit a vertical offset of a few inches. Two notable exceptions are the Picacho earth fissure, which has more than 2 feet of vertical offset at many places along its 10-mile length, and the fissure near Chandler Heights, which has about 1 foot of vertical offset.

DID YOU KNOW?

Subsidence has occurred in basins with large water-level declines, but the relationship between the magnitude of water-level decline and subsidence varies between and within basins. Representative profiles show that subsidence is greater near the center of basins, where the aggregate thickness of fine-grained sediments is generally greater (USGS, 2013).

Fissures can undercut and damage infrastructure and present a hazard to the public. Hazards associated with earth fissures (Table 12.1) are generally more local and include damage to homes and buildings, roads, dams, canals, sewer and utility lines, as well as providing a conduit for contaminated surface water to rapidly enter groundwater aquifers.

TABLE 12.1

Hazards Directly Associated with Earth Fissures

- Cracked or collapsing roads
- Broken pipes and utility lines
- Damaged or breached canals
- Cracked foundation/separated walls
- Loss of agricultural land
- Livestock and wildlife injury or death
- Severed or deformed railroad track
- Damaged well casing or wellhead
- Disrupted drainage
- Contaminated groundwater aquifer
- Sudden discharge of ponded water
- Human injury or death

Source: Adapted from Pew (1990) and Bell and Price (1993).

DID YOU KNOW?

In agricultural areas, fissures and fissure gullies are often obscured by culti-vation. Reactivation of fissures can recur, only to be obscured again by cul-tivation. In some cases, farmers periodically fill fissures with soil and other materials because the gully formation processes are persistent. Such fissures are commonly known only to the farmers who cultivate the fields.

THE BOTTOM LINE

In order to arrest, control, and mitigate the effects of land subsidence in Arizona, importation of water for consumptive use and groundwater recharge, retirement of some farmlands, and water-conservation measures have been put in place. These measures have resulted in cessation of water-level declines in many areas and the recovery of water levels in some areas. However, some basins are still experiencing subsidence, because much of the aquifer-system compaction has occurred in rela-tively thick aquitards. It can take decades or longer for fluid pressures to equilibrate between the aquifers and the full thickness of many of these thick aquitards. For this reason, both subsidence and its abatement have lagged pumping and recharge. A glimmer of hope exists from data at the borehole extensometer near Eloy, where water levels have recovered more than 150 feet and compaction has decreased markedly.

NOTE

1 Based on USGS (2013) Carpenter, M.C., South-Central Arizona *in* Land Subsidence in the United States. U.S. Geological Survey, Tucson, AZ.

REFERENCES

Bell, J.W. and Price, J.G. (1993). Subsidence in Las Vegas Valley, 1980-91-final project report, Reno: Nevada Bureau of Mines and Geology Open-File Report 93-4.

Holzer, T.L. (1980). Research at the U.S. Geological Survey on Faults and earth fissures associated with land subsidence. Washington, DC: USGS.

Johnson, N.M. (1980). The relation between ephemeral stream regime and earth fissuring in south-central Arizona: Tucson, University of Arizona, M.S. thesis, 158 p.

Laney, R.L., Raymond, R.H., and Winikka, C.C. (1978). Maps showing water-level declines, land subsidence, and earth fissures in south-central Arizona: U.S. Geological Survey Water-Resources Investigations Report 78-83, 2 sheets, scale 1:125,000.

Pew, T.L. (1990). Land subsidence and earth-fissure formation caused by groundwater withdrawal in Arizona. A review. In *Groundwater Geomorphology: The Role of Subsurface Water in Earth-Surface Process and Landforms*. Eds. Higgins, C.G. and Coates, D.R. Boulder, CO: Geological Society of America, pp. 218–233 (Special Paper 252).

Robinson, G.M. and Peterson, D.E. (1962). Notes on earth fissures in southern Arizona: U.S. Geological Survey Circular 466, 7 p.

Schumann, H.H. (1995). Land subsidence and earth fissure hazards near Luke Air Force Base, Arizona: U.S. Geological Survey Subsidence Interest Group Conference, Edwards Air Force Base, Antelope Valley, California, November 18–19, 1992, [Prince, K.R., Galloway, D.L., and Leake, S.A., eds], Abstaces and Summary: U.S. Geological Open-File Report 94-532, pp. 18–21.

Schumann, H.H. and Cripe, L.S. (1986). Land subsidence and earth fissures caused by groundwater depletion in southern Arizona, U.S.A.: International Symposium on Land Subsidence, 3rd, Venice, 1984, [Proceedings, Johnson, A.I., Laura, C., and Ubertini, L., eds], International Association of Scientific Hydrology Publication 151, pp. 841–851.

Schumann, H.H. and Poland, J.F. (1970). Land subsidence, earth fissures, and groundwater withdrawal in south-central Arizona, U.S.A: International Association of Scientific Hydrology Publication 88, pp. 295–302.

Strange, L (1983). Research at the U.S. Geological Survey on Faults and earth fissures associated with land subsidence. Washington, DC: USGS.

USGS. (2013). *Land Subsidence in the United States. Circular 1182*. Washington, DC: U.S. Department of the Interior, U.S. Geological Survey.

13 A Bolt from the Blue

INTRODUCTION

In order to gain a fundamental understanding of land subsidence presently occurring in the United States, it is important to know the lay of the land (so to speak)—from the earliest time to the present. This is challenging because none of us were around to witness the formation of Earth's landforms shaped by forces other than human hands—and considering the tumultuous nature of early Earth-shaping processes, not being around during those periods has been and continues to be beneficial for many forms of life, especially for us.

The key point here is that in order to comprehend and to appreciate land subsidence and relative sea-level rise in a particular region of the United States, it is helpful to have knowledge of what is known about the region's formation, development, and history—keep in mind that much of what we know is based on educated guesswork, based on knowledge and experience. The region discussed in this and the following chapters is the Lower Chesapeake Bay region (aka Tidewater or Hampton Roads), and its land subsidence problem, its relative sea level issues, and its mitigation efforts are reviewed and correlated.

So, let's begin with the beginning (as best we have discerned it).

TIMELINE 33,000,000 BCE[1]

They were all gone. They had been gone for approximately 32 million years; actually, a bit more than 32 million years before the present timeframe referred to herein. We would later know that they were actually here on the planet because we discovered their preserved skeletal remains, and discoveries continue here and there at various locations on Earth. When the remains are discovered and carefully preserved, they are often displayed to the delight and wonder of onlookers.

Don't you just love them?

Them?

Yes. We will get to them.

For the moment consider that many experts say (or at least speculate via an educated guess) that it was either a combination of volcanic activity, asteroid impact (creating the Chicxulub crater in Yucatan; about the size of Staten Island), and/or climate change that effectively ended 76% of life on Earth 65 million years ago. Others speculate that it might have been only one of these events that brought the mighty rulers of the Earth, the terrible lizards (aka dinosaurs) to their terminal end. Well, you can take your pick of the actual causal factor(s); the fact is that much of the life as known during that time period was extinguished in what is identified as the K–T event, but recently renamed the K–Pg for the Cretaceous–Paleogene extinction event—and also known as (and the title most often thought of, if thought about at all):

DOI: 10.1201/9781003461265-15

the fifth massive extinction event on Earth. Importantly, meteors (small parts of asteroids) have been recorded throughout history.

For the purpose of this book we will assume, speculate, guess, or believe (take your pick) that it was the meteor that created the Chicxulub crater, the 125-m hole off the coast of the Yucatan Peninsula, deep beneath the depths of the Gulf of Mexico. The meteor, or meteorite (i.e., a meteor that survives burning up in the atmosphere and reaches land is called a meteorite), was responsible for the largest mass extinction event in history. The impact was so intense that it started wildfires hundreds of miles away from the impact site. So much ash, sulfur, and fragmented debris were flung into the air that the Sun was blotted out. The atmospheric disturbance was so great that Earth sat in perpetual darkness for months, creating a very long and unexpected winter that changed the way of the planet forever. Approximately 75% of life on Earth (including the major dominant species, dinosaurs) died off at this time. Note that mammals survived this apocalyptic event by being small and warm-blooded. And keep in mind that when an animal's chief predators are no longer around, the chance of maintaining life and procreation is enhanced exponentially. Moreover, the lack of predators following the meteor impact led to the rise of evolution of all mammals alive today, including humans.

BACK TO 33,000,000 BCE

Before getting to the point of why this discussion begins with an era long past, it is important to point out that this book is a geologically based presentation of facts as we know them or as we think we know them. To begin this discussion, we need to begin at the beginning—the first link in the chain.

About 35 million years ago, a large comet or meteorite (actually it was a bolide—explained later) slammed into the western Atlantic Ocean on a shallow shelf, creating the Chesapeake Bay impact crater. At that catastrophic moment in Earth's history, it was as if Mother Nature had reached into her endless glove (the universe) and grabbed hold of that fiery, flaming, scorching mass and wound up her arm and threw Earth a massive knuckle ball. Yes, it was an arching, white hot knuckle ball at least two times brighter than Earth's moon. It was at least two miles wide; from the northwest horizon, it crossed paths with Earth at more than 76,000 miles per hour, more than 1260 miles per minute—roughly 21 miles per second. It moved too quickly to be heard and its white-hot light would have blinded had it not killed before optic nerves could signal brain matter (Tennant and Hall, 2001). This was the age, the time, and, in some cases, the world of the prehistoric sharks, whales, tiny camels, modern ungulates, bats, sea cows, eagles, pelicans, quail, and vultures. Were all these lifeforms present when Mother Nature's fireball impacted Earth?

Probably not.

But will we never really know who was there and who was not there—humans (as we call them) weren't even?

Probably not.

What we do know is those who were present died in a wink—for those who love dinosaurs (remember, don't we all?) not to worry, mighty T-Rex, nasty Utahraptor, and that affable, turkey-/chicken-sized Velociraptor (who would gobble you and me up in an instant) were nowhere to be seen; they had all perished at least 30 million years earlier during the K–T event—mass extinction. Some might call the Chesapeake Bay bolide impact the ultimate bolt from the blue, the ultimate shock and awe event; however, because the event was instantaneously deadly to all those within impact-affected areas, it would be better stated to say that the event was a whole bunch of shock (to minerals, especially quartz) with very little awe. When death is instant, who is around, at least in the immediate area, to be awed?

So, what about the crater?

Good question.

The crater is now covered by Virginia's central to outer Coastal Plain sediments and the Lower Chesapeake Bay. The Chesapeake Bay impact crater is a 56-mile-wide, complex peak-ring crater with an inner and outer rim, a relatively flat-floored annular trough, and an inner basin that penetrates the basement to a depth of at least 1.2 miles (Powers and Bruce, 2000).

Let's fast-forward to the bolide's impact result as determined today. For millennia, humans had no knowledge that this event had occurred or of the literal impact it would have on southeastern Virginia, on the formation of Chesapeake Bay, and the cause of many present-day local problems with groundwater quality and land subsidence; it was not until a handful of modern-day Sherlock Holmeses, with Dr. C.W. Poag in the lead, figured it out. Through intuition, bore-holing operations, application of scientific protocols, and a lot of common sense, they determined that at the end of the Eocene Mother Nature's fiery curveball impacted the coastline of what today is known as Virginia's Chesapeake Bay region.

DID YOU KNOW?

Descriptions of the location and geometry of the Chesapeake Bay impact crater are based on correlation of lithostratigraphic and biostratigraphy data from cores and well cuttings, borehole geophysical logs, and seismic reflection data (Powars and Bruce, 2000).

Today we call this event the Chesapeake Bay meteor incident. Those scientists in the know, however, understand that the term meteor is best replaced with the term bolide. Well, that brings us to the point and need of explaining the term bolide and many other relevant meteorite terms. So, let's do just that.

METEORITE TERMS (USGS, 1998; PWNET, 2016)

Asteroid—A rocky body orbiting the sun, usually greater than 100 m in diameter. Most asteroids orbit between Mars and Jupiter.

Bolide—Randall (2015) in her bestselling book *Dark Matter and the Dinosaurs* defines bolide as "an object from space that disintegrates in the atmosphere" (p. 127). USGS (1998) points out that there is no consensus on the definition of a bolide; however, it uses the term to mean an extraterrestrial body in the 1- to 10-km

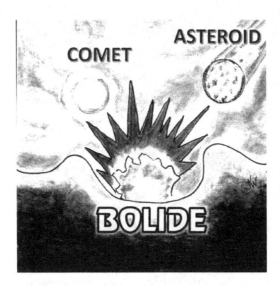

FIGURE 13.1 Bolide impact. Source: Adaptation from USGS (1998). Illustration by Spellman and Welsh (2019).

size range, which impacts the Earth at velocities of literally faster than a speeding bullet (Mach 75 = 20 to 70 mg/s), explodes upon impact, and creates a large crater.

The bottom line is that bolide is a generic term used to imply that we do not know the precise nature of the impacting body (and we do not)—whether it is a rocky or metallic asteroid or any icy comet, for example (see Figure 13.1). Note that in this book we call a bolide an extraterrestrial body of some size and of some composition that impacts Earth and creates a crater; for our purposes, it is the descriptor "impact crater" that is the key word.

Cape Charles, Virginia—The location of a huge peaking impact crater whose center is located near this Eastern Shore Virginia town (see Figure 13.2).

Central peak—A small mountain that forms at the center of a crater in reaction to the force of the impact.

Chesapeake Bay Bolide—This is one of the largest known impact structures found in North America. This event has been dated at about 35–33 million years ago during Eocene Epoch of the Cenozoic Era. (*Note*: For the purpose of simplicity and continuity in this text, we assume bolide impact occurred 35 million years ago.) The nature of the impact substantially affected the geology of the Atlantic continental crust and is suspected to affect the nature and quality of groundwater in southeastern Virginia. Estimated at 90 km (~55.8 miles), the crater may be about 1.3 km (~0.81 miles) deep.

Crater—The result of a bolide body impacting the surface of another planetary body. The resulting explosion leaves a round hole, or crater.

Eocene Epoch—This time period occurred 58–33.8 million years ago. This time period is marked by the emergence of mammals as the dominant land animals. The fossil record reveals many mammals quite unlike anything we have today. However, there were increasing numbers of forest plants, freshwater fish, and insects, much like those today. In fact, the term Eocene means "dawn of the recent."

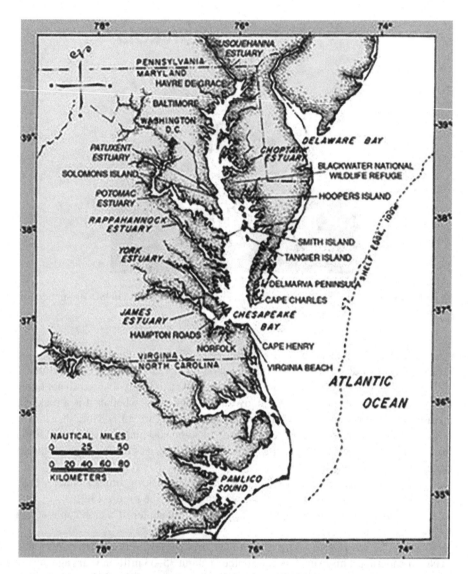

FIGURE 13.2 Chesapeake Bay area. Adapted from USGS (1998).

Ejecta—This is the debris that shoots out of the impact site when a crater forms.

Fault fracture/ground fissure—This is caused either by tectonic movement or impact events.

Floor—This is the bottom part of an impact crater. It can be flat or rounded and is often lower than the surrounding surface of the planet or moon.

Impact breccia—This is a rubble sediment that contains a mix of debris resulting from an impact event.

Iridium—This is a very hard and brittle metal, atomic number 77, often associated with meteorite impacts.

Mass—This is the measure of an object's inertia, that is, how heavy it is. Mass is different from weight, which measures the gravitational force on an object.

Meteor—This is a bright streak of light in the sky caused by a meteoroid or small icy particle entering Earth's atmosphere. It is also known as a "shooting star." Meteor showers occur when Earth passes through debris left behind by an orbiting comet.

Meteorite—This consists of small rocky remains of meteoroids that survive a fiery journey through Earth's atmosphere and land on Earth.

Metric units—1 km = 0.621 miles; 1 m = 3.28 feet; 4000 km^3 = 960 cubic miles.

Micrometeorite—This is a very small meteorite with a diameter of less than 1 mm.

Ray bright—These are lines of debris projecting from the edges of craters.

Rim—This is the highest point along the edge of a crater hole.

Rubble bed—Jumbled sediments and aged dated fossils that are associated with the Exmore beds of the Chesapeake impact structure.

Shocked materials—These are minerals, especially quartz, that show the result of tremendous forces, such as those found in impact events that alter and distort the normal optical qualities of a quartz crystal.

Tektites—These are millimeter- to centimeter-size glass beads (see Figure 13.3) derived from sediments melted by a bolide impact.

Tsunami—This is a very large ocean wave usually associated with underwater earthquakes or volcanic eruptions. Tsunamis may also be associated with large meteorite impacts in the oceans.

Wave material—This is material left among the trajectory of a meteor after the head of the meteor has passed.

Wall—This refers to the sides of the bowl of a crater.

With regard to the bowl of the crater and the basin created by the bolide impact event, the inner basin includes a central uplift surrounded by a series of concentric valleys and ridges.

FIGURE 13.3 Tektites. From the collection of and photo by F.R. Spellman.

A contour tracing of seismic reflection data, including basement data down to 6.0 seconds two-way travel time, shows the seismic "fingerprint" of a bowl-shaped zone of intensely shocked basement rocks down to about 3.5 seconds two-way travel time (about 33,000–37,0000 feet; 6.2–7 miles). The outer rim of the crater traverses the lower York–James Peninsula (Powars and Bruce, 2000).

DID YOU KNOW?

In this text, the work of Dr. C. Wylie Poag is referenced often. This is only fitting due to his flagship work on investigating and finding the Chesapeake Bay bolide crater. Therefore, it is only fitting that Dr. Poag, a senior research scientist with the United States Geological Survey, be highlighted here by sharing a short version of his biography. Dr. Poag's research emphasizes the integration of subsurface geophysical, geological, and paleontological data to reconstruct the stratigraphic framework and depositional history of the Atlantic and Gulf Coast margins of the United States. His 30-year geological career includes experience as a petroleum explorationist, a university professor, and a project coordinator for the National Science Foundation's Deep Sea Drilling Project. D. Poag has published more than 200 abstracts, articles, and books. A recent highlight of Dr. Poag's research has been the identification of the largest impact craters in the United States buried beneath the lower part of Chesapeake Bay and its surrounding peninsulas (USGS, 1997).

"THE HORRIBLE GASH"

Earth-changing events occur almost daily; they are ongoing processes that are beyond human control: we can't control earthquakes; we can't prevent or stop iceberg calving; we can't prevent volcano eruption or collapse; we can't, in most cases, prevent landslides, avalanches, mega floods; and we can't prevent meteor strikes. While it is true that today we can witness or review news and film reports about such events, it is also true that millions of these events occurred in ancient times, before humans were even thought of. Because many of these ancient events occurred millions and even a billion or more years ago, today many are not visible because the master architect, Mother Nature, continues to remove, degrade, eliminate, or cover up past Earth-shaping and Earth-changing events. Only through discovery, research, intuition, and, in some cases, common sense are we able to discern Earth-changing events of the past. This is certainly the case with the discovery of the Chesapeake Bay crater.

Here is what we know or think we know now about the creation of the Chesapeake crater and its subsequent formation of the Chesapeake Bay region as we know it now.

It was certainly an out-of-the-blue event. However, most experts just refer to it as a bolide blasting into Earth and into the shallow sea that covered Virginia from Cape Henry to Richmond; thus, this certainly can be classified as an out-of-the-blue Earth-changing event. It certainly changed things in southwest to upper Virginia. The Chesapeake Bay bolide exploded with more force than the combined nuclear arsenal of today's world powers (Tennant and Hall, 2001). The impact cracked the Earth as deep as 7 miles. The bolide caused the creation a crater 85 miles wide, creating a flash of evaporating ocean water (millions of gallons evaporated instantly; millions more were hurled 60 miles into the atmosphere) and a volume of ejected bedrock that may have risen in a towering column about 30 miles high (National Geographic, 2001). Most of the debris fell back into the crater. Some "shocked quartz" and tiny glass beads—"tektites" (see Figure 13.3)—were scattered as far as New Jersey. The bolide acting like a giant drill or jackhammer and unimpeded by ocean ripped through almost a mile of sediment and sand and penetrated the 600-million-year-old granite bedrock and pounded, milled, pulverized, and minced it. Huge chunks, boulders, and grains of solid earth were propelled upward. Keep in mind that when the bolide was drilling its way into Earth, the front end slowed down a bit but the back end was still flying at supersonic speed. The bedrock, at the front end, and the back end squeezed together like a humongous sponge, and then rebounded with a vengeance. The bedrock splintered and massive faults split open. In the chaos of searing heat and ferocity, the bolide vaporized, leaving a crater 55+ miles wide with a network of fractures and fissures spread more than 40 miles beyond its rim.

As the rocks blasted out into the air, they were ignited by friction and subsequently sparked firestorms for hundreds of miles—the surrounding area was literally an inferno of unimaginable proportions. And then the heavy hand of Mother Nature using her gravity grabbed and reclaimed the boulders and water from the sky and returned them to the gaping crater. We all know that when we toss a small stone into a river pool concentric circles ripple outward as the stone drops through to pool bottom. For a brief instant, we are struck by the obvious: the stone sinks to the bottom, following the laws of gravity. Eventually the ripples die away, leaving as little mark as the usual human lifespan creates in the waters of the world, and then disappears as if it had never been. Well, this is a scene, happening, or experience many of us are familiar with in the calmness and normality of our lives. Imagine, however, the bolide and its ejected materials rolling out, creating swells, forming concentric circles, headed across the ocean to Greenland, Europe, and the East Coast of the United States. Those swells raced at extraordinary speeds across the ocean floor, and they rose with the land and exploded. Even the Blue Ridge Mountains felt the impact from tsunamis thousands of feet high (Tennant and Hall, 2001). And then came the run back, adhering to the time-worn axiom applied within the realm of the bounds of Earth and based on scientific fact that what goes up will eventually come back. And come back it did, with a settling. Remember, there are very few substances, if any, that are or can be more destructive than water on the move. All that tsunami water that ran up the Blue Ridge (and other places) and loosened house-sized boulders and other objects rushed back carrying tons of rock, soil, trees, and wildlife of that time and filled the gigantic empty crater as deep as the Grand Canyon and the rest went off to the sea.

And then, oh it was so quiet, so peaceful, the surrounding area steaming, barren; it was over and there was no one around to remember that the impact had happened at all. Unlike the stone thrown in a pond with ripples that form and disappear in rings and leave no evidence that they were ever formed, the Chesapeake Bay bolide was somewhat different; so it can be said again: it was like the common river pool momentarily disturbed by a pebble thrown into its surface in that its bolide-struck Atlantic coastline was still, quiet, at peace—however, below its surface, in the murky depths, it left scars and the effects of its occurrence, but it took time for us to discover this, some of which we are still discovering today and hopefully more tomorrow.

You probably noticed that it was mentioned above that the bolide struck the Atlantic coastline and not Chesapeake Bay. This was the case, of course, because at the time of impact there was no Chesapeake Bay or Eastern Shore; it was certainly not a coastline formation as we know it today. The Atlantic Ocean at that time hugged the area near what is now known as Richmond, Virginia, roughly in the path of Interstate-95 today. During the late Eocene and at the time of the bolide impact, the area was covered by ocean (sea level was much higher than it is today due to the climate being much warmer than it is at present). The Chesapeake Bay itself did not form until after the Wisconsin glaciation ice sheet melted 18,000 years ago. As described in a story in the Richmond Times-Dispatch (2005):

> The waves nearly overlapped the Blue Ridge Mountains before washing back into the horrible gash, and then covered the super-heated water beneath a thick blanket of debris, rock and sediment. Over time, as new geologic formation settled, it set the stage for Virginia's baffling groundwater system, with its pockets of salty groundwater. USGS geologist Wylie Poag, another co-discover of the bay's ancient depression, has called it "probably the most dramatic geological event that ever took place in the Atlantic margin of North America."

Over the last 35 million years, erosion has deposited sediments on top of the water, and shifts in the path of the Susquehanna River have formed the peninsula of the Eastern Shore. Today, the impact crater is buried under 1500 feet of sand, silt, clay, and gravel, with the center of the crater underneath the Delmarva Peninsula (GSA, 2009). Most people know what an impact crater looks like. When we look up to Earth's moon, it is quite evident to us (or so we think is the case) what an impact crater looks like. Also, the image derived from Meteor Crater in Arizona, some 38 miles east of Flagstaff, is the archetypical example of what cratering experts call a *simple* crater. It is a shallow, bowl-shaped excavation, 1 km in diameter, with an upraised subcircular rim, and is extraordinarily well preserved. Craters wider than 10 km are classified as complex craters, because they exhibit additional features. Complex craters are different than the simple bowl-shaped craters because the object that created it hit hard and fast enough to melt the rock and splash it tall in the center like a skyscraper, where it hardened. Like simple craters, the outer margin of complex craters is marked by a raised rim. Inside the rim is a broad, flat, circular plain, called the annual trough. Large slump blocks fall away from the center's outer wall and slide out over the floor of the annular trough toward the crater center. The inner edge of the annular trough is marked by either a central mountainous peak, a ring of peaks (a peak ring), or both. Inside the peak ring is the deepest part of the crater, called the inner basin. The Chesapeake Bay crater has all the characteristics of a peak-ring crater and is said to look like an upside-down sombrero, with its upturned outer rim, a trough, and then a high peak in the center (see Figure 13.4).

Before we move onto the effects of the Chesapeake Bay bolide strike, we need to make an important point clear. Namely and specifically, the Chesapeake Bay bolide did not create Chesapeake Bay.

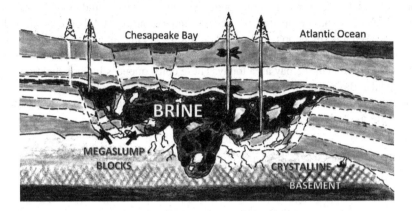

FIGURE 13.4   Adaptation of USGS illustration of the upside-down sombrero-shaped Chesapeake Bay bolide impact crater (1997). Illustration by F.R. Spellman and Kat Welsh.

No?

Yes, no.

Here is what we know: the Bay is nowhere near 35 million years old. The fact is as late as 18,000 years ago, the Bay region was dry land.

Dry land?

Yes.

It was the last ice sheet that created what we know of Chesapeake Bay today. This last great ice was at its maximum over North America, and sea level was about 200 m lower than at present. This sea level exposed the area that is now the bay bottom and part of the continental shelf. With sea level this low, the major East Coast rivers had to cut narrow valleys across the region all the way to the shelf edge. About 10,000 years ago, however, the ice sheets began to melt rapidly, causing sea level to rise and flood the shelf and the coastal river valleys—these flooded valleys became the major modern estuaries, such as Delaware Bay and Chesapeake Bay. Note that the rivers of the Chesapeake region converged at a location directly over the buried crater. In short, the impact crater did not create the Chesapeake Bay, but instead created a long-lasting topographical depression, which, after the end of the latest Ice Age, helped determine the eventual location of Chesapeake Bay.

EFFECTS OF THE CHESAPEAKE BAY BOLIDE IMPACT

One might think that after the bolide hit, water blasted out of the area would be sucked back in along with whatever debris was nearby—including sediments laid down over 35 million years scattered in a jackstraw clutter and crushed debris—and all this would be enough to fill the crater's enormous gorge.

Well, not exactly.

Although it is true that water and debris were sucked back into the chasm, it is also true that it was not enough to eliminate the low spot in the floor of what is now the Chesapeake Bay.

Keep in mind that we did not know until recently that the crater existed on the floor of Chesapeake Bay and the surrounding area. Dr. Poag's (and others') discovery of the giant crater has completely revised our understanding of Atlantic Coastal Plain evolution. What many people in the Hampton Roads (Tidewater) Region do not know is that Poag's studies revealed several consequences of the ancient cataclysm that still affect citizens in the bay area today. These consequences include location of the Chesapeake Bay, river diversion, ground instability, disruption of coastal aquifers, and land subsidence. It is also important to point out that the effects of the bolide impact are multiple, and these are briefly discussed below.

LOCATION OF CHESAPEAKE BAY

Thirty-five million years ago, the Chesapeake Bay did not exist. In fact, as late as 18,000 years ago, the bay region was dry land; And again, the last great ice sheet was at its maximum over North America, and sea level was about 200 m lower than at present. This sea level exposed the area that is now the bay bottom and continental shelf. Because sea level was so low, the major East Coast rivers had to cut narrow valleys across the region all the way to the shelf edge. About 10,000 years ago, however, the ice sheets began to melt rapidly, causing sea level to rise and flood the shelf and the coastal river valleys. The flooded valleys became the major modern estuaries, such as Delaware Bay and Chesapeake Bay. To come to the point, the impact crater created a long-lasting topographic depression, which helped determine the eventual location of Chesapeake Bay (USGS, 2016).

RIVER DIVERSION

The rivers of the Chesapeake region converged at a location directly over the buried crater. Some might think the convergence of these rivers as merely coincidence.

Is it?

The short answer: no, it is not coincidence. Notice that in Figure 13.5 the important river channels in the area change course significantly just after they cross the rim of the buried bolide crater. These channels are actually successive buried ice-age channels of the ancient Susquehanna River (formed from 450,000 years ago to 20,000 years ago). Combined with seismic evidence that past impact units sag and thicken over the crater, the river diversion indicates that the ground surface over the crater remained lower than the areas outside the crater for 35 million years.

The question is, why?

Why does the Rappahannock River flow southeastward to the Atlantic and in contrast the York and James Rivers make sharp turns to the northeast near the outer rim of the crater?

What is the answer?

Well, the course of the York and James Rivers in the lower bay region are the result of the ongoing influence of differential subsidence over the bolide crater.

Differential subsidence?

Yes. Absolutely. Two factors cause subsidence in the region. First, subsidence is the result of loading during the past 35 million years since the impact. Secondly,

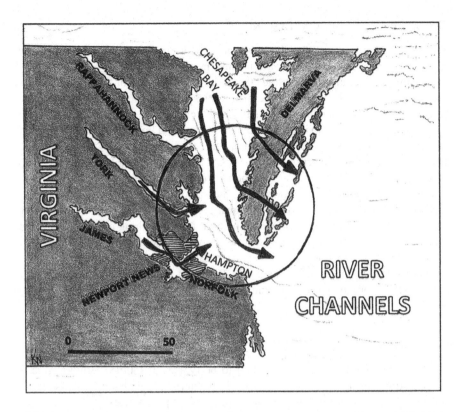

FIGURE 13.5 River channels Chesapeake Bay region. Source: Adapted from USGS (1997). Illustration by F. R. Spellman and Kat Welsh.

subsidence is also due to compaction of the breccia; that is, rock composed of broken fragments of minerals or gravel cemented together by a fine-grained matrix. The crater breccia is 1.2 km thick and was deposited as water-saturated sandy, rubble-bearing, non-jelled jello-like slurry. The sediment layers surrounding it were already partly consolidated, so the mushy breccia would compact much more rapidly under its subsequent sediment load than the surrounding strata.

You may be asking what all this has to do with the local bay rivers? The combination of the two factors detailed above produced a subsidence differential, causing the land surface over the breccia to remain lower than the land surface outside the crater. Therefore, the river valleys covered the crater and were located in those particular places when rising sea level flooded them. In short, the impact crater created a long-lasting topographic depression, which helped predetermine the eventual location of Chesapeake Bay (USGS, 1997). Finally, it is important to point out that one of the main focuses of this book and the SWIFT initiative is land subsidence in the Chesapeake Bay area; this will be discussed in detail later.

For now the important point to know and to remember is the continued influence of differential subsidence over the crater.

Ground Instability due to Faulting

Seismic profiles across the crater show many faults that cut the sedimentary beds above the breccia and extend upward toward the bay floor. The current resolution of our seismic profiles allows us to trace the faults to within 10 m of the bay floor. These faults represent another result of the differential compaction and subsidence of the breccia. As the breccia continues to subside under the load of post-impact deposits, it subsides unevenly due to its viable content of sand and huge clasts. This eventually causes the overlying beds to bend and break and to slide apart along the fault planes. These faults are zones of crustal weakness and have the potential for continued slow movement, or sudden large offsets if reactivated by earthquakes.

Some might ask why it is important to know about the faults and their location. It is important for us to know in detail the location, orientation, and amount of offset of these compaction faults because of the potential for the faulting to separate adjacent sides of the confining unit over the saltwater reservoir. If this occurred, it could allow the salty water to flow upward and to contaminate the freshwater supply.

Using the seismic profiles on hand, Dr. Poag has identified and is mapping more than 100 faults or fault clusters around and over the crater, which reach to or near the bay floor (USGS, 2016).

Disruption of Coastal Aquifers

The hydrogeological framework thought to be typical of southeastern Virginia, in cross section, consists of groundwater aquifers alternating with confining beds. The aquifers are mainly sand, which contain water-filled pore spaces between the sand grains. The pore spaces are connected, which allows the water to flow slowly though the aquifers. The confining beds are mainly clay beds, which have only very fine pores, and these are poorly interconnected, which greatly retards or prevents the flow of water. Before we knew about the Chesapeake Bay crater, this framework of alternating aquifers and confining units was applied to models of groundwater flow and water quality assessments in the Lower Chesapeake Bay region.

Based on core samples, researchers have determined that in the crater area itself the orderly stack of aquifers seen outside the crater does not exist; instead, they were truncated and excavated by the bolide impact. In place of those aquifers, there is now a single huge reservoir with a volume of 4000 km³. That's enough breccia to cover all of Virginia and Maryland with a layer 30 m thick. But the most startling part is that this huge new reservoir does not contain fresh water like the aquifers it replaced; the pore spaces are filled with briny water that is 1.5 times saltier than normal seawater. This water is too salty to drink or to use in industry (USGS, 2016). It is interesting to note that for decades geohydrologists and others in the Hampton Roads region scratched their collective heads and wondered why in locations away from the crater water wells yielded good-quality fresh water suitable for potable purposes; however, whenever wells were drilled within the crater ring or close to it salty water was all that could be found.

DID YOU KNOW?

The parameters for saline water are as follows:

- *Fresh water*—Less than 1000 ppm
- *Slightly saline water*—From 1000 to 3000 ppm
- *Moderately saline water*—From 3000 to 10,000 ppm
- *Highly saline water*—From 10,000 to 35,000 ppm
- *Ocean water*—About 35,000 ppm of salt
- *Chesapeake Bay crater water*—About 1.5 times more salt than normal seawater

The presence of this hypersaline aquifer has some practical implications for groundwater management in the lower bay region. For example, we need to know how deeply buried the breccia is in order to avoid drilling into it inadvertently and contaminating the overlying freshwater aquifers. Its presence also limits the availability of fresh water. On the Delmarva Peninsula, over the deepest part of the crater, only the aquifers above the breccia are available for fresh water. The crater investigation shows that we need to be especially conservative of groundwater use in the area (USGS, 2016).

Land Subsidence

Land subsidence and the potential for land rebound provided by injecting treated wastewater to drinking water quality into the Potomac Aquifer are the focus of this book. Later much will be said about this topic and how it is related to the Chesapeake Bay bolide impact crater and its effect, along with relative sea-level rise occurring at the present time. For now it is important to point out that there is growing evidence that accelerated land subsidence is reflected in the geology and topography of the modern land surfaces around the bolide crater. The breccia is 1.3 km thick and was deposited as water-saturated, sandy, rubble-bearing slurry (like concrete before it hardens). The sediment layers surrounding the crater, on the other hand, were already partly consolidated, and so the mushy breccia compacted much more rapidly under its subsequent sediment load than the surrounding strata. The compaction differences produce a subsidence differential (i.e., the difference in subsidence between two points on the crater), causing the land surface over the breccia (due to breccia compaction) to remain lower than the land surface over sediments outside the crater.

During Dr. Poag's investigation, he and his team observed that the boundary between older surface rocks and younger surface rocks coincides with the position and orientation of the crater rim on all three peninsulas that cross the rim. The older beds have sagged over the subsiding breccia, and the younger rocks have been deposited in the resulting topographic depression. The topography also reflects the differential subsidence. The Suffolk Scarp and the Ames Ridge are elevated landforms (10–15 m high) located at, and oriented parallel to, the crater rim.

Crater-related ground subsidence may also play a role in the high rate of relative sea-level rise documented for the Chesapeake Bay region. One of the locations of the highest relative sea-level rise is a Hampton Roads (the lower part of the James River, located over the crater rim).

LAND SUBSIDENCE[2]

From the Lower Chesapeake Bay region (Hampton Roads or Tidewater) of Virginia to San Francisco Bay/Delta to the Florida Everglades and from upstate New York to Houston, people are dealing with a common problem in these diverse locations—vanishing land as a result of land subsidence due to the withdrawal of groundwater. Vanishing land due to subsidence is not an isolated problem: an area of more than 17,000 square miles in 45 states, an area roughly the size of New Hampshire and Vermont combined, has been directly affected by land subsidence (USGS, 2013). More than 80% of the identified subsidence in the nation is a consequence of our exploitation of underground water. Moreover, it seems certain that the increasing development of land and water resources threatens to exacerbate existing land-subsidence problems and initiate new ones. This chapter focuses on three principal processes causing land subsidence: the compaction of aquifer systems because of groundwater withdrawal, the oxidation of organic soils, and the collapse of cavities in carbonate and evaporite rocks. Additionally, in this chapter we point out the value of applying science and engineering innovations used in effectively mitigating or limiting damages from land subsidence. One thing is certain: scientific understanding is critical to the formulation of balanced decisions about the management of land and water resources. When scientific information is presented in plain English and in a conversational approach and when obstacles are not present, understanding can flow like unimpeded groundwater.

By the way, did you know that land subsidence helps explain why the region has the highest rates of sea-level rise on the Atlantic Coast of the United States. An important fact to keep in mind is that land subsidence and rising water levels combine to cause what is known as *relative sea-level rise*.

DID YOU KNOW?

Water budgets provide a means for evaluating availability and sustainability of a water supply. A water budget simply states that the rate of change in water stored in an area, such as a watershed, is balanced by the rate at which water flows into and out of the area. Keep in mind that water slowly flows into aquifers via infiltration. If you ever visit and hike the Narrows in Zion National Park in Utah (highly recommended), there is a waterspout at the lodge that when opened pours water that took an estimated 3000 years to flow from the top of the canyon to the tap. Having personally consumed gallons of this water, I can confirm that it is the tastiest water I have ever consumed.

GROUNDWATER WITHDRAWAL

As mentioned, permanent subsidence can occur when water stored beneath the Earth's surface is removed by pumping. The reduction in fluid pressure in the pores and cracks of aquifer systems, especially in unconsolidated rocks, is inevitably accompanied by some deformation of the aquifer system. Because the granular structure—the so-called "skeleton"—of the aquifer system is not rigid, but more or less complaint, a shift in the balance of support for the overlying material causes the skeleton to deform slightly. Both the aquifers and aquitards that constitute the aquifer system undergo deformation, but to different degrees. During the typically slow process of aquitard drainage (when the irreversible compression or consolidation of aquitards occurs) is when almost all the permanent subsidence takes place (Tolman and Poland, 1940). This concept, known as the aquitard-drainage model, has formed the theoretical basis of many successful subsidence investigations.

DID YOU KNOW?

Studies of subsidence in the Santa Clara Valley in California established the theoretical and field application of the laboratory-derived principle of effective stress and theory of hydrodynamic consolidation to the drainage and compaction of aquitards (Tolman and Poland, 1940; Poland and Green, 1962; Green, 1964; Poland and Ireland, 1988).

EFFECTIVE STRESS

The principle of effective stress was first proposed by Terzaghi (1925). For our purpose in this book, "effective" means the calculated stress that was effective in moving soil and/or causing displacements. According to this principle, when the support provided by fluid pressure is reduced, such as when groundwater levels are lowered, support previously provided by the pore-fluid pressure is transferred to the skeleton of the aquifer system, which compresses to a degree. On the other hand, when the pore-fluid pressure is increased, such as when groundwater recharges the aquifer system, support previously provided by the skeleton is transferred to the fluid and the skeleton expands. In this way, the skeleton alternatively undergoes compression and expansion as the pore-fluid pressure fluctuates with aquifer-system discharge and recharge. When the load on the skeleton remains less than any previous maximum load, the fluctuations create only a small elastic deformation of the aquifer system and small displacement of land surface. This fully recoverable deformation occurs in all aquifer systems, commonly resulting in seasonal, reversible displacements in land surface of up to 1 inch or more in response to the seasonal changes in groundwater pumpage (USGS, 1998).

PRECONSOLIDATION STRESS

The maximum level of past stressing of a skeletal element is termed the precon-solidation stress. Stated differently, preconsolidation stress is the maximum effec-tive vertical overburden stress that particular soils has sustained in the past. When the load on the aquitard skeleton exceeds the preconsolidation stress, the aquitard skeleton may undergo significant, permanent rearrangement, resulting in irrevers-ible compaction. Because the skeleton defines the pore structure of the aquitards, this results in a permanent reduction in pore volume as the pore fluid is "squeezed" out of the aquitards into the aquifers. In confined aquifer systems subject to large-scale overdraft, the volume of water derived from irreversible aquitard compaction is essentially equal to the volume of subsidence and can typically range from 10% to 30% of the total volume of water pumped. This represents a one-time mining of stored groundwater and a small permanent reduction in the storage capacity of the aquifer system. Alternative names of the preconsolidation stress are preconsoli-dation pressure, pre-compression stress, pre-compaction stress, and preload stress (Dawidowski and Koolen, 1994).

AQUITARDS ROLE IN COMPACTION

In recent decades, increasing recognition has been given to the critical role of aqui-tards in the intermediate- and long-term response of alluvial systems to groundwater pumpage. Aquitard systems play an important role in compaction. In many such systems, interbedded layers of silt sand clays, once dismissed as non-water yielding, comprise the bulk of the groundwater storage capacity of the confined aquifer sys-tem. This is the case based on their substantially greater porosity and compressibility and, in many cases, their greater aggregate thickness compared to the more transmis-sive, coarser-grained sand and gravel layers (USGS, 2013).

Aquitards are less permeable than aquifers. Thus, the vertical drainage of aquita-rds into adjacent pumped aquifers may proceed very slowly and thus lag far behind the changing water levels in adjacent aquifers. The lagged response within the inner portions of a thick aquitard may be largely isolated from the higher-frequency sea-sonal fluctuations and more influenced by lower-frequency, longer-term trends in groundwater levels. Because the migration of increased internal stress into the aqui-tard accompanies its drainage, as more fluid is squeezed from the interior of the aquitard, larger and larger internal stresses from the interior of the aquitard, larger and larger internal stresses propagate farther into the aquitard.

DID YOU KNOW?

Responses to changing water levels following several decades of groundwater development suggest that stresses directly driving much of the compaction are somewhat insulated from the changing stresses caused by short-term water-level variations in the aquifers.

When the preconsolidation stress is exceeded by the internal stresses, the compressibility increases dramatically, typically by a factor of 20–100 times, and the resulting compaction is largely nonrecoverable. At stresses greater than the preconsolidation stress, the lag in aquitard drainage increases by comparable factors, and concomitant compaction may require decades or centuries to approach completion. The theory of hydrodynamic consolidation (Terzaghi, 1925)—an essential element of the *aquitard drainage model*—describes the delay involved in draining aquitards when heads are lowered in adjacent aquifers, as well as the residual compaction that may continue long after drawdowns in the aquifers have essentially stabilized. Numerical modeling based on Terzaghi's theory has successfully simulated complex histories of compaction observed in response to measure water-level fluctuations (Helm, 1975).

DID YOU KNOW?

Hydrocompaction—compaction due to wetting—is a near-surface phenomenon that produces land-surface subsidence through a mechanism entirely different from compaction of deep, overpumped aquifer systems.

THE VANISHING OF HAMPTON ROADS

When the name Hampton Roads is mentioned, many people shake their heads and say, "What?"

When the name Hampton Roads is roughly familiar to some people, they ask, "Do you mean the body of water ... or do you mean the location?"

"Answer, please?"

Well, actually the term Hampton Roads is the name of both the body of water in Virginia and the surrounding metropolitan region in southeastern Virginia and northeastern North Carolina. The land area is also known as Tidewater.

In this text, when we refer to Hampton Roads we are referring to both the water body and the land region as one because land subsidence and relative sea-level rise in the area pertain to both. With regard to the total area, it is comprised of 527 square miles (1364 km²) and is made up of nine major cities: Norfolk, Virginia Beach, Chesapeake, Newport News, Hampton, Portsmouth, Suffolk, Poquoson, and Williamsburg; as a combined statistical area, it also includes Kitty Hawk, Elizabeth City, North Carolina. The entire area has a population of over 1.7 million. With regard to the body of water known as Hampton Roads, it is one of the world's largest natural harbors. It incorporates the mouths of the Elizabeth River, Nansemond River, and James River with several smaller rivers and empties into the Chesapeake Bay—a treasured estuary—near its mouth leading to the Atlantic Ocean.

HAMPTON ROADS: SEA-LEVEL RISE

Of all the potential impacts of natural (cyclical) or human-induced climate change, a global rise in sea level appears to be the most certain and the most dramatic.

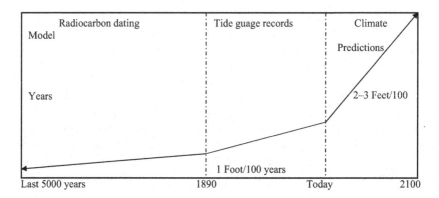

FIGURE 13.6 Sea-level rise in Hampton Roads region past and present. Source: USFWS (1995).

As shown in Figure 13.6, for the last 5000 years the rate of sea level rise was only 3 feet per 1000 years. In the Chesapeake Bay region, the relative rise in sea level has been about 1 foot during the last 100 years (Figure 13.6). While scientists view this rapid rate as possibly a temporary acceleration, many of these same scientists believe that it signals a new trend in response to global warming. The point is if the rate of rise accelerates in the near future as projected, it could have serious repercussions for Chesapeake Bay.

Because water levels are measured relative to the land, and as stated earlier, relative sea-level rise in the Chesapeake Bay region has two components: global water level increase and land subsidence. Worldwide or eustatic seal-level rise is caused by water released from melting glaciers and thermal expansion of seawater. Both are related to global warming and have amounted to a 6-inch rise in the last century.

MAJOR PHYSICAL EFFECTS OF SEA-LEVEL RISE

With increased global temperatures, global sea-level rise will occur at a rate unprecedented in human history (Edgerton, 1991). Changes in temperature and sea level will be accompanied by changes in salinity levels. For example, a coastal freshwater aquifer is influenced by two factors: pumping and mean sea level. In pumping, if withdrawals exceed recharge, the water table is drawn down and saltwater penetrates inland. With mean sea level, the problem occurs if sea level rises and the coastline moves inland, reducing aquifer area. Additional problems brought about by changes in temperature and sea level are seen in tidal flooding, oceanic currents, biological processes of marine creatures, in runoff and landmass erosion patterns, and saltwater intrusion.

The most important direct physical effects of sea-level rise is on a coastal beach system. At current rates of sea-level rise of 1–2 mm/year, significant coastal erosion is already produced. Two major factors contribute to beach erosion. First, deeper coastal waters enhance wave generation, thus increasing their potential for overtopping barrier islands. Second, shorelines and beaches will attempt to establish new equilibrium positions according to what is known as the *Bruun rule*; these adjustments will include a recession of shoreline and a decrease in shore slope (Bruun, 1962, 1986).

Major Direct Human Effects of Sea-Level Rise

Along with the physical effects of sea-level rise, in one way or another, directly or indirectly, accompanying effects have a direct human side, especially concerning human settlements and the infrastructure that accompanies them: highways, airports, waterways, water supply and wastewater treatment facilities, landfills, hazardous waste storage areas, bridges, and associated maintenance systems. Sea-level rise could also cause intrusion of saltwater into groundwater suppliers (Edgerton, 1991).

To point out that this infrastructure will be placed under tremendous strain by a rising sea level coupled with other climatic change is to understate the possible consequences. Indeed, the impact on infrastructure is only part of the direct human impact. For example, there is widespread agreement among scientists that any significant change in world climate resulting from warming or cooling will (1) disrupt world food production for many years, (2) lead to a sharp increase in food prices, and (3) cause considerable economic damage.

Just how much of a rise in sea level are we talking about? According to USEPA (1995), "if the experts on whom we relied fairly represent the breadth of scientific opinion, the odds are fifty-fifty that greenhouse gases will raise sea level at least 15 cm by the year 2050, 25 cm by 2100, and 80 cm by 2200" (p. 123).

LAND SUBSIDENCE IN HAMPTON ROADS[3]

This section concentrates specifically on the Hampton Roads area and in particular the Lower Chesapeake Bay region. Within the Lower Chesapeake Bay region, the continuing appearance of dead zone problems, nutrient pollution, sediment contamination, and sea life decline are without a doubt serious problems that persist to garner the attention of officials responsible for monitoring and managing the health of Chesapeake Bay. As serious as these problems are, it is the ongoing rise in relative sea-level problem that is beginning to impact Hampton Roads and offers a future that can best be described as foreboding and quite wet unless certain mitigation practices are put into place, not tomorrow but at the present and continuing into the future. Part of the relative sea-level rise problem is due to land subsidence, and it is this problem that is addressed in the section.

So, let's get to it.

Local or isostatic factors contribute to *relative* sea-level rise in Hampton Roads, Virginia, through subsidence or sinking of the land. In the Chesapeake Bay area, subsidence of land is due to both geological factors and excessive withdrawal of groundwater, which has amounted to 6 inches in the last 100 years at rates of 0.039–0.189 inches per year (1.1–4.8 mm/year). Consequently, there has been a relative increase in sea level in the Chesapeake Bay area of 1 foot in the last century. More specifically, land subsidence in the region is the result of flexing of the Earth's crust from glacial isostatic adjustment in response to glacier formation and melting. In addition, more than half of the observed subsidence is the result of the aquifer system in Hampton Roads that has been compacted by extensive groundwater extraction at rates of 1.5–3.7 mm/year. This helps explain why the southern Chesapeake Bay region has the highest rate of sea-level rise on the Atlantic Coast of the United States (Zervas, 2009). Because the communities in the region must grapple with flooding problems that lead to disappearance of existing land by combination of rising sea levels and subsiding land, all of which will continue to worsen in the future, it is important to understand and potentially manage land subsidence.

Data indicate that land subsidence has been responsible for more than half the relative sea-level rise measured in the Hampton Roads area; it also suggests that the problem will be ongoing in the future. This is bad news for those residing in the area. Land subsidence is a serious issue because the increased flooding has and will continue to have important economic, environmental, and human health consequences for the heavily populated and ecologically important southern Chesapeake Bay region.

As explained earlier, land subsidence is the sinking or lowering of the land surface. In the United States, most land subsidence is caused by human activities (Galloway et al., 1999). Earlier we described land subsidence problems in the western region of the country. And after recognition of the problems associated with groundwater drawdown and resulting land subsidence, the areas affected set up monitoring networks and ultimately adopted new water-management practices to prevent or arrest land subsidence.

Experience has shown that rates and locations of land subsidence change over time, so accurate measurements and predictive tools are needed to improve understanding of land subsidence. Although rates of land subsidence are not as high on the Atlantic Coast as they have been in the Houston-Galveston area or the Santa Clara Valley, land subsidence is important because of the low-lying topography and susceptibility to sea-level rise in the southern Chesapeake Bay region.

WHY LAND SUBSIDENCE IS A CONCERN IN THE CHESAPEAKE BAY REGION

In the Chesapeake Bay region, increased flooding, wetland and coastal ecosystem alteration, and damage to infrastructure and historical sites are all the result of land subsidence. This problem is not a new one; it is well known to regional planners who have gained understanding of what land subsidence is, that is, why, where, and how fast it is occurring, now and in the future.

LAND SUBSIDENCE CONTRIBUTES TO RELATIVE SEA-LEVEL RISE

Land subsidence contributes to the relative sea-level rise that has been measured in the Chesapeake Bay. Tidal-station measurements of sea levels, however, do not distinguish between water that is rising and land that is sinking—the combined elevation changes are termed *relative sea-level rise*. Global sea-level rise and land subsidence increase the risk of coastal flooding and contribute to shoreline retreat.

As relative sea levels rise, shorelines retreat and the magnitude and frequency of near-shore coastal flooding increase. This is particularly a problem in Norfolk,

Virginia (downtown area and Ocean View District), where during a coastal storm event and corresponding high tide, downtown Norfolk streets flood; at times, it can flood to several feet. Although land subsidence can be slow, its effects accumulate over time; this has been an expensive problem in the Norfolk and other parts of the southern Chesapeake Bay region. Analysis by McFarlane (2012) found that between 59,000 and 176,000 residents living near the shores of the southern Chesapeake Bay could be either permanently inundated or regularly flooded by 2100. This estimate was based on the 2010 census data, using the spring high tide as a reference elevation and assuming a 1-m relative sea-level rise. Damage to personal property was estimated to be $9 billion to $26 billion, and 120,000 acres of economically valuable land could be inundated or regularly flooded, under these same assumptions. Historic and cultural resources are also vulnerable to increased flooding from relative sea-level rise in the southern Chesapeake Bay, particularly at shoreline sites near tidal water, such as the 17th-century historic Jamestown site.

It should be pointed out that the shoreline area in southern Hampton Roads is not the only area prone to flooding. Land subsidence can also increase flooding in areas away from the coast in low-lying areas such as Franklin, Virginia. The city of Franklin is about 60 road miles west of Hampton Roads. The Blackwater River Basin, which encompasses Franklin and other local areas, can be subject to increased flooding as the land sinks. In fact, Franklin and the counties of Isle of Wight and Southampton have experienced large floods in years (Federal Emergency Management Agency, 2002). Land subsidence may be altering the topographic gradient that drives the flow of the river and possibly contributing to the flooding.

Wetland and marsh ecosystems in low-lying coastal areas are sensitive to small changes in elevation (Cahoon et al., 2009). Salt marshes, which are widespread in the southern Chesapeake Bay region, are dependent on tidal dynamics for their existence. Small changes in either land or sea elevations can alter sediment deposition, organic production and plant growth, and the balance between fresh water and seawater (Morris et al., 2002). The effects of sea-level rise on tidal wetlands are numerous and already apparent in local wetlands. Some of these effects include:

- Shoreline erosion
- Habitat loss
- Changes in tidal amplitude
- Landward migration of tidal waters
- Landward migration of habitats
- More frequent inundation
- Changes in plant and animal species composition
- Changes in tidal flow patterns
- Migration of estuarine salinity gradients
- Changes in sediment transport

Although sea-level rise has one of the most direct effects on tidal wetlands, shoreline environments also are affected by land subsidence. When land subsides, it subjects shorelines to increased waver action, increasing erosion and wash-over. This type of damage is happening in the Chesapeake Bay because of relative sea-level rise (Erwin et al., 2011; Kirwan and Guntenspergen, 2012; Kirwan et al., 2012). Major changes in the coastal and marine ecosystem of the southern Chesapeake Bay are expected to be caused by relative sea-level rise (Spellman and Whiting, 2006); these changes will likely be more severe if land subsidence continues.

Land subsidence can damage infrastructure: buildings, bridges, canals, water and wastewater treatment plants, electrical substations, communication towers, pipes, and other components that make up a region's infrastructure can be damaged from relative groundwater rise or from differential settling in areas with high subsidence gradients (Galloway et al., 1999). As land sinks and sea level continues to rise, groundwater levels rise toward the land surface in coastal areas, which can cause problems for subterranean structures, septic fields, buried pipes and tanks and cables, and infrastructure not designed for elevated groundwater levels. Storm and wastewater interceptor lines in urban areas are vulnerable because land subsidence can alter the topographic gradient driving the flow through the sewers, causing increased flooding and more frequent sewage discharge from combined sewer overflows.

AQUIFER COMPACTION

When groundwater is pumped from the Potomac aquifer system in southern Chesapeake Bay, pressure decreases. The pressure change is reflected by water levels in wells, with water levels decreasing as aquifer-system pressure decreases. As water levels decrease, the aquifer system compacts causing the land surface to subside (see Figure 13.7). Water levels have decreased over the entire Virginia Coastal Plain and in the Potomac aquifer, which is the deepest and thickest aquifer in southern Chesapeake Bay region and supplies about 75% of groundwater withdrawn from the Virginia Coastal Plain aquifer system (Heywood and Pope, 2009).

Three factors determine the amount of aquifer-system compaction: water-level decline, sediment compressibility, and sediment thickness. If any of these three factors increase in magnitude, then the amount of aquifer-system compaction and land subsidence increases. Because all three of these factors vary spatially across the southern Chesapeake Bay region, rates of load subsidence caused by aquifer-system compaction also vary spatially across the region.

The Virginia Coastal Plain aquifer system consists of many stacked layers of sand and clay. Although groundwater is withdrawn primarily from the aquifers (sandy layers), most compaction occurs in confining units and clay lenses, the relatively impermeable layers sandwiched between and within the aquifers (Pope and Burbey, 2004). The compression of the clay layers is mostly nonrecoverable, meaning that, if groundwater levels later recover and increase, then the aquifer system does not expand to its previous volume and the land surface does not rise to its previous elevations (Pope and Burbey, 2004). Konikow and Neuzil (2007) estimated that 95% of the water removed from storage in the Virginia Coastal Plain aquifer system between 1891 and 1980 was derived from the confining layers.

The timing of aquifer-system compaction is also important. After groundwater levels drop, compaction can continue for many years or decades. When groundwater is pumped from an aquifer, pressure decreases in the aquifer. The pressure decrease

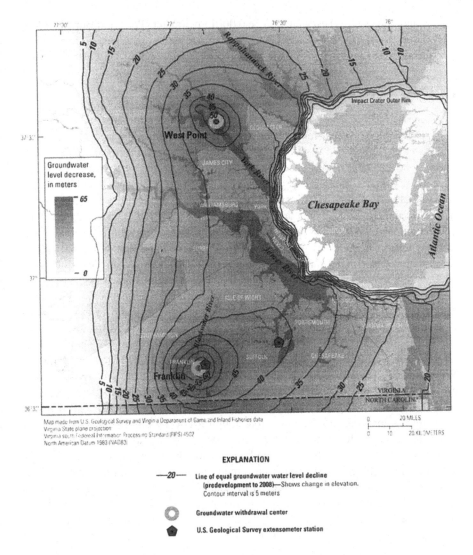

FIGURE 13.7   Groundwater level decreases from 1990 to 2008. Source: USGS (2013); modified from Heywood and pope (2009).

then slowly propagates into clay layers that are adjacent to or within the aquifer. As long as pressure continues to decrease in the clay layers, compaction continues (USGS, 2013).

The layered sediments of the Virginia Coastal Plain aquifer system range in grain size from very fine (silts and clays) to coarse (sand and shell fragments) (McFarland and Bruce, 2006). Based on the hydrogeologic framework of McFarland and Bruce (2006) and Heywood and Pope (2009), confining layers outside the bolide impact crater occupy about 16% of the total aquifer-system thickness, an average of 100 m out of the total average thickness of 619 m. These continuing layers have high specific storage (compressibility) estimated to be 0.00015 per meter (Pope and Burbey, 2004). Clay layers overlying and within the Potomac aquifer are compressing as aquifer pressure decrease migrates vertically and laterally from pumping wells (USGS, 2013).

GLACIAL ISOSTATIC ADJUSTMENT

The last ice age occurred about 16,000 years ago, when great sheets of ice covered much of Earth's Northern Hemisphere. Though the ice melted long ago, the land once under and around the ice is still rising and falling in reaction to its ice burden. This ongoing movement of land is called glacial isostatic adjustment or postglacial rebound. Here's how it works: imagine that you are lying down on a soft mattress and then get up from the same spot. You will likely see an indentation in the mattress where your body had been, and a puffed-up area around the indentation where the mattress rose. Once you get up, the mattress takes a little time before it releases back to its original shape. Earth is constantly on the move. We can't always see the movement unless we witness an earthquake or its aftermath or a volcanic eruption like Mount St. Helens' event in 1980 or land fissure and sinkholes opening in the Earth around us (NOAA, 2015).

The Virginia Coastal Plain aquifer consists of layered sediments overlaying crystalline bedrock. Bedrock is not solid and unyielding, but actually flexes and moves in response to stress. Bedrock in the mid-Atlantic region is moving slowly downward in response to melting of the Laurentide ice sheet that covered Canada and the northern United States during the last ice age (Sella et al., 2007; Boon et al., 2010). When the ice sheet still existed, the weight of the ice pushed the underlying Earth's crust downward and, in response, areas away from the ice sheet were forced upward (called glacial forebulge; see Figure 13.8). The southern Chesapeake Bay region is in the glacial forebulge area and was forced upward by the Laurentide ice sheet. The ice sheet started melting about 18,000 years ago and took many thousands of years to disappear entirely. As the ice melted and its weight was removed, glacial forebulge areas, which previously had been forced upward, began sinking and continue to sink today. Again, this movement of the Earth's crust in response to ice loading or melting is called glacial isostatic adjustment. Data from GPS and carbon dating of marsh sediments indicate that regional land subsidence in response to glacial isostatic adjustment in the southern Chesapeake Bay region may have a current rate of about 1 mm/year (Engelhart, 2010; Engelhart et al., 2009). This downward velocity rate is uncertain and probably not uniform across the region.

THE BOTTOM LINE

This chapter has set the first row of foundation blocks for the material to follow. Specifically, this chapter described the late Eocene period in the Virginia Coastal Plain area when the formerly quiescent geological regime was dramatically transformed when a bolide struck in the vicinity of the Delmarva Peninsula. This consequential event produced the following principal consequences (USGS, 1998):

- The bolide carved a roughly circular crater twice the size of the state of Rhode Island (~6400 km^2) and nearly as deep as the Grand Canyon (1.3 km deep).

- The excavation truncated all existing groundwater aquifers in the impact area by gouging ~4300 km^3 of rock from the upper lithosphere, including Proterozoic and Paleozoic crystalline basement rocks and Middle Jurassic to upper Eocene sedimentary rocks.
- A structural and topographic low formed over the crater.
- The impact crater may have predetermined the present-day location of Chesapeake Bay.
- A porous breccia lens, 600–1200 m thick, replaced local aquifers, resulting in groundwater ~1.5 times saltier than normal sea water.
- Long-term differential compaction and subsidence of the breccia lens spawned extensive fault systems in the area, which are potential hazards for local population centers in the Chesapeake Bay area.

Note: In Part 3 of this book, subsidence mitigation is discussed with particular emphasis on recharging aquifers with wastewater treated to drinking water standards to halt (hopefully) subsidence. Thus, in Chapter 14, advanced treatment of municipal wastewater using environmental and engineered buffers is discussed in order to provide basic, foundational information for understanding Hampton Roads Sanitation District's Sustainable Water Initiative For Tomorrow (SWIFT)—basically an engineered water cycle—presented in Chapter 15.

NOTES

1 Much of the information presented here is based on F.R. Spellman's *SWIFT* (2021), Lanham, Maryland: Bernan Press.
2 Based on information from USFWS (1995). Vanishing Lands. Accessed 12/20/22 @ http://www.fws.gov/s/amer/Vanishinglands.Sealevel.
3 Based on information from Eggleston and Pope. USGS (2013) Circular 2013–1992.

REFERENCES

Boon, J.D., Brubaker, J.M., and Forest, D.M. (2010). Chesapeake Bay land subsidence and sea-level change-An evaluation of past and present trends and future outlook: Virginia Institute of Marine Science Special Report 425 in Applied Marine Science and Ocean Engineering, 41 p., plus appendixes. Accessed from https://web.vims.edu/GreyLit/VIMS/stramsoe425.pdf.

Bruun, P. (1962). Sea level rise as a cause of shore erosion. *Proceedings of the American Society of Engineers and Journal Waterways Harbors Division*, v. 88, pp. 117–130.

Bruun, P. (1986). Worldwide impacts of sea level rise on shorelines. In *Effects of Changes in Stratospheric Ozone and Global Climate*, vol. 4. New York: UNEP/EPA, pp. 99–128.

Cahoon, D.R., Reed, D.J., Kolker, A.S., Brinson, M.M., Stevenson, J.C., Riggs, S., Christian, R., Reyes, E., Voss, C., and Kunz, D. (2009). Coastal wetland sustainability. In *Coastal Sensitivity to Sea-Level Rise-A Focus on the Mid-Atlantic Region, A Report by the U.S. Climate Change Science Program and the Subcommittee on Global Change Research.* Eds. Titus, J.G., Anderson, K.E., Cahoon, D.R., Gesch, D.B., Gill, S.K., Gutierrez, B.T., Thieler, E.R., and Williams, S.J. Washington, DC: U.S. Environmental Protection Agency, U.S. Climate Change Science Program Synthesis and Assessment Product 4.1, pp. 57–72, Chapter 4.

Dawidowski, J.B. and Koolen, J.J. (1994). Computerized determination of the preconsolidation stress in compaction texting of field core samples. *Soil and Tillage Research*, v. 31,
no. 2, pp. 277–282.

Edgerton, L. (1991). *The Rising Tide: Global Warming and World Sea Levels*. Washington,
DC: Island Press.

Eggleston, J. and Pope, J. (2013). Land Subsidence and Relative Sea-Level Rise in the Southern
Chesapeake Bay Region. USGS Circular 2013-1392. Reston, VA. Accessed 12/21/16 @
https://doi.org/10.3133/cir1392.

Engelhart, S.E. (2010). Sea-Level Changes along the U.S. Atlantic Coast: Implications for
Glacial Isostatic Adjustment Models and Current Rates of Sea-Level Change. Publicly
Accessible Penn Dissertations, 407. Accessed from https://repositioryupenn.edu/
edisserations/407.

Engelhart, S.E., Horton, B.P., Douglas, B.C., Peltier, W.R., and Tornqvist, T.E. (2009). Spatial
variability of late Holocene and 20th century sea-level rise along the Atlantic coast of the
United States. *Biology*, v. 37, pp. 1115–1118.

Erwin, R.M., Brinker, D.F., Watts, B.D., Costanzo, G.R.., and Morton, D.D. (2011). Islands at
bay-Rising seas, eroding islands, and waterbird habitat loss in Chesapeake Bay, USA.
Journal of Coastal Conservation, v. 15, pp. 51–60.

Federal Emergency Management Agency. (2002). *Flood Insurance Study of Franklin, Virginia,
Community 510060 (Revised September 4, 2002)*. Federal Emergency Management
Agency, Washington, DC , 16 p.

Galloway, D.I., Jones, D.R., and Ingebritsen, S. E., eds. (1999). Land Subsidence in the United
States: U.S. Geological Survey Circular 1182, 177 p. Accessed from https://pubs.usgs.
gov/circ/circ1182/.

Green, J.H. (1964). Compaction of the aquifer system and land subsidence in the Santa Clara
Valley, California: U.S. Geological Survey water-Supply Paper 1779-T, 11 p.

GSA. (2009). Chesapeake Bay impact structure: Postimpact sediments. In *ICDP-USGS Deep
Drilling Project in the Chesapeake Bay Impact Structure: Results from the Eyreville
Core Holes*. The Geological Society of America: Washington, DC, pp. 112–118.

Helm, D.C. (1975). One-dimensional simulation of aquifer system compaction ear Pixley,
Calif., part 1. *Constant Parameters: Water Resource Research*, v. 11, pp. 465–478.

Heywood, C.E. and Pope, J.P. (2009). Simulation of groundwater flow in the Coastal Plain
aquifer system of Virginia: U.S. Geological Survey Scientific Investigations Report
2009-5039, 115 p. Accessed from https://pubs.usgs.gov/sir/2009/5039/.

Kirwan, M.L. and Guntenspergen, G.R. (2012). Feedback between intimidation, root production, and shoot growth in a rapidly submerging brackish marsh. *Journal of Ecology*, v.
100, no. 3, pp. 760–770.

Kirwan, M.L., Langley, J.A., Guntenspergen, G.R., and Megonigal, J.P. (2012). The impact of
sea-level rise on organic matter decay rates in Chesapeake Bay brackish tidal marshes.
Biogeosciences Discussions, v. 9, no. 10, pp. 14689–14708.

Konikow, L.F. and Neuzil, C.E. (2007). A method to estimate groundwater depletion from
confining layers. *Water Resources Research*, v. 43, no. 7, p. W07417, 15 p.

McFarland, E.R. and Bruce, T.S. (2006). The Virginia Coastal Plain hydrogeologic framework:
U.S. Geological Survey Professional Paper 1731, 118 p, 25 pls. Accessed from https://
pubs.water.usgs.gov/pp1731/.

Morris, J.T., Sundareshwar, P.V., Nietch, C.T., Kjerfve, B., and Cahoon, D.R. (2002). Responses
of coastal wetlands to rising sea level. *Ecology*, v. 83, no. 10, pp. 2869–2877.

National Geographic. (2001). Chesapeake Bay Crater Offers Clues to Ancient Cataclysm.
November 13. Accessed from https://newsnationalgeographic.com/news/2001/1113_
chesapeakecrater.html.

NOAA. (2015). What is glacial isostatic adjustment? National Oceanic and Atmospheric Administration. Accessed from https://oceanserive.noass.gov/facts/galicial-adjustment.html.

Poland, J.F. and Green, J.H. (1962). Subsidence in the Santa Clara Valley, California-a progress report: U.S. Geological Survey Water-Supply Paper 1619-C, 16 p.

Poland, J.F. and Ireland, R.L. (1988). Land subsidence in the Santa Clara Valley, California, as of 1982: U.S. Geological Survey Professional Paper 497-F, 61 p.

Pope, J.P. and Burbey, T.J. (2004). Multiple-aquifer characteristics from single borehole extensometer records. *Ground Water*, v. 42, no. 1, pp. 45–58.

Powars, D.S. and Bruce, T.S. (2000). *The Effects of the Chesapeake Bay Impact Crater on the Geological Framework and Correlation of Hydrogeologic Units of the Lower York-James Peninsula, Virginia.* Washington, DC: USGS. Accessed 09/29/19 @ https://pubs.usgs.gov/pp/p1612/powars.html.

PWNET. (2016). The Impact Crater. Prince William NET. Accessed from https://meteor.pwnet.org/impact_event/impact_crater.htm.

Randall, L. (2015). *Dark Matter and the Dinosaurs.* New York: Harper Collins.

Richmond Times Dispatch. (2005). Drill explores blast: Research seeks insight into explosion that carved huge crater under the Chesapeake. September 8. No longer online.

Sella, G.F., Stein, S., Dixon, T.H, Craymer, M., James, T.S., Mazzotti, St., and Dokka, R.L. (2007). Observation of glacial isostatic adjustment in "stable" North America with GPS. *Geophysical Research Letters*, v. 34, no. 2, p. 1.02306, 6 p.

Spellman, F.R. and Whiting, N. (2006). *Environmental Science and Technology: Concepts and Applications.* Boca Raton, FL: CRC Press.

Tennant, D. and Hall, M. (2001). The Chesapeake Bay Meteor: A mystery, meteors and one man's quest for the truth. *The Virginian-Pilot*, June 24.

Terzaghi, K. (1925). Principles of soil mechanics, IV-Settlement and consolidation of clay. *Engineering New-Record*, v. 95, no. 3, pp. 874–878.

Tolman, C.F. and Poland, J.F. (1940). Ground-water infiltration, and ground-surface recession in Santa Clara Valley, Santa Clara County, California. Transactions American Geophysical Union, v. 21, pp. 23–24.

USEPA. (1995). *The Probability of Sea Level Rise.* Washington, DC: Environmental Protection Agency.

USFWS. (1995). Vanishing Lands. Accessed 12/20/22 @ https://www.fws.gov/s/amer/Vanishinglands.Sealevel.

USGS. (1997). Location of Chesapeake Bay. Accessed from https://woodshole.er.usgs.gov/epubs/bolide/location_of_bay.html.

USGS. (1998). *The Chesapeake Bay Bolide.* Washington, DC: US Geological Survey.

USGS. (2013). *Land Subsidence in the United States. Circular 1182.* Washington, DC: U.S. Department of the Interior, U.S. Geological Survey.

USGS. (2016). The Chesapeake Bay Bolide Impact: A New View of Coastal Plain Evolution: Fact Sheet 049-98. Accessed from https://pubs.usgs.gov/fs/fs/49-98/.

Zervas, C. (2009). Sea level variations of the United States, 1854-2006: National Oceanic and Atmospheric Administration Technical Report NOS CO-OPS 053, 76 p., plus appendixes. Accessed from https://tidesandcurrents.noaa.gov/publications/Tech_rpt_53.pdf

Section III

Land Subsidence Mitigation

14 Yuck Factor Exaggerated
Fluence (Radiant Exposure) = Rate of UV Dose

FIXING A PROBLEM

If you find a flat tire on your vehicle, you change the tire (hopefully with an available spare). If your roof leaks, you fix the leak or have it fixed. If your cloth washing machine fails to operate, you have it repaired or replaced. These are everyday common problems (i.e., when things break or wear out or don't function as expected) that all of us, sooner or later, experience. And for these aforementioned problems, recognition of the problem(s) is relatively simple (however, not necessarily inexpensive). Simple problems but again they are also easily recognizable, for example: One way or another, we discover we have a flat tire; when inside our home and raindrops plop on our heads and everything else we know we have a problem, and when our cloth washers do not operate, we know we have a problem. Again, identifying problems with their result is generally simple; moreover, fixing many of these problems is also simple.

Simple is simple but not all is simple. Take all those insidious things going on all around us, daily. Cancer can be insidious sometimes and often is; we find out about the disease after it manifests itself on us or inside of us. Take for example, asbestosis and mesothelioma that do not show their sinister attributes until after a latency period of 10–20 years (or more or less), and by the time the disease raises is life-threatening characteristic(s).

Now the question is what does cancer and asbestos diseases have to do with land subsidence?

Good question.

Slow-moving landslides, landslides, volcanic eruptions, sinkholes, and subsidence are often insidious in that their internal acts and processes are hidden, making us unaware of their presence, that is, until they make their sometimes-deadly presence apparent and always destructive in one way or several others. Again, when these events take place, they are sometimes deadly because their ultimate coming out (so to speak) can be catastrophic like cancer or some other hidden, deadly disease.

So, what is the point, you ask?

The point is that whether the causal factors and results are brought about by natural processes or human activities is one thing, but the other thing (most important thing) is we often have no clue as to Earth's structural mechanizations going on underground, under the Earth's surface. Yes, we do not have X-ray vision, and therefore, we witness the results of subsidence (in particular) when they occur and cause damage to infrastructure, buildings, and wells; decreased aquifer capacity

DOI: 10.1201/9781003461265-17

FIGURE 14.1 The withdrawal of groundwater near Lucerne Lake (dry) in Mojave Desert, California, has caused the land to subside, with the results being the formation of fissures on the landscape. *Source:* Public Domain USGS photo, Credit Loren Metzer, USGS. Accessed 10/23/23 @ https://www.usgs.gov/special-topics/water-science-school/science/land-subsidence.

and water storage; landscape fissures (see Figure 14.1); changes in elevation and slope of waterway and drainage systems; coastal flooding and erosion; and habitat loss and loss of diversity. When these results are discovered, recognized, and identified, we must understand the root cause and determine a solution—taking action to solve the problem.

Okay, subsidence occurs and makes its arrival visible, the question is so what then?

The what then is to take whatever action deemed necessary to correct the subsidence problem. We can simply bulldoze a load of fill into whatever depression, fault, fissure, or sinkhole that appears and then grade it and walk away.

Is the problem fixed?

Well, that depends upon one's point of view. However, filling in a sinkhole is one thing but how about several acres of land subsidence?

As stated earlier, most of land subsidence is caused by groundwater withdrawal. Obviously, a bulldozer attempting to fill in acres where subsidence has occurred or is occurring is not a practical solution to the problem.

Simply, no problem is solved until the root cause of the problem is determined and then mitigated by whatever means is necessary, available, and practical, and so forth. Attacking land subsidence due to groundwater withdrawal must be identified, understood, and a plan of action put in place to mitigate subsidence. This can only be accomplished by gaining understanding of hydraulic and mechanical properties affecting groundwater and aquifer system compaction.

Okay, so far so good. But how do we solve land subsidence caused by groundwater withdrawal? Well, first we can stop pumping of groundwater in an affected area.

Second, water can be pumped back into the aquifer in an effort to stop aquifer compaction, reduce or freeze subsidence in its sliding tracks, so to speak and, in some cases, to inject enough water to increase the supply of groundwater supplies.

Okay, sounds like a winner, right?

Hold on, pumping water into aquifers to mitigate subsidence is a great idea, but only if water is available.

How much water is available in the Southwestern United States to replace that which has been withdrawn over the years? You would have better luck in Las Vegas than in trying to find more water in the region.

So, we know we need to pump water into several aquifers in different locations in the United States but where do we get the water?

How about reusing (reclaimed) wastewater, pumping our depleted aquifers with wastewater to aid in the halting of subsidence?

Well, what about the so-called yuck factor of using toilet water and other wastewater to replenish aquifers and eventually use this same wastewater as a source of potable water?

Personal note: While teaching undergraduate and graduate students at Old Dominion University in Norfolk, Virginia, I discussed the use of wastewater as a supplement to potable drinking water. While explaining how this can and is being done in certain locations, one of my students raised her hand and I acknowledge her to speak and she said, "The yuck factor of drinking wastewater is so high. I couldn't bring myself to even take a sip."

My reply: "Obviously you have never been thirsty ... dying of thirst ... willing to do anything to quench your thirst... and besides ... the truth be told we drink wastewater everyday thanks to the water cycle and *de facto* reuse—again, the fact is we drink recycled wastewater every day."

Anyway, let's look at the so-called yuck factor involved with drinking wastewater (treated to drinking water standards, of course). Let's begin at a good place with that great mythical hero, Hercules, arguably the world's first environmental engineer, who was ordered to perform his 5th labor by Eurystheus to clean up King Augeas' stables. Hercules, faced with a mountain of horse and cattle waste piled high in the stable area had to devise some method to dispose of the waste; he did. He diverted a couple of river streams to the inside of the stable area so that all the animal waste could simply be deposited into the river streams: out of sight out of mind. The waste simply flowed downstream. Hercules understood the principal point in pollution control technology that is pertinent to this very day and to this discussion, that is, *dilution is the solution to pollution* (including proper treatment to drinking water standards).

When people say they would never drink toilet water, they have no idea what they are saying. As pointed out in my textbook, *The Science of Water,* 4th edition, the fact is we drink recycled wastewater every day. In Hampton Roads, lower Chesapeake Bay region, for example, Hampton Roads Sanitation District (HRSD—often considered the premier wastewater treatment operation on the Globe)—HRSD's wastewater treatment plants (WWTPs) outfall (discharge) treated water to the major rivers in the region. Many of the region's rivers are sources of local drinking water supplies. Even local groundwater supplies are routinely infiltrated with surface water inputs, which, again, are commonly supplied by treated wastewater (and sometimes infiltrated by raw sewage that is accidentally spilled).

My compliments to Mr. Henifin, former General Manager of HRSD, who stated in a recent local newspaper article (2020) that he would be first to drink the treated wastewater effluent from the unit treatment processes at York River Treatment Plant (and drink it, he did). My only contention with his statement is that because of Mother Nature's Water Cycle, the one we all learned about in grade school, we have been drinking toilet water all along. I have yet to find anything yucky about it or its taste.

DID YOU KNOW?

The Environmental Protection Agency (EPA) does not require any type of reuse. Generally, states maintain primary regulatory authority (i.e., primacy) in allocating and developing water resources. Some states have established programs to specifically address reuse, and some have incorporated water reuse into their existing programs—and this is what this book is all about.

NOTHING NEW UNDER THE TAP

Depending upon your place of residence the drinking water from your tap may contain reclaimed wastewater that has been indirectly provided via surface water augmentation, groundwater recharge, and aquifer storage and recovery; this is nothing new under many of your taps. The following includes four examples where municipal wastewater for potable water reuse is practiced: Washington State, Nevada, Pennsylvania, and Virginia.

The following two things need to considered before describing wastewater reuse for potable service in the four selected states. *First,* note that USEPA does not have specific regulations regarding wastewater reuse for potable purposes. Instead, USEPA provides guidelines and acts as an authoritative reference on water reuse practices published in its *2017 Potable Reuse Compendium*—this document supplements state regulations and guidelines by providing technical information and outlining key implementation considerations. *Second,* in order to comprehend the material that follows it is important to define the terminology used, but keep in mind that each state has its own interpretation of guidelines provided by USEPA and so does each State Water Control Board (or other State Agency). USEPA, States, tribes, and local governments implement programs under the Clean Water Act (CWA) and the Safe Drinking Water Act (SDWA) to protect the quality of source waters and drinking water. The SDWA and the CWA provide a foundation from which states can further develop and support portable water reuse as they deem appropriate. As described in the 2017 *Potable Reuse Compendium*, the terminology associated with treating and reusing municipal wastewater varies both within the United States and globally. For example, some states and countries use the term "reclaimed water" and "recycled" interchangeably. By the same token, the terms "water recycling" and "water reuse" are often used synonymously. This book uses terms reclaimed water and water reuse.

The main point being made here and within the entire text is that potable water reuse provides another option for expanding a region's water resource portfolio.

Potable water reuse—the process of using treated wastewater for drinking water.

Planned potable reuse—the publicly acknowledged, intentional use of reclaimed wastewater for drinking water supply. Commonly referred to as to simply as *potable reuse*.

De facto (aka in reality) *reuse*—a situation where reuse of treated wastewater is practiced but is not officially recognized (e.g., a drinking water supply intake located downstream from a WWTP discharge point [i.e., outfall]).

Direct potable reuse—the introduction of reclaimed water (with or without retention in an engineered storage buffer) directly into a drinking water treatment plant (WTP). This includes the treatment of reclaimed water at an advanced wastewater treatment facility (AWTF) for direct distribution.

Indirect potable reuse (IPR)—deliberative augmentation of a drinking water source (surface water or groundwater aquifer) with treated reclaimed water, which provides an environmental buffer prior to subsequent use.

ENVIRONMENTAL AND ENGINEERED BUFFERS

In order to treat municipal wastewater to drinking water standards and use it to augment drinking water supplies, some type of environmental or engineered buffer is required within the advanced treatment train (see Figure 14.2).

So, what is an environmental and engineering buffer? As shown in Figure 14.2, an environmental and engineering buffer refers to an aquifer, wetland, or other body of water such as a river, stream, lake, or reservoir, that serves as an intermediate discharge and holding point within a potable reuse scheme. The environmental buffer receives treated water from an AWTF. Dilution, blending, and some contaminant removal through filtration (aquifers), photolysis (surface waters), or biological degradation can occur before IPR withdrawal (Spellman, 2021). Environmental buffers tend to dissociate the origin of the water (wastewater discharge, plant effluent, outfall) from the end point (drinking water). Note that environmental buffers create a

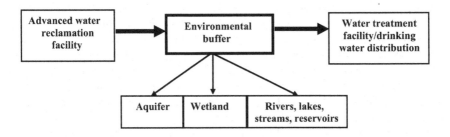

FIGURE 14.2 Environmental buffers in potable water reuse treatment schemes.

window of time in which the water enters into a natural environment (Khan, 2013). Also note that although environmental buffers can improve water quality, they are not universally required components in portable reuse projects, and they do not conform to controlled performance standards (Khan, 2013). Finally, keep in mind that some environmental buffers may degrade the quality of purified water—including risks of surface water contamination.

With regard to an environmental buffer's importance to public health largely it depends on the influent water quality (bad water in equals bad water out) and the buffer's specific characteristics. For instance, reclaimed water that undergoes advanced treatment upstream of the environmental buffer may have less stringent dilution (Hercules: dilution is the solution to pollution) and residence time requirements than reclaimed water that only undergoes filtration and disinfection (USEPA, 2017). Three types of environmental buffers, namely, aquifers, surface storage, and wetlands (with the emphasis on aquifers; later presented in Chapter 15), are discussed herein.

DID YOU KNOW?

Aquifer storage recovery (ASR; see Chapter 15) is a water resource management technology for actively storing water underground during wet periods for recovery when needed, usually during dry periods.

AQUIFER RECHARGE

As shown in Figure 14.2, aquifers can serve as subsurface buffers. In this approach, treated effluent is either diverted to surface spreading basis whereby infiltration occurs or used in more modern approaches such as rapid infiltration basins (RBIs; see Figure 14.3), vadose zone injection wells (VZM; aka dry wells or recharge shafts), infiltration trenches, or riverbank filtration to reach the water table.

Reclaimed water percolates through sediment until it reaches aquifer and blends with groundwater; it remains underground for a predetermined residence time before being extracted as a drinking water source (USEPA, 2017). The process of enhancing natural groundwater supplies using engineered conveyances to route water to an aquifer is known as managed aquifer recharge (MAR), often interchangeably referred to as enhanced aquifer recharge, and in a number of cases in near coastal areas, MAR has also helped to reduce saltwater intrusion. Purified water sent to a spreading basin undergoes a natural water treatment process known as soil aquifer treatment (SAT). As opposed to fully advanced treatment (discussed in Chapter 15), SAT does not result in the generation of a brine requiring disposal and provides additional pathogen removal, therefore making it a popular option for inland geographies. Lab analysis of soil columns can assess the removal of pathogens, such as Cryptosporidium (see Sidebar 14.1), through SAT.

Given a suitable aquifer (such as the Potomac Aquifer discussed in Chapter 15), SAT is considered the most economical potable reuse alternative. Note that the level

FIGURE 14.3 Artificial rapid infiltration basins outside Orlando, Florida. USGS Public domain photo accessed 10/27/23 @ https://www.usgs.gov/media/images/rapid-inflitriton-basins-recharge-groundwater-florida.

SIDEBAR 14.1   CRYPTOSPORIDIUM

Cryptosporidium (crip-toe-spor-ID-ee-um) is one of several single-celled protozoan genera in the phylum Apircomplexa (all referred to as coccidian). *Cryptosporidium* along with other genera in the phylum Apircomplexa develop in the gastrointestinal tract of vertebrates through all of their life cycle – in short, they live in the intestines of animals and people. This microscopic pathogen causes a disease called *cryptosporidiosis* (crip-toe-spor-id-ee-O-sis).

The dormant (inactive) form of *Cryptosporidium* called an oocyst (O-o-sist) is excreted in the feces (stool) of infected humans and animals. The tough-walled oocysts survive under a wide range of environmental conditions.

Several species of *Cryptosporidium* were incorrectly named after the host in which they were found; subsequent studies have invalidated many species. Now, eight valid species of *Cryptosporidium* (see Table 14.1) have been named.

Upton (1997) reports that *C. muris* infects the gastric glands of laboratory rodents and several other mammalian species, but (even though several texts state otherwise) is not known to infect humans, However, *C. parvum* infects the small intestine of an unusually wide range of mammals, including humans, and is the zoonotic species responsible for human cryptosporidiosis. In most mammals, *C. parvum* is predominately a parasite of neonate (newborn) animals. He points out that even though exceptions occur, older animals generally develop poor infections, even when unexposed previously to the parasite. Humans are the one host that can be seriously infected at any time in their lives, and only previous exposure to the parasite results in either full or partial immunity to challenge infections.

TABLE 14.1
Valid Named Species of
Cryptosporidium

Species	Host
C. baileyi	Chicken
C. felis	Domestic cat
C. meleagridis	Turkey
C. muris	House mouse
C. nasorium	Fish
C. parvum	House mouse
C. serpentis	Corn snake
C. wrairi	Guinea pig

Source: Adapted from Fayer et al. (1997).

Oocysts are present in most surface bodies of water across the United States, many of which supply public drinking water. Oocysts are more prevalent in surface waters when heavy rains increase runoff of wild and domestic animal wastes from the land, or when sewage treatment plants are overloaded or break down.

Only laboratories with specialized capabilities can detect the presence of *Cryptosporidium* oocysts in water. Unfortunately, present sampling and detection methods are unreliable. Recovering oocysts trapped on the material used to filter water samples is difficult. Once a sample is obtained, however, whether the oocyst is alive or whether it is the species *C. parvum* that can infect humans is easily accomplished by looking at the sample under a microscope.

The number of oocysts detected in raw (untreated) water varies with location, sampling time, and laboratory methods. WTPs remove most, but not always all, oocysts. Low numbers of oocysts are sufficient to cause cryptosporidiosis, but low numbers of oocysts sometimes present in drinking water are not considered cause for alarm in the public.

Protecting water supplies from *Cryptosporidium* demands multiple barriers. Why? Because *Cryptosporidium* oocysts have tough walls that can withstand many environmental stresses and are resistant to the chemical disinfectants such as chlorine that are traditionally used in municipal drinking water systems.

Physical removal of particles, including oocysts, from water by filtration is an important step in the water treatment process. Typically, water pumped from rivers or lakes into a treatment plant is mixed with coagulants, which help settle out particles suspended in the water. If sand filtration is used, even more particles are removed. Finally, the clarified water is disinfected and piped to customers. Filtration is the only conventional method now in use in the United States for controlling *Cryptosporidium*.

Ozone is a strong disinfectant that kills protozoa if sufficient doses and contact times are used, but ozone leaves no residual for killing microorganisms in the distribution system, as does chlorine. The high costs of new filtration or ozone treatment plants must be weighed against the benefits of additional treatment. Even well-operated WTPs cannot ensure that drinking water will be completely free of *Cryptosporidium* oocysts. Water treatment methods alone cannot solve the problem; watershed protection and monitoring of water quality are critical. For example, land use controls such as septic system regulations and best management practices to control runoff can help keep human and animal waste out of water.

Under the Surface Water Treatment Rule of 1989, public water systems must filter surface water sources unless water quality and disinfection requirements are met and a watershed control program is maintained. This rule, however, did not address *Cryptosporidium*. The USEPA has now set standards for turbidity (cloudiness) and coliform bacteria (which indicate that pathogens are probably present) in drinking water. Frequent monitoring must occur to provide officials with early warning of potential problems to enable them to take steps to protect public health. Unfortunately, no water quality indicators can reliably predict the occurrence of cryptosporidiosis. More accurate and rapid assays of oocysts will make it possible to notify residents promptly if their water supply is contaminated with *Cryptosporidium* and thus avert outbreaks. The collaborative efforts of water utilities, government agencies, health care providers, and individuals are needed to prevent outbreaks of cryptosporidiosis.

The Bottom Line: Field studies using actual *Cryptosporidium* are relatively rare due to low oocyst concentrations, even in raw surface waters. Instead, field studies typically utilize surrogates, such as bacterial spores or microspores, to assess log removal of pathogens through SAT (WRRF, 2014).

---

of treatment achieved through aquifer recharge depends on the quality of the feed water. Keep in mind that there always exists the possibility of water quality degradation; blending high-quality free water with groundwater that was exposed to municipal, agricultural, industrial, and natural contaminants can result in added treatment requirements when the water is extracted (WE&RF, 2016). ASR is a specific type of MAR practiced to augment groundwater resources and recover the water in the future for various uses (USEPA, 2012). In the United States, ASR is used frequently as a method for improving water availability during drought conditions, halting subsidence, and offsetting water shortages; ASR projects utilize underground injection and require an underground injection control (UIC) permit (USEPA, 2017). Most of the current ASR practices are utilized in non-potable water and wastewater reuse applications intended for irrigation, industrial, and urban landscape end uses; however, applications intended for irrigation, industrial, halt subsidence, and urban landscape end uses; however, ASR for IPR has recently increased in popularity, as discussed in Chapter 15.

SURFACE WATER STORAGE

Surface water storage occurs when reclaimed effluent is discharged into a lake, reservoir, or river. In this particular case, the receiving surface water blends with reclaimed water before being extracted and sent to the WTP. Surface water storage provides a mitigation response time in the event of process failure and can provide a level of treatment; however, the effectiveness of treatment depends on the water quality of the reclaimed effluent and the water quality and environmental conditions of the surface water (WE&RF, 2016),

SELECTED CASE STUDIES

In this section, four selected locations are described where treated municipal wastewater for potable water reuse is practiced in the United States: Pennsylvania, Washington, Nevada, and Virginia. For each of these practices, the technical basis is discussed along with types of planned potable reuse approved for use in each state, water reuse treatment category/type, and portable reuse specifications. Keep in mind that USEPA does not regulate what the states do in this area; instead, it provides guidance. Note that some states have guidance for producing Class A+ reclaimed water, but not all of them use it to produce potable drinking water—at the present time.

Case Study 14.1—Pennsylvania (Treated Municipal Wastewater for Potable Water Reuse; Source: DEP, 2012)

In Pennsylvania, potable water reuse applications include indirect potable reuse (surface water spreading and groundwater recharge). The source of water (treated municipal wastewater) is specified by the state as municipal wastewater. State terms are used in this write-up when discussing sources or uses of water that may differ from the Regulations and End-Use Specifications.

Note that at the technical level, Pennsylvania abides by potable water regulations and meets all applicable SDWA requirements and the National Pollutant Discharge Elimination System (NPDES permit—40 CFR § 122). Pennsylvania does not have statutes or regulations that specifically address the reuse of treated wastewater; however, water reuse in Pennsylvania is implemented through the technical guidance manual and Water Quality Management permits (Pennsylvania DEP, 2012). If the reclaimed water is used for direct injection, certain requirements of the EPA's Underground Injection Control Program (USEPA, 2023) may need to be met. The goal of federal regulations is to prevent contamination of underground sources of drinking water from the placement of fluids underground through injection wells. The UIC regulations do this by regulating the construction, operation, and closure of injection wells. Additionally, Pennsylvania indirect portable reuse guidelines require specific treatments, which vary by the class of reclaimed water. With regard to chemicals, Pennsylvania reuse guidelines require Class A+ finished water meets a less than 1 mg/L total organic carbon (TOC) concentration. The guidelines require that Class A reclaimed water has less than 10 mg/L TOC concentration. Note that the technical basis for the removal of microbial and chemical contaminants is not explicitly specified.

DID YOU KNOW?

In general, for Class A+ reclaimed water, the treatment requirements are secondary treatment, filtration, nitrogen removal treatment, and disinfection using oxidants, UV light, or other agents to kill or inactivate pathogenic organisms. Class A+ water is not required for a specific type of direct reuse but may be used for any Class A, B, or C reuse.

Pennsylvania DEP (2012) defines the following approved planned potable uses:

- Indirect potable reuse
  - Surface water spreading (Class A)—augmentation or recharge to potable (or non-potable) water aquifers via surface spreading.
  - Groundwater recharge.
  - Direct injection (Class A+)—augmentation or recharge to potable (or non-potable) water aquifers or saltwater intrusion barriers. In Pennsylvania, reclaimed water should be retained underground for a minimum of 12 months prior to withdrawal in aquifers used as drinking water sources (Pennsylvania DEP, 2012). In cases where UV light is used for disinfection, a dose of ≥ 50 mj/cm^2 [for fluence (radiant exposure) rate of UV dose] is required. In cases where chlorine is used for disinfection, a total chlorine residual of at least 1.0 mg/L should be maintained for a minimum contact time of 30 minutes of design average flow. Table 14.2 lists Pennsylvania's source of water type and specifications.

DID YOU KNOW?

Chlorine residual—the amount of chlorine (determined by testing) remaining after the demand is satisfied. Residual, like demand, is based on time. The longer the time after dosage, the lower the residual will be, until all of the demand has been satisfied. Residual, like dosage and demand, is expressed in mg/L. The presence of a *free residual* of at least 0.2–0.4 ppm usually provides a high degree of assurance that the disinfection of the water is complete. *Combined residual* is the result of combining free chlorine with nitrogen compounds. Combined residuals are also called chloramines. *Total chlorine residual* is the mathematical combination of free and combined residuals. Total residual can be determined directly with standard chlorine residual test kits (Spellman, 2020).

TABLE 14.2

Pennsylvania Potable Reuse Specifications

Recycle Water Class/Category	Source Water Type	Water Quality Parameter	Specification
Class A+ reclaimed water	Municipal wastewater	BOD	<2 mg/L (monthly average)
			5 mg/L (maximum)
		TOC	<1 mg/L
		Turbidity	≤0.3 NTU[c] (monthly average)
			1 NTU (maximum)
		Fecal coliform	<2.2 organisms per 100 mL (monthly average)
			23 organisms per 100 mL (maximum)
		TOX	<0.2 mg/L (monthly average)
		Total nitrogen	<10 mg/L (monthly average)
		All other primary and secondary drinking water contaminants	See SDWA regulations (40 CFR § 141)
		UV light design dose[a]	≥50 mj/cm^2
Class A reclaimed water	Municipal wastewater	BOD	<2 mg/L (monthly average)
			5 mg/L (maximum)
		TOC	<10 mg/L
		Turbidity	≤2 NTU (monthly average)
			5 NTU (maximum)
		Fecal coliform	<2.2 organisms per 100 mL (monthly average)
			23 organisms per 100 mL (maximum)
		TOX	0.2 mg/L (monthly average)
		Total nitrogen	<10 mg/L (monthly average)
		All other primary and secondary drinking water contaminants	SWDA (40 CFR§ 141)
		Total chlorine residual[b]	≥1.0 mg/L (maintained for a minimum contact time of 30 minutes at design average flow)
			>0.02 mg/L (at the point of reuse application)
		Dose[a]	≥100 mj/cm^2 (for granular media filtration)
			UV light design ≥80 mj/cm^2 (for porous membrane filtration)
			≥50 mj/cm^2 (for semipermeable membrane)

Source: Adapted from DEP (2012).

TOC, total organic carbon; BOD, biological oxygen demand; TOX, total organic halides; NTU, nephelometric turbidity unit.

[a] Only applies if UV light was used for disinfection.

[b] Only applies if chlorine was used for disinfection.

[c] NTU—indicates amount of turbidity in a waste sample.

DID YOU KNOW?

Fecal coliforms are used as indicators of possible sewage contamination because they are commonly found in human and animal feces. Although they are not generally harmful themselves, they indicate the possible presence of pathogenic (disease-causing) bacteria and protozoans that also live in human and animal digestive systems. Their presence in streams suggests that pathogenic microorganisms might also be present and that swimming in and/or eating shellfish from the waters might present a health risk. Since testing directly for the presence of a large variety of pathogens is difficult, time-consuming, and expensive, water is usually tested for coliforms and fecal streptococci instead. Sources of fecal contamination to surface waters include WWTPs, on-site septic systems, domestic and wild animal manure, and storm runoff. In addition to the possible health risks associated with the presence of elevated levels of fecal bacteria, they can also cause cloudy water, unpleasant odors, and an increased oxygen demand (Spellman, 2020).

Case Study 14.2—Washington (Treated Municipal Wastewater for Potable Water Reuse; Source: USEPA, 2023)

In the State of Washington, potable water reuse applications include direct potable reuse and indirect reuse (surface water augmentation, groundwater recharge, and aquifer storage and recovery). The source of water (treated municipal wastewater) is specified by the state as wastewater with a domestic wastewater component. In the case study, the write-up uses state terms when discussing sources or used of water that may differ from the Regulations and End-Use Specifications Explorer's terms.

Note that at the technical level, Washington abides by potable water regulations and meets all applicable SDWA requirements and the NPDES (permit—40 CFR § 122). Additionally, all microbial and chemical contaminants must meet Washington State drinking water standards (Wash. Admin. Code § 246-290-310). In Washington, no reclaimed water may be distributed or used without a state-issued reclaimed water permit (Wash. Admin. Code § 219-070, Wash. Rev. Code § 90-46). In Washington, reclaimed water permits for surface water discharges are issued as combined NPDES/Reclaimed Water permits (40 CFR § 122). Additionally, Washington IPR regulations require certain treatment requirements, which vary by class of recycled water. Note that the technical basis for the removal of microbial and chemical contaminants is not explicitly specified.

Washington Administrative Code § 173-219 defines the following approved planned potable use:

- Indirect potable reuse
  - Surface water augmentation (Class A or B)—the intentional use of reclaimed water for rivers and streams of the state or other surface water bodies, for the purpose of increasing volumes.

- Groundwater recharge—introduction of reclaimed water to groundwater aquifers including.
  - Indirect (Class A or B)—reclaimed water is introduced to groundwater through surface or subsurface infiltration or percolation, where the introduced water travels through an unsaturated vadose zone and the coming with groundwater of the state is not immediate.
  - Direct (Class A)—reclaimed water is released directly and immediately into groundwater of the state through direct injection or other means.
- Aquifer storage and recovery (Class A)—Recovery of reclaimed water stored in an aquifer.

Direct potable reuse (Class A+)—Introduction of reclaimed water into an existing water distribution storage or treatment system without an environmental buffer.

Note that in Washington, reclaimed water is defined as "water derived in any part from wastewater with a domestic wastewater compound that has been adequately and reliably treated, so that it can be used for beneficial purposes" (Wash. Rev. Code § 90-46). The various classes of water (Class B, Class A, and Class A+) are defined by their respective wastewater treatment requirements and applicable performance standards. The respective treatment requirements are briefly summarized for potable reuse applications:

- Class A treatment must meet Class B requirements (secondary treatment including biological oxidation and disinfection) and one of the following treatment process train requirements:
  - Biological oxidation, followed by coagulation, filtration, and disinfection, demonstrating at least a 4-log virus removal of inactivation.
  - Biological oxidation, followed by membrane filtration and disinfection, demonstrating at least 4-log virus removal or inactivation.
  - Combination of biological oxidation and membrane filtration via a membrane bioreactor followed by disinfection, demonstration at least a 4-log virus removal or inactivation.
  - An alternative treatment method that demonstrates to the satisfaction of the lead agency that provides for equivalent treatment and reliability.
- Class A+ reclaimed water requirements are "established on a case-by-case basis by health" and must have approval of the State Board of Health before it be beneficially used for direct potable reuse. Note that no facilities currently produce Class A+ reclaimed water in Washington. Table 14.3 lists Washington's source of water type and specifications.

DID YOU KNOW?

CFU, cfu, or Cfu is a unit which estimates the number of bacteria, fungi, viruses, etc. in a sample that are viable. MPN is the most probable number.

TABLE 14.3

Washington Potable Reuse Specifications

Recycle Water Class/Category	Source Water Type	Water Quality Parameter	Specification
Class A+ reclaimed water	Not specified		Must meet the treatment requirements for Class A reclaimed water and any additional treatment criteria determined necessary on a case-by-case basis by health for direct potable reuse
Class A reclaimed water	Not specified		Must meet the treatment requirements for Class B reclaimed water and any additional treatment criteria as specified
		Turbidity (coagulation/ filtration)	2 NTU (monthly average) 5 NTU (maximum)
		Turbidity (membrane filtration)	0.2 NTU (monthly average) 0.5 NTU (monthly average)
		Total coliform	2.2 MPN/100 mL or CFU/100 mL (7-day median) 23 MPN/100 mL or CFU/100 mL 100 mL (monthly average) Wash Admin. Code § 173-219
		Virus removal	Minimum 4-log virus removal before additional water treatment—if chlorine is used
		Total nitrogen	10 mg/L (monthly average) 15 mg/L (weekly average)
Class B reclaimed water	Not specified	Dissolved oxygen	Must be measurably present
		5-day BOD	30 mg/L (monthly average) 45 mg/L (weekly average)
		CBOD5	25 mg/L (monthly average) 40 mg/L (weekly average)
		TSS	30 mg/L (monthly average) 45 mg/L (weekly average)
		pH	6–9 6.5–8.5 (for groundwater recharge)
		Total coliform	23 MPN/mL or CFU/mL (7-day median) 240 MPN/mL or CFU/mL (sample maximum)
		Chlorine residual (if chlorine used as the disinfectant)	$\geq$1 mg/L
		Chemical contaminants and other parameters	Must meet SDWA regulations and state drinking water regulations

Source: Adapted from USEPA (2023).

SDWA, Safe Drinking Water Act; MPN, most probable number; BOD, biological oxygen demand; CBOD5, 5-day carbonaceous BOD; NTU, nephelometric turbidity unit; TOS, total suspended solids.

Case Study 14.3—Nevada (Treated Municipal Wastewater for Potable Water Reuse; Source: USEPA, 2023)

In the state of Nevada, potable water reuse includes indirect potable reuse (aquifer augmentation or recharge). Treated municipal water, the source of water, is specified by the state as municipal wastewater. The written description herein uses state terms when discussing sources or uses of water that may differ from the regulations.

Note that at the technical level, Nevada abides by potable water regulations and meets all applicable SDWA requirements and the NPDES (permit—40 CFR § 122). Additionally, all microbial and chemical contaminants must be treated and removed in order to meet requirements of (40 CFR § 141). Moreover, to use reclaimed water in Nevada, a state permit must be obtained pursuant to Nev. Admin Code § 445A.228 to 445.263 and Nevada Code § 445A.274 to 334A.280. Nevada indirect potable reuse regulations have specific treatment requirements for certain pathogens and chemicals.

Note that microbial (pathogen) log reduction values were derived assuming raw sewage maximum densities of densities of 10^5 culturable enteric viruses/L, 10^5 *Giardia lamblia* cysts/L, and 10^4 *Cryptosporidium* oocysts/L and a health-based target of less than 1 infection per 10,000 people per year. Risk-based calculations resulted in treatment requirements of 12-log enteric virus reduction, ten-log *Giardia lamblia* cyst reduction, and ten-log *Cryptosporidium* spp. Oocyst reduction. The required log removals must be demonstrated from the point where raw sewage enters a treatment works to the point of extraction from an aquifers for potable use (DEP, 2012).

Chemical contaminants must meet the maximum contaminant levels (MCLs) in SDWA regulations (40 CFR § 141), and certain secondary contaminants (including chloride, iron, magnesium, manganese, sulfate, and total dissolved solids) must meet the MCLs as set forth in Nevada's secondary standards for public water systems (DEP, 2012).

The type of planned potable water reuse approved for use in Nevada is indirect potable reuse, where reclaimed water is discharged into an aquifer for the purpose of augmentation or recharge of a drinking water source. Note that the reclaimed water travels through an environmental buffer before the reclaimed water is recovered into an extraction well for potable use. Reuse Category A+ is required for indirect potable reuse through injection wells or spreading basins. Table 14.4 lists Nevada's source of water type and specifications.

Case Study 14.4—Washington (Treated Municipal Wastewater for Potable Water Reuse)

In Virginia, potable water reuse applications include indirect potable reuse (surface water augmentation). Treated municipal wastewater, the source of water, is specified by the state as reclaimed water. The information provided in this case study uses state terms when discussing sources or uses of water that may differ from regulations.

Virginia defines reclaimed water as "water-resulting from the treatment of domestic, municipal or industrial wastewater that is suitable for a water reuse that would not otherwise occur" and specifically excludes gray water, harvested rainwater, and stormwater.

Source: USEPA (2023).

TABLE 14.4
Nevada Potable Reuse Specifications

Recycle Water Class/Category	Source Water Type	Water Quality Parameter	Specification
Category A+ indirect potable reuse	Municipal wastewater	Viruses (enteric)	12-log reduction
		Giardia lamblia	10-log reduction
		Cryptosporidium	10-log
		Total coliforms	≤2.2 CFU or MPN/100 mL (30 day geometric mean)
			≤23 CFU or MPN/100 mL (maximum daily number)
		BOD5	≤30 mg/L (30-day average)
		TSS	≤30 mg/L (30-day average)
		Chloride	≤400 mg/L
		Iron	≤0.6 mg/L
Category A+ indirect potable reuse	Municipal wastewater	Magnesium	≤150 mg/L
		Manganese	≤0.1 mg/L
		Sulfate	≤500 mg/L
		TDS	≤1000 mg/l

Source: Adapted from USEPA (2023).
MPN, most probable number; BOD, biological oxygen demand; BOD5, 5-day BOD; TSS, total suspended solids; TDS, total dissolved solids.

DID YOU KNOW?

Gray water—water that has been used for showering, cloth washing, and faucet uses. Kitchen sink and toilet water is excluded. This water has excellent potential for reuse as irrigation for yards.

Virginia approves the discharge of reclaimed water to surface waters for IPR through a Virginia-delegated NPDES permit (or VPDES permit). For any new IPR projects proposed after January 29, 2014, Virginia requires the project to generate Level 1 reclaimed water and meet additional requirements like Virginia's water quality standards and total maximum daily loads, the strictest of which could be followed if there is more than one standard for a pollutant.

Treatment requirements and performance standards are applied for the removal of microbial contaminants, chemicals, and other relevant indicators for indirect potable reuse and are summarized in Table 14.5. The microbial standards for Level I reclaimed water were derived from fecal coliform water quality standards for

TABLE 14.5
Virginia Potable Reuse Specifications

Recycle Water Class/Category	Source Water Type	Water Quality Parameter	Specification
Level 1+ additional requirements (indirect potable reuse)	Municipal wastewater	In addition to the Level 1 treatment requirements, any new IPR projects proposed after January 29, 2014, must meet additional requirements like Virginia's water quality standards (9 Va. Admin. Code § 25–260) and total maximum daily loads (Virginia DEQ, 2023). If there is more than one standard for the same pollutant across these there regulations, the strictest standard should be applied	
	Fecal coliform—must meet bacterial standards for fecal coliform, *E. coli*, or enterococci. All three are not required	≤14 colonies/100 mL (monthly geometric mean) ≥49 colonies/100 mL (corrective action threshold)	
	E. coli—must meet bacterial standards for fecal coliform, *E. coli*, or enterococci. All three are not required	≤11 colonies/100 mL (monthly geometric mean) ≥35 colonies/100 mL (corrective action threshold)	
	Enterococci—must meet bacterial standards for fecal coliform, *E. coli*, or enterococci. All three are not required	≤11 colonies/100 mL (monthly geometric mean) ≥24 colonies/100 mL (corrective action threshold)	
	Total residual chlorine—only applies if chlorine was used for disinfection	<1.0 mg/L (corrective action threshold)	
	pH	6.0–9.0 SU	
	BOD5 or CBOD5	≤10 mg/L (monthly average for BOD5) or ≤8 mg/L (monthly average for BOD5)	
	Turbidity (if UV radiation is used for disinfection, other turbidity standards may apply)	≤2 NTU (daily average of discreet measurements recorded over a 24-hour period) ≥5 NTU (corrective action threshold)	
Level 1+ additional requirements (indirect potable reuse)	Municipal wastewater	UV light design dose—if UV I is used for disinfection	≥100,000 µWsec/cm^2
		UV transmittance—only applies if UV is used for disinfection	≥55% at 254 nm

Source: Adapted from USEPA (2023).
Note: A lower UV disinfection dose may be authorized.
IPR, indirect potable reuse; BOD, biological oxygen demand; CBOD5, 5-day carbonaceous BOD; BOD5, 5-day BOD; NTU, nephelometric turbidity unit; SU, specific units.

shellfish propagation waters of Virginia. The reasoning is that surface water quality standards considered safe for the propagation of shellfish for human consumption should also be considered safe for uses that have the potential for public contact, as is the case for the reuse applications for Level 1 reclaimed water. The Level 1 reclaimed water standards for *E. coli* and enterococci were derived from the reclaimed water standard for fecal coliform using conversion factors. Note that the technical basis for developing the specifications and/or removal of chemicals and other relevant indicators are not explicitly specified.

THE BOTTOM LINE

Water reuse or water reclamation reclaims water from a variety of sources then treats and reuses it for beneficial purposes such as potable water supplies, agriculture and irrigation, industrial processes, environmental restoration, and groundwater replenishment. Although water reuse provides alternatives to existing water supplies and is used to enhance water, security, sustainability, and resilience, it is water reuse replacing groundwater in aquifers in an attempt to mitigate land subsidence that this book is all about.

The preceding chapters have all combined to set the stage for the final chapter wherein a technology, technique, procedure, and/or practice of arresting land subsidence that is presently ongoing is described.

REFERENCES

DEP. (2012). *Point and Nonpoint Source Management. Reuse of Treated Wastewater Guidance Manual 385-2188-002*. Harrisburg: Pennsylvania Department of Environmental Protection.

DEQ. (2023). *Water Quality*. Richmond, VA: Department of Environmental Quality.

Khan, S. (2013). Drinking Water through Recycling: The Benefits and Costs of Supplying Direct to the Distribution System. Report of Australian Academy of Technology Sciences and Engineering, Funded by the Australian Water Recycling Centre of Excellence through the Commonwealth Government's Water for the Future Initiative. Accessed 11/5/23 @ https://aste.org.au/Documents/Publications/Reports/Water/drinking-water-through-recycling-full-report.pdf.

Spellman, F.R. (2020). *Handbook of Water/Wastewater Treatment Operations*, 4th ed. Boca Raton, FL: CRC Press.

Spellman, F.R. (2021). *Sustainable Water Initiative for Tomorrow (SWIFT)*. Lanham, MD: Bernan Press.

Upton, S.J. (1997). *Basic Biology of Cryptosporidium*. Kansas State University, Manhattan, Kansas.

USEPA. (2012). *Drinking Water Standards and Health Advisories. PA/600/R-12/618*. Washington, DC: U.S. Environmental Protection Agency.

USEPA. (2017). Decision Support System for Aquifer Storage and Recovery (ASR) Planning, Design and Evaluation-Principles and Technical Basis. Accessed 11/4/23 @ https://cfpub.epa.gov/si/si_public_record_report.CFM?dirEntryld=35408.

USEPA. (2023). Water Reuse. Accessed 10/31/23 @ https://epa.gov/waterreuse.

WE & RF. (2016). *Development of Operation and Maintenance Plan and Training and Certification Framework for Direct Potable Reuse (DPR) Systems*. Reuse 13-13. Alexandria, VA: Water Environment and Reuse Foundation.

WRRF. (2014). *Risk Reduction for Direct Potable Water Reuse*. Alexandra, VA: Water Reuse Research Foundation.

15 Water Recharge[1]

AQUIFER REPLENISHMENT

The *Virginian-Pilot* newspaper, November 2, 2023, published an account by Cianna Morales with the front-page headline that reads "Project Gives Area Water A Recharge" (pg. 1, 4). The chapter points out that Hampton Roads Sanitation District (HRSD) headquartered in Virginia Beach, Virginia, and with the newspaper's statewide distribution subscribers living in the surrounding cities, towns, and counties in the Lower Chesapeake Bay area (also known as Hampton Roads or Tidewater) about the sanitation district's ongoing treatment of drinking water quality wastewater recharge project underway in Newport News. The chapter discusses HRSD's Newport News project where a team of workers used heavy machinery and welding to fit together vertical lengths of pipe, each about 2.5 feet in diameter.

Ok, so what's this project all about?

The project is all about recharging the Potomac Aquifer (underlying the Lower Chesapeake Bay region) with wastewater treated to drinking water quality standards. The article explains that the workers fed the pipe straight down into a hole—into a total of ten holes in Riverview Park near the James River Wastewater Treatment Plant (WWTP) in Newport News, Virginia. The pipe was fed down through several layers of silt, gravel, and clay, all the way to the Potomac Aquifer—roughly 1200 feet beneath the surface, not quite the length of the Empire State Building.

The wells are designed to inject water back into the aquifer, which is an essential source of freshwater in the region, and, like many aquifers described earlier in this book and across the nation, are in danger of being overdrawn.

Actually, HRSD's goal is multifaceted and is detailed in its Sustainable Water Initiatives for Tomorrow (SWIFT) project. The purpose of HRSD's SWIFT program entails "more" than just recharging the aquifer with wastewater treated to drinking water standards (treated to match existing groundwater chemistry) and injecting it back into earth.

So, what's the "more" about HRSD's SWIFT program?

Well, along with recharging all three levels, the upper, middle, and lower Potomac Aquifer (LPA), the treated to drinking water injections also serve to address environmental concerns, including conservation, saltwater intrusion, and land compaction.

Initially, SWIFT was simply thought of as an engineering practice that will (as is hoped), in the long run, arrest relative sea level issues contributed by substantial land subsidence and worldwide sea level rise. Note that clay and silt hold water but do not allow it to pass easily after reaching a state of saturation. Sands and gravels, which allow for the flow of water, compose aquifers that feed wells. The wells in Newport News will contribute 16 million gallons of water per day to the aquifer.

DOI: 10.1201/9781003461265-18

Sounds like a lot of water, but it is only a drop in the bucket—more than 155 million gallons of water is withdrawn from the aquifer every day in Southeast Virginia. Actually, most Tidewater uses surface water sources for potable water but the aquifer is an excellent emergency supply when needed.

A part of SWIFT's recharging efforts is intended to increase pressure in the aquifer, creating what is known as a "saltwater curtain." This curtain, a hydrological phenomenon, occurs when freshwater is pumped into the aquifer; it is the interface, the wall between freshwater in the aquifer and infiltrating saltwater.

Obviously, the pumping of wastewater to treated drinking water quality standards into the aquifer is an excellent way to maintain pressure and to prevent saltwater intrusion.

Putting water into the Potomac Aquifer has another beneficial application: It provides an alternative sink for nutrients that would otherwise be pumped into the Elizabeth River, James River, York River, Potomac River, and finally end up in Lower Chesapeake Bay ecosystem.

Another benefit of added water pressure is prevention of subsidence in the Lower Chesapeake Bay region. Approximately 25% of relative sea level rise is due to the land sinking from aquifer compaction. The original HRSD 2018 SWIFT well pumps 1 million gallons per day (MGD) of treated wastewater, and that to date (2023) has made a slight observable improvement in the land subsidence issue.

The main goal of the SWIFT project is to not only to arrest land subsidence in the Lower Chesapeake Bay region but to also replenish the Potomac Aquifer with much needed drinking water. Note that when we use up one resource and then another resource and then another, there is one critical resource mankind never runs out of—that one item that sustains us—in a word it is innovation, innovation, innovation (Spellman, 1996). Hopefully, we never stop innovating.

HAMPTON ROADS SANITATION DISTRICT (HRSD)

Its genesis was driven by oysters. No, not the genesis of the Chesapeake Bay; its genesis was driven by a heavy, unstoppable, all-knowing hand. HRSD, arguably the premier wastewater treatment district on the globe, became a viable governor-appointed state commission-monitored entity because of a significant decline in the oyster population in the Chesapeake Bay. As a case in point, consider that in the Hampton Roads region of the Chesapeake Bay in 1607 when Captain John Smith and his team settled in Jamestown, oysters up to 13 inches in size were plentiful—more than could ever be harvested and consumed by the handful of early settlers. And this population of oysters and other aquatic lifeforms remained plentiful until the population gradually increased in the bay region.

Over-harvesting of oysters by the increased numbers of humans living in the Chesapeake Bay region was (and might still be) a major issue with the decline of the oyster population. However, the real culprit causing the decline in the oyster population is pollution. Before the bay became polluted from sewage, sediment, and garbage disposal, oysters could handle natural pollution from stormwater runoff and other sources. Ninety years ago when there was a much larger oyster population than today, it is estimated that the large oyster population could filter pollutants from the bay and clean it in as little as 4 days. By the 1930s, however, the declining oyster population was overwhelmed by the increasing pollution levels.

For years and in many written accounts, the author has stated that pollution is a judgment call. That is, pollution as viewed by one person may not be pollution observed by another. You might shake your head and ask a couple of questions, "Pollution is a judgment call? Why is pollution a judgment call?" A judgment is based on an opinion; it is an opinion because people differ in what they consider to be a pollutant based on of their assessment of the accompanying benefits and/or risks to their health and economic well-being posed by the pollutant. For example, visible and invisible chemicals spewed into the air or water by an industrial facility might be harmful to people and other forms of life living nearby. However, if the facility is required to install expensive pollution controls, forcing the industrial facility to shut down or move away, workers who would lose their jobs and merchants who would lose their livelihoods might feel that the risks from polluted air and water are minor weighed against the benefits of profitable employment and business opportunity. The same level of pollution can also affect two people quite differently. Some forms of air pollution, for example, might cause a slight irritation to a healthy person but cause life-threatening problems to someone with chronic obstructive pulmonary disease like emphysema. Differing priorities lead to differing perceptions of pollution (concern at the level of pesticides in foodstuffs generating the need for wholesale banning of insecticides is unlikely to help the starving). No one wants to hear that cleaning up the environment is going to have a negative impact on them. The fact is public perception lags behind reality because the reality is sometimes unbearable.

HRSD is a political subdivision of the Commonwealth of Virginia with a service area that includes 17 counties and cities encompassing 2800 square miles in its southeastern Virginia service area (see Figure 15.1). HRSD's collection system consists of more than 500 miles of piping, 6–66 inches in diameter. HRSD possesses more than 100 active pumping operations that pump raw wastewater to 9 major treatment plants in Hampton Roads and 4 smaller plants in the Middle Peninsula. The combined capacity of HRSD facilities is 249 MGD.

Probably another question rumbling through the reader's brain matter at this point is "Has HRSD solved the pollution problem in Chesapeake Bay?" This question leads us to another question, "Have the oysters rebounded in quantity?" The answer to both these questions is yes, that is, to a point. On an ongoing, 24-7 basis, HRSD treats wastewater to a quality better than is contained in the James River, Elizabeth River, York River, and other river systems in the region. Those who have no knowledge of wastewater treatment, or of HRSD, and/or the conditions of the rivers in this region might have second thoughts about this statement. However, it is true. It is all about the human-made water cycle. In this case, we are talking about the urban water cycle.

DID YOU KNOW?

Artificially generated water cycles or the urban water cycles consists of (1) source (surface or groundwater), (2) water treatment and distribution, (3) use and reuse, and (4) wastewater treatment and disposition, as well as the connection of the cycle to the surrounding hydrological basins.

HRSD Service Area
A Political Subdivision of the Commonwealth of Virginia

Facilities include the following:

1. Atlantic, Virginia Beach
2. Chesapeake-Elizabeth, Va. Beach
3. Army Base, Norfolk
4. Virginia Initiative, Norfolk
5. Nansemond, Suffolk
6. Lawnes Point, Smithfield
7. County of Surry
8. Town of Surry

09. Boat Harbor, Newport News
10. James River, Newport News
11. Williamsburg, James City County
12. York River, York County
13. West Point, King William County
14. King William, KingWilliam County
15. Central Middlesex, Middlesex County
16. Urbanna, Middlesex County

Serving the Cities of
Chesapeake, Hampton,
Newport News, Norfolk,
Poquoson, Portsmouth, Suffolk,
Virginia Beach, Williamsburg and the
Counties of Gloucester,
Isle of Wight, James City,
King and Queen, King William,
Mathews, Middlesex, Surry* and York
*Excluding the Town of Claremont

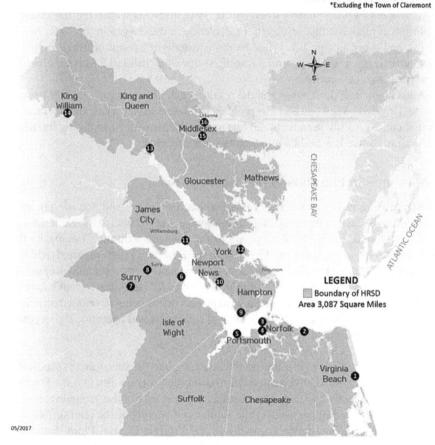

FIGURE 15.1 HRSD service area.

There is no question that the Chesapeake Bay is cleaner and that the sea life, including oysters are happier today because of the efforts of HRSD. The problem of making the bay cleaner is compounded by two factors. First, there are more than 300 WWTPs that outfall treated water to the Chesapeake Bay via its 9 major river systems and numerous tributaries. These treatment plants, separate and isolated from

HRSD's 13 plants, do the best they can to treat wastewater to a cleaner product (effluent) than the influent they received from various sources. However, some of these 300+ other plants treat only to primary treatment levels, and thus, their effluent is not as clean as secondary and tertiary plant effluent. Second, HRSD treats wastewater to a top-notch water quality level. However, treating wastewater to remove nutrients is a complicated and expensive undertaking. Biological nutrient removal and other nutrient removal technologies are available and in use in many locales, but the technology is expensive—expensive to the point where the treatment technology needed and used might overtax the ratepayers.

The Chesapeake Bay occasionally suffers from dead zones due to algal blooms. *Algal bloom* is a phenomenon whereby excessive nutrients within the bay cause an explosion of plant life that results in the depletion of the oxygen in the water needed by fish and other aquatic life. Algal bloom is usually the result of urban runoff (of lawn fertilizers, etc.). The potential tragedy is that of a "fish kill," where the bay life dies in one mass execution.

Algal bloom and dead zones and the resulting fish kill events are a major issue, of course. However, when you add this problem to relative sea level rise and land subsidence, it can be readily seen that the issue and problems with maintaining the health of the Chesapeake Bay and its inhabitants is multifaceted.

THE SOLUTION TO POLLUTION IN THE CHESAPEAKE BAY[2]

HRSD's mantra is as follows: Every problem has a solution. And that has proven to be true in most cases with many more environmental problems/issues that need solutions to be found, hopefully. With regard to the problems with the Chesapeake Bay, land subsidence, and relative sea level rise in the Hampton Roads region, HRSD has developed the innovative SWIFT program (a work in progress; a decadal project). Do not confuse the acronym SWIFT with the adjectives for fast, speedy, rapid, hurried, immediate, or quick. SWIFT is a long-term project that is being developed on a timeline that is set for installation of the technical equipment and operational procedures with a completion date of 2030.

What is HRSD's SWIFT program?

SWIFT is a program to inject treated wastewater into the subsurface; specifically, it is designed to inject treated wastewater to drinking water quality into the Potomac Aquifer. Injection of water into the subsurface is expected to raise groundwater pressures, thereby potentially expanding the aquifer system, raising the land surface, and counteracting land subsidence occurring in the Virginia Coastal Plain. In 2016, a pilot project site is under construction at the HRSD Nansemond WWTP in Suffolk, Virginia, to test injection into the aquifer system. HRSD has asked United States Geological Survey (USGS) to prepare a proposal for installation of an extensometer monitoring station at the test site to monitor groundwater levels and aquifer compaction and expansion.

SWIFT is designed to counter land subsidence at various locations in the Hampton Roads area of southern Chesapeake Bay where land subsidence rates of 1.1–4.8 mm/year have been observed (Eggleston and Pope, 2013; Holdahl and Morrison, 1974).

Injection of treated wastewater (treated to drinking water quality) is expected to counteract land subsidence or raise land-surface elevations in the region. Careful monitoring of aquifer system compaction and groundwater levels can be used to optimize the injection process and to improve fundamental understanding of the relation between groundwater pressures and aquifer system compaction and expansion.

There is more to HRSD's treated wastewater injection project, SWIFT, than just arresting or mitigating land subsidence and relative sea level rise in the Hampton Roads region. One of the additional goals of the project is to stop discharge of treated wastewater from seven of its plants that would mean 18 million pounds a year less of nitrogen, phosphorus and sediment outfalling into the bay. Assuming SWIFT works as designed, this is a huge benefit to the Chesapeake Bay in that it may help to prevent or reduce the formation of algal bloom dead zones. Not only would success as a result of treated wastewater injection benefit the bay but it would also be a huge benefit for the ratepayers at HRSD. To meet regulatory guidelines to remove nutrients from discharged treated wastewater would cost hundreds and millions of dollars and almost non-stop retrofitting at the treatment plants to keep up with advances in treatment technology and regulatory requirements. Another goal of HRSD's SWIFT project is to restore or restock potable groundwater supplies in the local aquifers. The drawdown of water from the groundwater supply has not only contributed to land subsidence but to a reduction of water available for potable use.

HRSD's planned restocking Hampton Roads groundwater supply with injected wastewater treated to potable water quality is not without its critics. The critics state that HRSD's wastewater injection project would contaminate potable water aquifers. For the critics and others, this is where the so-called yuck factor comes into play. The yuck factor, in this particular instance, has to do with the thought that groundwater for consumptive use will be contaminated basically with toilet water. This is the common view of many of the critics who feel HRSD's SWIFT project is nothing more than direct reuse of wastewater, that is, a pipe-to-pipe connection of toilet water to their home water taps.

What the critics and others do not realize is that we are already using and drinking treated and recycled toilet water. As far as HRSD's SWIFT project contaminating existing aquifers with toilet water, it is important to point out that this water is to be treated (and already is at Nansemond Treatment Plant in Suffolk, Virginia) to drinking water quality—to drinking water quality is the key phase here. This sophisticated and extensive train of unit drinking water quality treatment processes—treated wastewater that HRSD General Manager and several others drank right out of the process recently, and, by the way, they are doing just fine today, thank you very much—is discussed in detail later in the text. The bottom line is as follows: Statements about yuck factor involved in drinking treated toilet water are grossly overstated—drink up!

The SWIFT project includes construction of an extensometer monitoring station with the ability to accurately measure land-surface elevations, bedrock-surface elevations, and changes in aquifer system thickness. Monitoring of groundwater levels and aquifer system elastic response will benefit operation of the wastewater injection system at the Nansemond Treatment Plant in Suffolk, Virginia, and provide guidance of future wastewater injection facilities for the SWIFT project.

OBJECTIVES

The objectives of the SWIFT project are to

- Design and construct an extensometer station for collection of aquifer system thickness data with sub-millimeter accuracy for long-term (decadal) operation.
- Operate and maintain the station for at least 3 years to collect data describing groundwater levels, land-surface and bedrock vertical motion, and changes in aquifer system thickness.

THE NANSEMOND TREATMENT PLANT EXTENSOMETER PLAN

The HRSD extensometer station at the Nansemond Treatment Plant will be designed and constructed to produce reliable and accurate data describing aquifer thickness, groundwater levels, land-surface elevation, and bedrock for at least 20 years and likely for 40 years or more.

EXTENSOMETER STATION DESIGN

The USGS, from its National Research Drilling Program, has obtained unique and extensive experience in designing and installing extensometers in many locations around the United States, including California, Texas, and Louisiana. Based on this past experience, USGS recommended including the following features in the extensometer instrument:

- Casing slip joints to accommodate the vertical stress and strain that accompany aquifer compaction/expansion and that can cause casing failure.
- Casing centralizers and a counterweighted fulcrum at the surface to reduce friction and striking between the extensometer pipe and casing.
- Combined mechanical strain gauges and digital potentiometers to achieve both accuracy and long-term record continuity.
- Deep surface mount support piers for the reference table to exclude surface soil compaction and heaving from the aquifer compaction/expansion signals.
- Heavy steel beam frames to support surface equipment.

THE PLAN

Under the proposed plan and as discussed earlier, the USGS has constructed an extensometer station at HRSD's Nansemond wastewater treatment facility in Suffolk, VA. USGS Water Science Center personnel oversaw all aspects of construction and operation of the extensometer system, including the drilling and down-hole work. USGS also deployed its own drill rig, support equipment and drill crew. A staff geologist with the USGS and a geologist for the Virginia Department of Environmental Quality (DEQ) shared on-site geologist duties during the approximate 1 month of

FIGURE 15.2   USGS drilling rig at Nansemond Treatment Plant Suffolk, VA. Photograph by F.R. Spellman.

drilling (see Figure 15.2). Installation of surface reference frame structures and monitoring equipment was performed or overseen by USGS staff. The work lasted from July 2016 through August 2019. Drilling and construction occurred in Year 1, and the USGS operated the station in Years 1–3. Station operation can continue beyond Year 3 by mutual agreement.

The extensometer station was constructed within the fenced perimeter of the Nansemond WWTP in Suffolk, Virginia, and the final location was chosen jointly with HRSD. The final location was located approximately 500 feet from the injection test well, has paved road access, and has adequate space for equipment including a large drilling rig, delivery trucks, a generator, work lights, forklift, dumpsters, and backhoe. A roughly 60 feet × 100 feet area is used to house materials, a work area for geologists and drillers was provided, and a pit was dug for circulation of drilling fluids.

A permanent structure with a 20 feet × 20 feet footprint and a 10 feet height was built to house the top of the extensometer instrument and the associated monitoring equipment (see Chapter 4, Figures 4.3 and 4.4). It is an insulated metal building with a concrete floor and a window and door, but alternative exteriors can be considered later if desired by HRSD. The building will have locked windows and doors.

A cluster of four piezometers is highly recommended to allow monitoring of groundwater levels (pressures) in multiple aquifers. Each piezometer will be housed in a protective casing that protrudes about 3 feet above ground. These piezometers are permanent structures and are located near the extensometer building.

HRSD's SWIFT project proposes to add advanced treatment process to several of its facilities to produce water that exceeds drinking water standards and to pump this clean water into the ground and the Potomac Aquifer. This will ensure a sustainable source of water to meet current and future groundwater needs through eastern Virginia while improving water quality in local rivers and the Chesapeake Bay. The SWIFT project benefits include (HRSDa,b, 2016):

- Eliminates HRSD discharge to the James, York, and Elizabeth rivers except during significant storms
- Restores rapidly dwindling groundwater supplies in eastern Virginia upon which hundreds of thousands of Virginia residents and businesses depend
- Creates huge reductions in the discharge of nutrients, suspended solids, and other pollutants to the Chesapeake Bay
- Make available significant allocations of nitrogen and phosphorous to support regional needs
- Protects groundwater from saltwater contamination/intrusion
- Reduces the rate of land subsidence, effectively slowing the rate of sea level rise by up to 25%
- Extends the life of protective wetlands and valuable developed low-lying lands

POTOMAC AQUIFER

Be advised that HRSD is in the early stages of operation of its SWIFT initiative and lessons are being learned via this work in progress. Also, before HRSD commenced pumping up to 130 MGD into the Potomac Aquifer System (PAS) or as it will do more in the future into any other aquifer it—along with the expert assistance of CH2M, its primary consultant in this matter—HRSD first had to determine the feasibility of aquifer replenishment by recharging clean water, purified from the advanced treatment of WWTP effluent. In this section, a description of the essential elements of recharging clean water into the PAS at the following seven HRSD WWTPs: Army Base, Boat Harbor, James River, Nansemond, Virginia Initiative Plant (VIP), Williamsburg, and York River. Also, a determination of the capacity of individual injection wells at the seven WWTPs has been made or is being studied at the present time; moreover, a projection of the injection capacity within the existing site area of the seven WWTPs has or is being determined, and a characterization of the regional beneficial hydraulic response of the PSA to clean water injection has and is being determined.

The material presented in this section is based on data available in city/country, state and federal databases, reports, scientific papers, interviews, operators findings, and other literature to characterize the injection capacity of individual wells at each of the WWTPs. Injection well capacities and analytical mathematical modeling were used to estimate the injection capacity of each WWTP based on the plant's flow rate, property size, and the transmissivity of the underlying PAS aquifer.

THE POTOMAC FORMATION[3]

Given the elevated volume requiring disposal and the importance of minimizing the number of injection wells, the most suitable aquifer units are those that exhibit the highest production capacity. Furthermore, a thick, confining bed composed of impermeable materials like silt or clay should overlie the aquifer to prevent vertical migration of the injection fluid (injectate) into the surrounding aquifer units. Beneath the HRSD service area, the Cretaceous age, Potomac Formation meets these criteria. The Potomac Formation contains thick sand deposits, forming three discrete aquifer units. Although the modern convention developed by the USGS and the Virginia Department of Environmental Quality (VDEQ) is to group the three aquifers as one, named the Potomac Aquifer (McFarland and Bruce, 2006), locally and in this book, because they behave hydraulically as three distinct units, they must be and are examined separately. Accordingly, in this presentation, these units are treated as three distinct units referred to as upper Potomac Aquifer (UPA) zone, middle Potomac Aquifer (MPA) zone, and LPA zone (Laczniak and Meng, 1988; Hamilton and Larson, 1988). Each discrete aquifer in the HRSD service area is separated from adjacent aquifers by clay confining beds of measurable thickness, while a cumulative thickness of silt and clay units totaling several hundred feet overlies the top (UPA) aquifer unit of the PAS. Production wells screened in the PAS exhibit significantly greater pumping capacities than wells screened in other aquifers of the Virginia Coastal Plain (Smith, 1999). Some wells can pump at rates approaching 3000 gallons per minute (gpm) (4.3 MGD). In addition to confinement and production capacity, aquifers in the PAS exhibit deep static water levels, ranging from 80 to 180 feet below grade (fbg), providing available head for injection.

Recent USGS sedimentological studies suggest the HRSD service area, spanning the York-James Peninsula and Southeastern Virginia, is well situated regarding the quality of aquifers in the PAS (McFarland, 2013). PAS aquifer sands display greater thickness, coarser grain size, and better sorting in the HRSD service area than units in the northern Virginia Coastal Plain, or to the south in the northern North Carolina. As a result, aquifers exhibit excellent hydrologic coefficients (hydraulic conductivity, transmissivity) and, thus, more productive well capacity (HRSDa,b, 2016).

The PAS outcrops at the ground surface west of King William WWTP in King William County Virginia (see Figure 15.1). PAS is recharged by infiltrating precipitation. In the recharge area, the aquifers range in thickness from 70 (UPA) to 400 feet (LPA). Further downdip, recharge enters the PAS by leakage through overlying confining beds (Meng and Harsh, 1988).

Individual aquifers thicken and dip to the southeast toward the Atlantic coastline reaching thickness ranging from 170 feet (UPA) to approaching 1000 feet (LPA) at the coast (Treifke, 1973). These thicknesses comprise all sediments contained in the vertical section including discrete sand beds representing aquifer materials, and interleaving silt and clay lenses from intra-aquifer contained beds. Individual aquifers exhibit a strongly interbedded morphology consisting of thin to thick beds of sands, silts, and clays. Beneath Newport News (refer to Figure 15.1), the MPA consists of six discrete sand intervals. Obtaining maximum production (or injection) capacities from wells installed in layered aquifers such as the PAS requires extending the well screen assembly across the maximum thickness of aquifer sand. Screen assemblies can consist of multiple screens and blank sections (HRSDa,b, 2016).

INJECTION WELLS

Typically, when we think about a well, the image of a hole in the ground and some type of device to pump water to the surface for use or for storage. A well pumping water from the subsurface to the surface functions to provide whatever use it might be intended for. However, wells used for pumping are not what we are concerned with in this book. We are concerned with just the opposite, that is, wells that inject treated injectate and that do not pump fluids to the surface.

Injection wells are known as Class V wells. There are 22 types of Class V injection wells. The types of wells are shown in Table 15.1.

SUBSIDENCE CONTROL WELLS

It is the last type of Class V well listed in Table 15.1 (subsidence control wells) that is the focus of this presentation. Subsidence control wells are injections wells whose primary objective is to reduce or eliminate the loss of land-surface elevation due

TABLE 15.1

Class V Underground Injection Wells

Type of Injection Well	Purpose
Agricultural drainage wells	Receive agricultural runoff
Stormwater drainage wells	Dispose of rainwater and melted snow
Carwashes without undercarriage washing or engine cleaning	Dispose of wash water from car exteriors
Large-capacity septic systems	Dispose of sanitary waste through a septic system
Food processing disposal wells	Dispose of food preparation wastewater
Sewage treatment effluent wells	Used to inject treated or untreated wastewater
Laundromats without dry cleaning facilities	Dispose of fluid from laundromats
Spent brine return flow wells mine backfill wells	Dispose of mining byproducts
Aquaculture wells	Dispose of water used for aquatic sea life cultivation
Solution mining wells	Dispose of leaching solutions (lixiviants)
In situ fossil fuel recovery wells	Inject water, air, oxygen solvents, combustibles, or explosives into underground or oil shale beds to free fossil fuels
Special drainage wells	Potable water overflow wells and swimming pool drainage
Experimental wells	Used to test new technologies
Aquifer remediation wells	Use to clean up, treat, or prevent contamination of underground sources of drinking water
Geothermal electrical power wells	Dispose of geothermal fluids
Geothermal direct heat return flow wells	Dispose of spent geothermal fluids
Heat pump/air conditioning return flow wells	Re-inject groundwater that has passed through a heat exchanger to heat or cool buildings
Saline intrusion barrier wells	Injected fluids to prevent the intrusion of saltwater
Aquifer recharge/recovery wells	Used to recharge an aquifer
Noncontact cooling water wells	Used to inject noncontact cooling water
Subsidence control wells	Used to control land subsidence caused by groundwater withdrawal or over-pumping of oil and gas

to removal of groundwater providing subsurface support. Subsidence control wells are important to HRSD's SWIFT project. The goal is to inject treated wastewater to drinking water quality into the underground PAS to maintain fluid pressure and avoid compaction and to ensure that there is no cross-contamination between infected water and underground sources of drinking water. Thus, the injectate must be of the same quality or superior in quality as the existing groundwater supply.

INJECTION WELL HYDRAULICS

With regard to aquifer injection hydraulics, a well's injection capacity depends on its specific capacity and the pressure (head) available for injection, a function of the static head (water level) of the aquifer in which the well is screened. Specific capacity describes a well's yield per unit of head decrease (drawdown) in a pumping well or head increase (draw-up) in an injection well. Specific capacity is expressed in units of feet of drawdown/draw-up per unit of pumping/injection rate in gpm per foot (gpm/ft), respectively. When injection begins, the water level in the well rises as a function of the transmitting properties of the receiving aquifer and the well's efficiency (Warner and Lehr, 1981). While the transmitting character of the aquifer should remain stable over the service life of an injection well, the available head for injection will decline as injection recharges the aquifer causing the static water level to rise toward the ground surface (HRSDa,b, 2016).

In this discussion of the planning and study phase (and now the processing phase at Nansemond Treatment Plant, NATP) of the SWIFT project, it is important to differentiate between production and injection by referring to injection-specific capacity as injectivity. Moreover, for our purposes, here the evaluation of specific capacity of local production wells and its conversion to injectivity forms an important variable in determining the capacity of individual injection wells and ultimately the total injection capacity across the affected area and number of wells required at each WWTP.

DID YOU KNOW?

Specific capacity is one of the most important concepts in well operation and testing. The calculation should be made frequently in the monitoring of well operation. A sudden drop in specific capacity indicates problems such as pump malfunction, screen plugging, or other problems that can be serious. Such problems should be identified and corrected as soon as possible. *Specific capacity* is the pumping rate per foot of drawdown (gpm/ft), or

Specific capacity = Well yield ÷ drawdown

Problem: If the well yield is 300 gpm and the drawdown is measured to be 20 feet, what is the specific capacity?

Solution:

Specific capacity = 300 ÷ 20
Specific capacity = 15 gpm per ft of drawdown

In determining injectivity for wells screened in specific aquifer units, specific capacity values were evaluated in the three Potomac Aquifer units with respect to their proximity to HRSD's WWTPs. The VDEQ supplied a database of production wells that contained a total of 98 wells screened in either the UPA, MPA, or LPA, with some wells screened across sands in two of the three aquifers. In these wells, a technique was employed to determine the specific capacity contributed by each aquifer to the well, based on the length of the screen penetrating the aquifer, divided by the total length of the well screen. It proved necessary to separate the aquifers across individual wells for the less-represented LPA. Because of its greater depth and water quality, production wells rarely penetrated all the sand units in the LPA, often screening sand intervals in both the MPA and UPA.

Across the study area, the specific capacity of production wells screening the UPA and MPA averaged 35.5 and 32.4 gpm/ft, respectively, essentially equaling each other, while wells screened in the LPA exhibited a 40% lower value of 21 gpm/ft. Wells screening the LPA, as identified by the VDEQ, were only located around the Franklin Paper Mill and Franklin City, Virginia area. Wells screened in the UPA and MPA were better represented across the study area.

As stated previously, the available head for injection represents an important factor in determining injection capacities. Multiplying the well's injectivity by the available head for injection provides the injection capacity of the well. As a driver for this study, local industrial and residential development has resulted in elevated pumping, drawing water levels in aquifers of the PAS downward at rates averaging 1–2 feet per year (McFarland and Bruce, 2006). The declining water levels represent greater available head for injection. However, once injection commences, water levels should rebound with a corresponding loss in available head (HRSDa,b, 2016).

INJECTION OPERATIONS

Because of the deep static water levels at most HRSD WWTPs and available head for injection in feet at each site for each level of the Potomac Aquifer (see Table 5.2), HRSD could inject water under gravity conditions or pressurized conditions. Under gravity conditions, HRSD would fit a foot valve to the base of the injection piping to reduce head and maintain a positive pressure in the injection and well-header piping. This design facilitates better control of injection rates compared with cascading water down the injection piping, inducing a vacuum through the system. If HRSD decides to inject under gravity conditions, it would not be necessary to seal the injection head, well-header piping, and associated fittings. Instead, the design requires monitoring injection levels rising in the well's annular space to prevent it from topping the ground surface.

DID YOU KNOW?

"No owner or operator shall construct, operate, maintain, convert, plug, abandon, or conduct any other injection activity in a manner that allows the movement of fluid containing any contaminant into underground sources of drinking water, if the presence of that contamination may cause a violation of any primary drinking water regulation under 40 CFR part 142 or may otherwise adversely affect the health of persons" (40 CFR 144.121).

TABLE 15.2

Summary of Available Head for Injection (feet) in PAS at HRSD's Treatment Plants

Wastewater Treatment Plant	PAS Elevation	PAS Aquifer	Available Head for Injection (feet)
Army Base	11	UPA	107.28
		MPA	101.28
		LPA	101.28
Virginia Initiative Plant	8.5	UPA	109.78
		MPA	106.28
		LPA	101.28
Nansemond	22.5	UPA	123.78
		MPA	106.28
		LPA	106.28
Boat Harbor	4	UPA	100.28
		MPA	91.28
		LPA	96.28
James River	18	UPA	107.28
		MPA	81.28
		LPA	86.28
York River	5.5	UPA	86.78
		MPA	76.28
		LPA	76.28
Williamsburg	58.5	UPA	125.78
		MPA	67.28
		LPA	76.28
Rate of water-level decline	0.92 ft/yr		

Source: Adapted from CH2M (2016).

Note: ft, feet; WWTP, wastewater treatment plant; PAS, Potomac Aquifer System; VIP, Virginia Initiative Plant.

An option to injecting under gravity conditions could entail sealing the injection head, well-header piping, and associated fittings, while allowing the injection head to rise above the elevation of the ground surface by maintaining a positive pressure in the annular space of the well. With greater available head for injection, HRSD could achieve higher injection rates, while anticipating risking regional water levels inherent with injecting large volumes of water.

The operator is responsible for ensuring annular pressures stay below an established threshold. Elevated annular pressures can stress the sand filter pack surrounding the well screen by lifting it and initiating the formation of channels that connect the well screen to the surrounding formation materials. Limiting the injection pressure to 10 pounds per square inch (psi) in the well's annular space precludes damage to the filter pack while making available another 23 feet of head for injection.

Regular maintenance is required to operate injection wells to their maximum capacity whenever they are screening fine sandy materials like those found in the PAS. In aquifer storage and recovery (ASR) wells, screening the PAS, of similar aquifers at Chesapeake, Virginia, and in North Carolina, Delaware, and New Jersey, fine sandy aquifers have proven susceptible to clogging from total suspended solids (TSS) entrained in the recharge water in the recharge water (McGill and Lucas, 2009). Even high-quality treated water can contain some amount of TSS that accumulates (clog) in the screen and pores spaces of sand filter pack and formation. Clogging reduces the permeability around the screen, filter pack, and aquifer proximal to the well (wellbore environment), resulting in higher injection levels while reducing injection capacity.

ASR wells operating in these states are equipped with conventional well pumps, allowing periodic well back-flushing during recharge. Back-flushing entails temporarily shutting down recharge and turning on the well pump for a sufficient time to remove fine-grained materials from the wellbore environment to the ground surface.

By adopting the approach in maintaining ASR wells by installing a pump in each of HRSD's injection wells for back-flushing, it will slow the accumulation of fine-grained materials in the wellbore environment. To generate sufficient energy required to effectively remove solids from the wellbore, the capacities of back-flushing pumps should equal or exceed the injection rates planned for the well.

DID YOU KNOW?

The type and quality of injectate and the geology affect the potential for endangering an underground supply of drinking water. The following examples illustrate potential concerns:

- If injectate is not disinfected, pathogens may enter an aquifer. Some states allow injection of raw water and treated effluent. In these states, the fate of microbes and viruses in an aquifer is relevant.
- When water is disinfected prior to injection, disinfection byproducts can form in situ. Soluble organic carbon should be removed from the injectate before disinfection. If not, a chlorinated disinfectant may react with the carbon to form contaminating compounds. These contaminants include trihalomethanes and haloacetic acids.
- Chemical differences between the injectate and the receiving aquifer may create increased health risks when arsenic and radionuclides in the geologic matrix interact with injectate having a high reduction–oxidation potential.
- Carbonate precipitation in carbonate aquifers can clog wells when the injectate is not sufficiently acidic (USEPA, 2016).

Even with back-flushing, progressive clogging can occur and exceed the ability of back-flushing to maintain injection wells near their maximum capacity. Each WWTP should contain a sufficient number of wells to compensate for removing wells from service for rehabilitation, without compromising the facility's injection capacity.

INJECTION WELL CAPACITY ESTIMATION

In the planning and study stage, the following steps were employed in estimating the capacities of injection well in the PAS at each HRSD WWTP (HRSDa,b, 2016):

- Estimate specific capacity at individual wells (Specific capacity = well yield ÷ drawdown)
- Organize specific capacities by aquifer
- Separate well screens spanning two aquifers and calculate specific capacity for both
- In short screen assemblies, normalize screen length to 100 feet
- Convert specific capacity to injectivity
- Calculate available head for injection from USGS 2005 synoptic study
- Combine injectives of UPA and MPA, or MPA and LPA in a single injection well
- Average available head for injection across UPA and MPA, or MPA and LPA in a single injection well
- Add 23 feet of head to available injection head to account for maintaining a pressure of 10 psi in annular space of well
- Multiply injectivity by available head for pumping to obtain injection well capacity
- Practically limit injection well capacity to 3 MGD (2100 gpm)
- To estimate number of injection wells per WWTP, divide the plant's effluent rate by injection well capacity
- To facilitate periodic maintenance, add one injection well for every five, per WWTP

ESTIMATING SPECIFIC CAPACITY AND INJECTIVITY

The VDEQ database of wells and their locations was obtained and evaluated according to their location relative to HRSD's WWTPs at Army Base, Boat Harbor, James River, Nansemond, VIP, Williamsburg, and York River. The database contained static water levels, pumping levels, and stable production rates of most wells, enabling the calculation of specific capacity. Moreover, the VDEQ database identified the PAS unit spanned by each screen interval in each well. Several wells featured over five well screen intervals. As previously described, maximizing well screen length increases a well's production capacity. Accordingly, many larger-capacity productized wells were equipped with multiple screen intervals and, in many cases, screening more than one PAS unit (HRSDa,b, 2016).

Specific capacities were calculated for each well and then grouped according to the aquifer unit(s) in which the well was screened. Specific capacities for wells with screens spanning two aquifers resulted in developing two specific capacities for single well. Screen assemblies in these wells usually spanned the UPA and MPA, or MPA and LPA. Individual intervals were grouped according to the aquifer which they screened. In wells spanning several aquifers, the specific capacity for a discrete aquifer was then estimated by taking the length of screen spanning the aquifer and dividing it by the total screen length as follows:

$$\text{Total SC of well} \times (\text{SL Aquifer 1/TSL of well}) = \text{SC Aquifer 1}$$

where
  SC = specific capacity
  SL = screen length
  TSL = total screen length

HRSD should design their injection wells with the screens penetrating an entire aquifer's sand thickness to maximize injection capacity. Note, however, that many of the production wells studied in this project used shorter screen assemblies that only partially penetrated the aquifers in which they were installed. Accordingly, to make the estimates compatible with fully penetrating injection wells, the specific capacity for a shortened screen assembly was normalized to a well with 100 feet of screen by applying the following equation:

$$\text{SC Aquifer 1} = (100 \text{ feet of screen/SL})$$
$$= \text{SC of aquifer normalized to 100 feet of screen.}$$

One hundred feet of screen was recognized as an average total for assemblies fully penetrating the UPA, MPA, and LPA. Specific capacities for wells exhibiting screens extending over 100 feet across a specific aquifer unit of the PAS were accepted as representative and applied without modification (HRSDa,b, 2016).

Based on observations from ASR wells, which function in both injection and pumping modes of operation, the specific capacity and the injectivity of wells installed in the same aquifer usually vary slightly (Pyne, 2005). In a production well, water migrates toward a potentiometric head lowered by pumping in the wellbore. Water moves through an environment constrained by the size and heterogeneity of the porous media into the well. Upon entering the pumping well, the water is no longer impeded by porous media.

By comparison, injected water is driven down the wellbore by elevated head against the resistance of the wellbore environment. As a result of the greater resistance to flow, the injectivity of an ASR well often falls 10%–50% less than its specific capacity (Pyne, 2005). To re-create this important relationship for the study, converting specific capacity to injectivity involved multiplying specific capacity by a factor of 0.5, an average of the ratio of injectivity to specific capacity in Atlantic Coastal Plain injection type wells (Pyne, 1995).

Available Head for Injection

The available head for injection also played an important role in determining the capacity of an injection well. The available head in each aquifer at individual WWTPs was determined by obtaining the most recent general view of the whole (synoptic) water-level information. The US Geodetic Survey last measured synoptic water levels separated by aquifer unit of the PAS, in 2005 (VADEQ, 2006b). More recent work considers all aquifers of the PAS together.

Note that with progressively declining water levels attributed to over-pumping, potentiometric heads measured in 2005 cannot accurately match septic conditions in 2014. To adjust potentiometric levels to 2015 conditions, hydrographs were examined to quantify the annual conditions in water levels (VADEQ, 2006a). The evaluation revealed that potentiometric levels have declined an average of 0.9 feet per year since 2005 in the UPA, MPA, and LPA. The annual rate of decline in potentiometric levels was multiplied by 10 years and then applied to the potentiometric head from 2005, projecting elevations to 2014.

In estimating the depth of water, the projected potentiometric level for 2014 in each aquifer unit was subtracted from elevation of the land surface at each WWTP. A constant head of 23 feet, equal to injecting under a pressure of 10 psi in the well's annual space, was added to the depth of the water level to obtain the available head for injection (HRSDa,b, 2016).

Flexibility for Adjusting Injection Well Capacities

Maximizing the capacity of individual injection well required screening two aquifer units, either the UPA and MPA, or the MPA and LPA in each well. Because of potential hydraulic inefficiencies inherent with the difference in screen elevations, wells screening the UPA and LPA were not considered for this evaluation. This is important because it provides HRSD with some flexibility for adjusting injection well capacities while installing the injection well-fields. Upon installing the initial injection wells at a WWTP, HRSD could elect to screen all three PAS units in a single well or revert to screening one aquifer if capacities fail to meet, or significantly exceed expectations, respectively.

To obtain the injection well capacity, the final steps in estimating injection well capacities were deterring by adding the injectivities of two aquifer units in a well and averaging the available head for injection between the two aquifers. Then, the resulting injectivity was multiplied by the average available head for injection.

Number of Injection Wells Required

Elevated well capacities that exceeded reasonable constructability practices resulted from the contamination of high injectivities from merging screens and large available heads for injection. Constructing the relatively deep wells of sufficient diameter to inject at rates approaching 8 MGD (5600 gpm) or equipping the well with a back-flush pump capable of the same rate are likely impractical. To reduce well casing, screen, and pumps to dimensions consistent with local well drilling capabilities and operation (such as equipment available and electrical service), a capacity threshold of 3 MGD was set for the injections wells.

The number of injection wells at each WWTP was determined by dividing the plant's effluent rate by the injection well capacities. At large plants, injection wells were split evenly between UPA and MPA, and MPA and LPA combinations. These well totals and aquifer combinations at each WWTP formed the basis for the initial mathematical modeling runs discussed in the modeling section of this work. To accommodate removing wells form service for rehabilitation, one well was added for each five. At plants with smaller effluent flows, on well was added to any total less than five.

HRSD's VIP and Nansemond Treatment Plant required the largest number of injection wells to replenish the aquifer because of their higher effluent rates and relatively low injectivity in the MPA. Lower injectives also appear in the UPA adjacent to the VIP.

Production wells supporting the mapping of specific capacity in the MPA for this project were not present within a 5-mile radius of the Nansemond and Boat Harbor Plants. Regionally, specific capacity values in the MPA appear to increase to the southeast, with the exception of two production wells located west of Nansemond and Boat Harbor. To maintain a conservative approach to the project, low-specific capacity values imparted by these wells were maintained in estimating injection well capacity at Nansemond and Boat Harbor. By comparison, large-specific capacity values in the UPA and MPA around the James River, York River, and Williamsburg Plants resulted in elevated injectivities, which yielded elevated hypothetical injection capacities despite the relatively shallow available head for injection (HRSDa,b, 2016).

AQUIFER INJECTION MODELING

Hydrologists, hydrogeologists, and groundwater experts in other professional fields soon learn that groundwater flow models are simplified representation of often highly complex hydrogeological flow systems. Modeling tools are well suited for analyzing aquifer injection experiments. Generally, incorporating as much available hydrogeologic information as possible into the formulation of the conceptual and numerical models of the flow system is advantageous. This is the approach used by HRSD and its consultant in modeling for HRSD's SWIFT project. Hydrogeologic information takes many forms, including maps that show outcropping surfaces of geologic units and faults, cross-sections derived from geophysical surveys and wellbore information that show the likely subsurface location of geologic units and faults, maps of water table levels, independent point well data, and maps showing the hydraulic properties of the subsurface materials. HRSD and its consultant used this information to classify the geologic units into hydrogeologic units, which are convenient units to define hydrologic properties (Anderman and Hill, 2000). The modeling employed in the SWIFT project helps to project or estimate the capacity of injections of injectate and other important parameters used in this project.

Estimating the capacity of individual injections at each WWTP along with a preliminarily determining of the number of wells at each facility comprises one of several key elements of HRSD's SWIFT; the significance of these determinations can be seen when the goal is to dispose of nearly 130 MGD into the PAS. The modeling executed in this section tests the hydraulic interference between injection wells located within the boundaries of each WWTP property. This evaluation will identify whether individual WWTP properties are sufficiently large to contain the projected number of wells required to dispose of the projected effluent volumes.

MATHEMATICAL MODELING

Estimates of injection well capacity and the appropriate number of wells assigned at each WWTP described earlier did not account for hydraulic interference between wells in the same aquifer. With well screens combining the UPA and MPA or MPA and LPA, hydraulic interference in the MPA will exert the greatest influence on local injection levels. WWTPs that are situated on smaller properties will cause an increase in hydraulic interference; they require smaller inter-well spacing for fitting the number of injection wells required to dispose of effluent.

Mathematical modeling techniques were used in quantifying the interference between injection wells located at the WWTPs and rebounding water levels in the aquifers receiving effluent. In this section, analytical groundwater flow modeling is used to evaluate local groundwater mounding at individual WWTPs while injecting effluent (injectate) over 50 years.

DID YOU KNOW?

Hydraulic engineers and others are quite familiar with mathematical modeling. With the continuing advancements in computer technology and development of advanced computer engineering programs, engineers rely more and more on mathematical modeling. A mathematical model is an abstract model that uses mathematical language to describe the behavior of a system. Eykhoff (1974) defined a mathematical model as a representation of the essential aspects of an existing system (or a system to be constructed), which presents knowledge of that system in usable form.

GROUNDWATER FLOW MODELING

In evaluating potentiometric levels in the UPA, MPA, and LPA analytical groundwater flow modeling was applied at each injection well at the seven WWTPs. The modeling study extends the determination of individual injection well capacities by testing the injection rates under the spatial conditions unique to each WWTP property. Although the injection capacities of individual wells may appear feasible at a WWTP based on the head available for injection and the aquifer transmissivities, hydraulic interference between multiple wells can drive injection levels higher in inverse proportion to the available spacing between wells.

At smaller WWTP sties, interfering wells could cause injection levels to exceed 23 feet above the ground surface, the maximum threshold, established for injection head at individual wells. Mitigating elevated injection levels can entail screening all three PAS in a single well and/or reducing injection rates sufficiently to lower levels below the site injection elevation threshold. Lowering injection rates so that heads fall below site thresholds effectively limits flow injection rates lower than the projected 2040 target flows.

Accordingly, groundwater flow models were customized according to property size and the projected number of wells required at each WWTP. The computer program CAPZONE (Bair et al., 1992) was applied to conduct the analytical groundwater flow modeling at each WWTP.

CAPZONE is an analytical flow model that can be used to construct groundwater flow models of two-dimensional flow systems characterized by isotropic and homogeneous confined, leaky-confined, or unconfined flow conditions. CAPZONE computes drawdowns at the intersections of a regularly spaced rectangular grid produced by up to 100 wells using either Equation 15.1 for a confined aquifer developed in 1935 (while working for USGS) or the Hantush–Jacob equation for a leaky-confined aquifer (see Equation 15.2) (Bair et al., 1992). Unlike the numerical mathematical techniques employed by models like MODFLOW (which comprises simple algebraic equations that a computer cycles through multiple iterations to solve the flow equation), CAPZONE directly solves the differential flow equation. Subsequently, CAPZONE provides a more exact and conservative solution that models relying on numerical methods. However, analytical groundwater flow models offer less flexibility in simulating the heterogeneous conditions exhibited in natural systems, including multiple layers, variable grid spacing, spatially varying transmissivity, and boundary conditions. At the scale of a single WWTP, where neither the USGS nor DEQ has characterized heterogeneity beyond single wells, CAPZONE offers a reasonable method for simulating the hydraulic response to injection (or pumping) in the PAS.

The Theis equation (1935) is simply

$$s = \frac{q}{4\pi T} W(u) \tag{15.1}$$

$$u = \frac{r^2 S}{4 T t}$$

where
  s = drawdown (change in hydraulic head at a point since the beginning of the test)
  u = a dimensionless time parameter
  Q = the discharge (pumping) rate of the well
  T and S = are the transmissivity and storativity of the aquifer around the well
  r = distance from the pumping well to the point where the drawdown was observed
  t = the time since pumping began (seconds)
  $W(u)$ = well function

The Hantush–Jacob well function for leaky-confined aquifers is abbreviated $w(u, r/B)$. The Hantush–Jacob equation can be written in compact notation as follows:

$$s = \frac{Q}{4\pi T} w(u, r/B) \tag{15.2}$$

where
  s = drawdown
  Q = pumping rate
  T = transmissivity

In practice, CAPZONE produces drawdowns/draw-ups that are then subtracted from water levels that form either a uniform or non-uniform hydraulic gradient. Thus, the analyst can designate a hypothetical gradient of one based on an observed water-level distribution (non-uniform). The non-uniform options have proven particularly useful for injection well-field analyses at sites where a potentiometric surface exhibits irregularities or deflections that could potentially alter the potentiometric surface geometry.

At HRSD's individual WWTPs, a uniformed hydraulic gradient representing the site's position in the regional potentiometric surface (Figure 15.3) for the PAS developed by USGS was input to CAPZONE. In this approach, the regional gradient was considered locally at each WWTP in estimating ambient groundwater flow direction and hydraulic gradient. In this approach, the regional gradient was considered locally at each WWTP in estimating ambient groundwater flow direction and hydraulic gradient.

In applying CAPZONE, the boundaries of the simulated area were defined, and then the area was divided into a grid. The grid and cell dimensions for CAPZONE

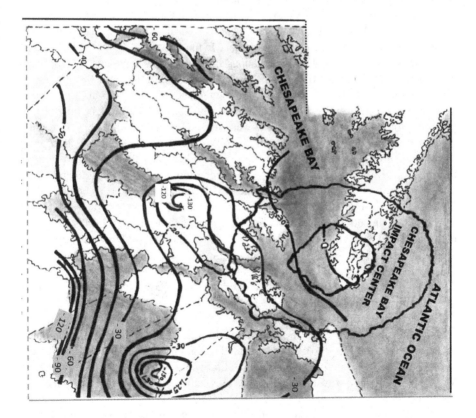

FIGURE 15.3 Potentiometric surface map of the Potomac Aquifer. Adapted from USGS (2015) in HRSD (2016). Illustration by F.R. Spellman and Kathern Welsh.

were unique to each WWTP, depending on the size of the site. The grid contained up to 75 columns and 75 rows, with a grid node spacing range from 11 to 200 feet at the Boat Harbor and York River WWTPs, respectively.

Other inputs included coefficients of transmissivity from the results of aquifer testing conducted at production wells in the vicinity of each WWTP and a storage coefficient of 0.0001, typical for a confined aquifer. Hydraulic gradients averaged around 0.0001 feet per year foot (ft/ft), with varying directions of groundwater flow based on the site's position within the USGS potentiometric map.

Other inputs included coefficients of transmissivity from the results of aquifer testing conducted at production wells in the vicinity of each WWTP and a storage coefficient of 0.0001, typical for a confined aquifer. Hydraulic gradients averaged around 0.0001 feet per year foot (ft/ft), with varying directions of groundwater flow based on the site's position within the USGS potentiometric map.

A unique static groundwater elevation was entered for each WWTP site and modified slightly depending on whether injection wells were screened across two or three aquifer units of the PAS. As described by the assessment of local-specific capacity, injection wells were first simulated to screen the two adjoining aquifer units of the three aquifers (UPA and MPA, and MPA and LPA) comprising the PAS. This approach entailed adding the transmissivity of the two aquifers together and obtaining an average static water elevation for the two aquifers (USEPA, 2016).

For wells screening the UPA and MPA, and the MPA and LPA discrete simulations were conducted. As the MPA received effluent whether it was combined with the UPA or LPA, to effectively simulate well interference in the aquifer, each simulation involved all the wells. As an example, in simulations involving wells screening the UPA and MPA, wells screening the MPA and LPA received one half of the total effluent flow.

To obtain the number of injection wells used in each simulation, maximum injection rates were held at 3 MGD per well and divided into the effluent flow HRSD projected for 2040. The injection wells were spaced (as much as possible) at roughly equal distance around the perimeter of the WWTP. Care was taken to avoid locating wells on existing structures. Locations were not evaluated for the practicality of positioning wells on lawns, parking lots, along fence lines, or other questionable areas that might host an injection well. Consistent with transient model runs conducted by USGS, CAPZONE simulations were set for a 50-year duration.

Simulated injection elevations were compared against the WWTP's threshold elevation in ft mean sea level, as defined by the ground surface elevation plus 10 psi (23 feet). Simulated injection levels from two aquifers exceeding the threshold elevation, which indicated that two aquifers could not facilitate effluent flows for the site. Accordingly, the simulation was run again combining all three aquifers of the PAS in each injection well at the WWTP.

Similar to the approach described previously, the transmissivities were added for each aquifer and static water-level elevations were averaged. In case the simulated injection levels continued to exceed threshold elevations, the effluent flow rates were reduced in each well until a solution was found where injection rates fell below the WWTP's designated threshold elevation. This approach resulted in determination the sustainable effluent rate for the site (HRSDa,b, 2016).

After the model runs that resolved the sustainable number of wells and effluent rates were completed, sensitivity testing was conducted to quantify the uncertainty in

input parameters used in obtaining the model solutions. Sensitivity testing was conducted under conditions prevalent at the HRSD's York River Plant. Additional test
was conducted at the York River Plant to investigate the relationship between well
spacing and interference. This testing proved particularly important to station with
elevated projected effluent flows, or smaller stations that could not support the number of wells to inject effluent at the projected 2040 rates. At these stations, HRSD
could potentially locate injection wells at offsite locations at distances sufficient to
lower the effects of hydraulic interference.

Modeling Results

After the model grid was set up, injection wells were located around the perimeter of each WWTP site, maximizing the number of wells given the site constraints.
Through iterations of well layouts and injection rates the 2040 projected demands
were tested at each WWTP. Table 10.5 shows the modeling results for each WWTP,
including the maximum injection rates for sites that did not meet the 2030 or 2040
demand projections within the existing boundaries.

Army Base Treatment Plant

The Army Base WWTP model was able to meet the 2040 projected demands of 12
MGD using four wells. Two wells were screened in the UPA/MPA, with the other
two set in the MPA/LPA, with all four-injecting effluent at 3 MGD. The injection
head elevation reached a maximum of 9 feet MSL within the UPA/MPA, falling
several feet below the threshold elevation of 35 feet MSL. Depending on the aquifer
used, the maximum draw-up found at the property boundary is approximately 61–68
feet above static conditions.

Boat Harbor Treatment Plant

Seven injection wells spaced between 200 and 600 feet apart and screening all three
PAS aquifers achieved an injection rate total 14, falling short of the 16 MGD tar-
geted 2040 flow projections. At 14 MGD, the injection level MGD has the maximum threshold value of 28 feet MSL. The total injection rate was constrained by
the location and layout of the plant. The adjacent highway and the harbor limit the
space available for wells, placing some wells at distances as close as 200 feet apart.
If HRSD can find locations outside the WWTP boundaries, increasing well spacing to greater than 600 feet, six injection wells should prove sufficient to meet the
2040 flow projections. The maximum draw-up found at the property boundary fell
approximately 93 feet above static conditions.

James River Treatment Plant

The James River WWTP model was able to meet the 2040 projected flows of 15
MGD using five wells. Unlike the Army base model, the James River model required
that the wells screened all three aquifers in the PAS. Using the three PAS aquifers, the injection head elevation fell below the threshold elevation (42 feet MSL)

by almost 3 feet. The maximum draw-up found at the property boundary totaled approximately 98 feet above static conditions.

NANSEMOND TREATMENT PLANT

The 2040 projections for the Nansemond WWTP represented the second highest of any of the WWTP sites texted, reaching 28 MGD. The results of the model include use well screening at all three PAS aquifers to inject a maximum rate of 24 MGD but fell short of reaching the 2040 projections. At 24 MGD, the injection head elevation remained 2 feet below the threshold elevation (46.5 feet MLS). The large number of wells required and the limited space available led to the 4 MGD shortfall. The adjacent river and marsh limit the space available within the Nansemond WWTP site for locating injection wells. The maximum draw-up found at the property boundary was approximately 122 feet above static conditions.

VIRGINIA INITIATIVE PLANT

The VIP's 2040 projections were the highest of all of the WWTP sites tested, requiring 33 MGD. The model included 14 UPA/MPA/LPA wells spread out across parcels north and east of the WWTP, comprising the golf course. The maximum attainable injection rate reached only 21 MGD, falling short of the 2040 projections. Meeting the projected 2040 flow of 33 MGD will require locating wells in offsite locations. The maximum projected draw-up found at the property boundary equaled approximately 101 feet above static conditions. In this scenario, draw-up was obtained from the boundary of the WWTP with the surrounding open space.

WILLIAMSBURG TREATMENT PLANT

The Williamsburg Treatment Plant, like the Army Base Treatment Plant, was able to meet 2040 projections using five wells split between the UPA/MPA and MPA/LPA. For Williamsburg simulations, two wells were set in the UPA/MPA, with the other three in the MPA/LPA. Given the high threshold elevation at this site (82.5 feet MSL), the WWTP was able to exceed the 2040 demands (13 MGD).

At 15 MGD, injected through five wells, the injection elevation reached 53 feet MSL, well below the threshold elevation for the site. The maximum draw-up found at the property boundary totaled approximately 77–93 feet above static conditions, depending on the aquifer combination (UPA/MPA or MPA/LPA).

YORK RIVER TREATMENT PLANT

The York River Treatment Plant also successfully met the 2040 projections but required using all three PAS aquifers. The model was able to achieve 15 MGD (over 2040 projections of 14 MGD) using six UPA/MPA/LPA wells, reaching a simulated injection elevation of 29.5 feet MSL, matching the threshold value (29.5 feet MSL). The maximum draw-up found at the property boundary fell approximately 79 feet above static conditions.

SENSITIVITY OF AQUIFER PARAMETERS

Whenever mathematical models are used as was the case in a phase of HRSD's SWIFT project, there is uncertainty in the inputs applied to get an output. Because of uncertainty in inputs and outputs used in mathematical modeling, a sensitivity analysis is called for, that is, it should be part of the entire process. With regard to the SWIFT project, a sensitivity analysis quantifies the doubt in a calibrated or predicted solution caused by uncertainty in the estimates of the aquifer parameters, injection stresses, and groundwater elevations. Basically what a sensitivity analysis accomplishes is a process of recalculating outcomes under alternative assumptions and has a range of various purposes (Parnell, 1997), including

- In the presence of uncertainty, it tests the strength of the results of the model.
- Amplified understanding of the relationships between input and output variables in a system to model.
- Further research can reduce uncertainty by identifying the model inputs that cause significant uncertainty in the output.
- Encountering unexpected relationships between inputs and outputs can be accomplished by searching for errors in the model.
- Model simplification.
- Increasing and enhancing the communication and the links from modelers and decision makers via using persuasion and straight talk.
- Employing Monte Carlo filtering to find regions in the space in input factors to optimum criterion.
- Knowing the sensitivity of parameters saves time by ignoring non-sensitive ones (Bahremand and De Smedt, 2008).
- To develop better models, important connections between observations, model inputs, and predictions or forecasts must be identified (Hill et al., 2015; Hill and Tiedeman, 2007).

In HRSD's SWIFT program, the sensitivity of aquifer parameters was performed on the scenario that simulated injecting 14 MGD at the York River WWTP site. The York River site was chosen for selectivity analysis because it represents a site that accommodated the 2040 flows but required using all three PAS aquifers.

Additionally, changes in pumping stresses and groundwater elevations were tested on a single well, eliminating interference from multiple wells. Finally, two wells were simulated to measure interference at varying distances.

For this sensitivity analysis, characterizing the uncertainty of the modeled solution, input values for transmissivity, storativity, injection rate, simulation duration, and the groundwater elevations were systematically adjusted to assess haw the changes affected groundwater elevations beneath the WWTP site. To quantify the evaluation, the maximum head generated from a sensitivity run was compared against the head form the original modeled solution.

The sensitivity analysis for aquifer transmissivity was performed by changing one parameter value at a time by −50% and +50% of the original parameter. The storage coefficient, injection rates, and groundwater elevations were tested by increasing and decreasing the values incrementally, not on a percent basis.

TRANSMISSIVITY

For purpose of clarity, in this text, transmissivity is defined as the capacity of a rock to transmit water under pressure. The coefficient of transmissibility is the rate of flow of water, at the prevailing water temperature, in gallons per day, through a vertical strip of the aquifer 1 foot wide, extending the full saturated height of the aquifer under a hydraulic gradient of 100%. A hydraulic gradient of 100% means a 1-foot drop in head in 1 foot of flow distance.

With regard to HRSD's SWIFT project, the transmissivity applied to the combined units of the UPA, MPA, and LPA at York River Treatment Plant was increased and decreased from the value used in the modeled solution. Values used in the sensitivity analysis ranged from 101,200 to 303,600 gpd/ft. The model was more sensitive to decreasing than increasing transmissivity by a factors approaching three times. Reducing transmissivity by 50% increased the maximum groundwater elevation at York River by 77 feet over the modeled solution of 29.5 feet MSL. Conversely, increasing the transmissivity value 50% decreased the mounding by only 27 feet from the modeled solution.

STORAGE COEFFICIENT

Storage coefficient is the volume of water released from storage in a unit prism of an aquifer when the head is lowered a unit distance. In the HRSD SWIFT model, the aquifer storage coefficient was increased and decreased from the value used in obtaining the modeled solution (0.0005). Storage coefficient used in the sensitivity analysis ranged from 0.00005 to 0.005. Reducing the storage coefficient by an order of magnitude increased the maximum head across the site by 13.6 feet. Conversely, increasing the storage coefficient by an order of magnitude decreased the maximum injection level by 13.5 feet. As the modeled solution fell close to the threshold elevation, adjusting the storage coefficient can have significant effects on whether the UPA/MPA/LPA can except the 2040 injection rate.

INJECTION RATES

Sensitivity of changes to injection rates was texted by comparing the maximum injection levels of a single well at different injection rates. Injection rates within the sensitivity analysis ranged from 0 (static conditions) to 4 MGD. The model was almost identically sensitive to increases and decrease in injection rate. Reducing the rate from 3 to 2 MGD reduced the maximum injection head by 8.3 feet while increasing the rate to 4 MGD increased the injection level by 8.2 feet.

SIMULATION DURATION

The duration of injection activity was adjusted to determine the model's sensitivity to changes of this parameter. An increase of 50% (75 years) over the original simulation (50 years) resulted in a rise in the maximum head value of 2.45 feet. Decreasing the simulation duration by 50% (25 years) lowered the maximum head value by 4.04 feet.

STATIC WATER LEVELS

The static water level was set at −56 feet MSL in the model solution. The model solution, with 14 MGD of injection into the UPA/MPA/LPA, reached a maximum groundwater elevation of 29.5 feet MSL. The static water level was raised and lowered 10 feet for the sensitivity analysis. The model appeared slightly more sensitive to an increase in the groundwater elevation than a decrease. Increasing the static water level to −46 ft MSL increased the maximum head at the well by 10.1 feet, decreasing the static water level to −66 ft MSL, reduced the maximum head by 9.9 feet.

WELL INTERFERENCE

To measure the effect of multiple wells injecting in proximity to each other, two wells were simulated with injection rates of 3 MGD under the subsurface conditions encountered at the York River WWTP for 50 years. The distance between wells was changed incrementally, while the maximum groundwater elevation was recorded at each well and at the midpoint between the wells. The well spacings tested ranged from 500 to 2000 feet. Injection heads at each well ranged from −30.9 feet MSL with only Well 1 injecting to −12 feet MSL when spaced 500 feet apart. Between 500 and 3000 feet of spacing, the maximum groundwater elevations in the wells varied about 4 feet. Interferences at Well 1 ranged from almost 19 to 15 feet and that of Well 2 spaced from 500 to 3000 feet away. At the midpoint between the wells, the head values ranged from −14.8 to −23.1 feet MSL and distances ranged from 500 to 3000 feet. The interference in the PAS changed by 8.3 feet or slightly less than 2 feet for every 500 feet of separation between the wells. The large amount of interference (greater than 15 feet) caused by a single nearby injection well even at relatively large spacings (3000 feet) appears consistent with the results of the modeling at other WWTPs. WWTPs carrying large injection rats require many wells, each well increasing the injection levels at other wells and in the aquifer (HRSDa,b, 2016).

HAMPTON ROADS REGION GROUNDWATER FLOW

The Virginia Coastal Plain Model (VCPM), a SEAWAT groundwater model, was employed to evaluate the hydraulic response of the PAS to injection operations at HRSD's seven WWTPs (Heywood and Pope, 2009; Langevin et al., 2008). This section presents the results of simulating injection at individual WWTPs and in scenarios with all seven WWTP injection simultaneously.

SEAWAT exemplifies a three-dimensional, variable density groundwater flow and transport model developed by the USGS based on MODFLOW and MT3DMS (Modular Three-Dimensional Multispecies Transport Model for Simulation). The VCPM groundwater model encompasses all of the coastal plain within Virginia and parts of the coastal plain in northern North Carolina and southern Maryland. The original VCPM was updated for use in the DEQ well permitting process and is now called VAHydro-GW. The VAHydro-GW model is discretized into 134 rows, 96 columns, and 60 layers. The majority of the model cells are square with the horizontal edges measuring 1 mile. The upper 48 model layers are 35 feet thick each. Layer thicknesses for the lower model layers increase to 50 feet after layer 48 (top to bottom) and then 100 feet beneath layer 52.

The model simulates potentiometric water levels in 19 coastal plain hydrogeologic units. The water levels are simulated for each year from 1891 through 2012, based upon historic pumping records. The VAHydro-GW also simulates water levels for 50 years beyond 2012. These water levels are based upon two scenarios: the total permitted scenario and the reported use scenario. The total permitted scenario simulates water levels for 50 years beyond 2012 by using the May 2015 total permitted withdrawal rates established for withdrawal permits issued by the DEQ together with the estimates for non-permitted (domestic wells, wells in Maryland and North Carolina, wells within unregulated portions of Virginia) withdrawals based upon 2012 estimated use. The total permitted scenario represents the estimated water levels 50 years into the future if all permittees within the coastal plain were to pump at their authorized maximum withdrawal rates for the duration of the 50-year period.

The reported use scenario simulates water levels for 50 years using pumping rates reported in 2012, for wells permitted by the DEQ and estimates for non-permitted withdrawals based upon 2012 estimated use. For most large permitted systems (greater than 1 MGD), reported pumping rates fall well below their total permitted diversion. The reported use simulation represents the best available estimate of water levels within the coastal plain aquifers over the next 50 years, if pumping were to continue at the currently reported pumping rates for the permitted wells within the coastal plain.

Virginia regulations have established limits on the amount of drawdown allowed as a result of permitted pumping within the coastal plain. The "critical surface" is defined as the surface that represents 80% of the distance between the land surface and the top of the aquifer. Individual model cells where simulated potentiometric water levels fall below the critical surface are referred to as "critical cells." Both the reported use and total permitted simulations show areas of the coastal plain for the Potomac, Virginia Beach, Aquia, Piney Point, and Yorktown-Eastover aquifers, where the predicted water levels at the end of the 50-year simulation end below the critical surface for those aquifers.

For any new or renewing permitted withdrawal, DEQ performs a technical evaluation which involves adding the proposed facility to the total permitted simulation. As a major criterion for permit issuance, the facility cannot create new critical cells in any aquifer due to their proposed withdrawal. The critical cells simulated at the end of the reported use simulation are not used for permit evaluation or issuance but represent a more plausible estimate of areas where water levels have lowered to crucial levels.

MODEL INJECTION RATES

VAHydro-GW row and column values were assigned to seven of HRSD's proposed injection WWTPs by using the well locations (latitude and longitude) to plot the position on a Geographic Information System coverage of the VAHydro-GW finite-difference grid. Each facility was simulated as a single point of injection and consequently assigned to only one row and one column. As explained earlier, each model cell is square with each cell edge measuring one mile. As a result, the rates injected through any number of wells are dependent on the individual WWTP. Because of course grid dimensions, the analysis is not intended to evaluate the number of injection wells that are required to dispose of injectate at each facility.

For the initial modeling, the well screen length for each WWTP was assumed to measure between 300 and 350 feet, thus screening across multiple layers of the VAHydro-GW model. Because the VAHydro-GW module utilizes the Hydrogeologic-Unit Flow (HUF) package, the model layers are independent of the hydrogeologic units. The HUF package is an alternative internal flow package that allows the vertical geometry of the system hydrogeology to be defined explicitly within the model using hydrogeologic units that can be different than the definition of the model layers. With regard to the model, a model layer may contain multiple hydrogeologic units. In order to ensure that simulated water levels were not artificially influenced by the Potomac confining unit, each injection well was assigned to the uppermost VAHydro-GW model layer filled by the PAS. The remainder of each injection well screen was assigned to lower adjacent model layers.

MODELING DURATION

In addition to modeling each WWTP operating individually (with the exception of the York River injection facility), all of the proposed facilities were modeled simultaneously at a combined flow of 114.01 MGD. The York River Treatment Plant was not included in the combined simulations because the facility lies with the outer rim of the Chesapeake Bay Bolide Impact Crater. As simulated in the VAHydro-GW, the horizontal hydraulic conductivity for the PAS at the cells within the bolide impact crater equals 0.0001 feet per day. As a result, simulated heads mounded to unrealistically high values when modeling injections at the York River Treatment Plant.

The individual WWTP scenarios and the combined WWTP scenario were simulated by adding the proposed injection rates to the total permitted and reported use simulations outlined previously, at the beginning of the 50-year predictive portion of those simulations (year 2013). The reported use and total permitted scenarios were also executed before adding the injection facilities to establish "baseline" conditions. This presentation refers to these scenarios as the "reported use baseline" and "total permitted baseline" simulations.

The model runs represent the DEQ's preferred metric for determining the beneficial impacts, if any, of proposed pumping/injection scenarios. The difference between water levels from an injection simulation and water levels from a baseline simulation represent the benefits, or recovery (rebound), resulting from the injection. The results of the injection and baseline simulations were compared at two points, 10 and 50 years into the predictive portion of the simulations.

RESEARCH FINDINGS

This section summarizes the findings and conclusions drawn from research, investigation, and modeling procedures conducted by HRSD, USGS, and CH2M for HRSD's SWIFT project. These summarized procedures and findings are based on the evaluation of injection well rates, WWTP injection capacities, and the hydraulic response of the PAS beneath the HRSD service area to injection operations. In addition, these bottom line conclusions are from HRSD's (2016b) *Sustainable Water Recycling Initiative: Groundwater Injection Geochemical Compatibility Feasibility Evaluation. Report 1.* Virginia Beach, VA. HRSD. Compiled by CH2M Newport News, VA, and are summarized as follows:

- The transmissivities and available head for injection in the PAS beneath each HRSD WWTP appear to support individual injection well capacities ranging between 3 and 8 MGD.
- In adhering to practical well design standards (such as borehole and casing diameter and pumping capacities), injection capacities were capped at 3 MGD for this evaluation.
- To account for maintenance necessary for injection wells screened in sandy aquifers, one additional injection well was added in every five required to meet the effluent disposal rate at each WWTP.
- Accordingly, the number of injection wells ranged from 5 at Army Base and Williamsburg to 17 at the VIP.
- Analytical groundwater flow modeling indicated that of the seven WWTP sties tested, Army Base, James River, Williamsburg, and York River were able to meet the 2040 projected demands, within the site boundaries.
- Only Army Base and Williamsburg met the demands using the original two-aquifer approach.
- Conditions at the James River and York River required screening all three of the PAS aquifers to meet the 2014 demands.
- Sensitivity testing at the York River Treatment Plant revealed that the modeled solution appeared sensitive to all parameters tested (transmissivity, storage coefficient, injection rates, simulation duration, and static water levels).
- The modeled solution exhibited the greatest sensitivity to changes in transmissivity. Changes in static water level resulted in increasing or decreasing the modeled head by the magnitude in the change of the water level.
- An evaluation of hydraulic interference between two wells at the York River WWTP revealed significant interference (greater than 15 feet) between wells spaced even 3000 feet away.
- The results of the analytical modeling showed that hydraulic interference exerts significant influence over the feasibility for replenishing the aquifer at the project 2040 rates.
- Injection was successfully simulated using the VCPM at each WWTP except York River. Because of its location inside the outer rim of the Chesapeake Bay Bolide Impact Crater, the VCPM simulates very low coefficients of hydraulic conductivity for the PAS beneath the York River Treatment Plant.
- Injection at each WWTP resulted in removing most of the critical cells and region-wide, recovering water levels in the PAS.
- Water levels in all injection scenarios resulted in simulated water levels that exceeded the land surface across the HRSD service area.

MIXING NATIVE GROUNDWATER AND INJECTATE

Ted Henifin's jaw-dropping, eyebrow-raising idea was first proposed in 2015, and last month, the sanitation district [HRSD] general manager kicked off its pilot phase to stop what some scientists have called a nightmare in super slow motion (Darryl Fears, *The Washington Post*, October 20, 2016).

"To stop a nightmare in slow motion…?" Are you kidding me? Is this the real message? Shouldn't Henifin dodge the challenge or the criticism that accompanies his jaw-dropping, eyebrow-raising project? Weren't there critics who mumbled these same words and doubts when da Vinci had the audacity to be far-thinking, when Newton worked his calculus, when the Wright Brothers attempted to fly like a bird, when Einstein developed his theories, when General Patton marched his army against the bad guys, and when Jonas Salk developed the vaccine for polio? Two things seem certain to me; critics abound by the mega-millions, while innovators, risk-takers, far thinkers, people with grit, and people with backbone and vision like Henifin and his HRSD associates are almost as rare as the Dodo Bird; moreover, it is only the innovator who thinks outside the stovepipe; the rest go up the chute, undampered (Spellman, 2017).

Bear with me as I present a very simplistic view of what this chapter is all about. Envision two 1-liter glass beakers. In one of these beakers, we fill it half-full of clean, safe drinking water. In the other beaker, we fill it half-full of salty sea water. A normal person who wants to quench his or her thirst would obviously prefer to drink from the beaker of clean, safe fresh drinking water; he or she would leave the beaker of salty sea water alone. Now, if we take either the beaker of clean, safe drinking water or the beaker of salty sea water and pour one into the other we have obviously mixed the two different solutions. The question now becomes is the new 1-L mixture of clean, safe fresh drinking water and salty sea water something that any of us would want to drink? The truth is some would use this mixture to gargle with. I have seen this done; have you? Anyway, the point I am making here is that by our action of mixing clean, safe drinking water with salty sea water we have changed or adulterated the mixture.

HRSD does not intend to mix clean, safe fresh drinking water with salty sea water or any other contaminant. No, the intent is not to adulterate native groundwater in any way. Instead, the intent is to inject, replenish, and recharge the Potomac Aquifer's native groundwater supply with purified, safe water from the advanced treatment of wastewater effluent. In order to do this, to ensure the injection of treated wastewater is of the same quality as the native groundwater contained in the Potomac Aquifer HRSD and its consultant (CH2M) conducted a feasibility study. This feasibility study evaluated the geochemical compatibility of recharging clean water (injectate), native groundwater, and injectate interactions with minerals in the PAS aquifers. Three discrete injectate chemistries originate from advanced water treatment processes (AWTPs) including reverse osmosis (RO), nanofiltration (NF), and biologically activated carbon (BAC).

The focus of the study was on a single WWTP where conditions (such as geography, flow, geology, injectate quality, and groundwater quality) best represent the HRSD system. This was based on a large number of permutations involved with comparing three injectates with native groundwater chemistries from the three PAS aquifers (i.e., UPA, MPA, and LPA) beneath HRSD's seven WWTPs and then applying the injectate chemistries to aquifer minerals in the three aquifers.

HRSD's York River Treatment Plant was selected for this evaluation and for the subsequent pilot study. In addition to displaying fairly representative conditions, the property surrounding the York River Treatment Plant is sufficiently spacious to accommodate at WWTP upgraded with AWTP and an injection well-field.

HRSD'S WATER MANAGEMENT VISION[4]

As mentioned earlier, the Hampton Roads region is faced with a variety of future challenges related to the management of the region's water supply and receiving water resources. These challenges involve a combination of technical, financial, and institutional complexities that invite the exploration of using non-traditional approaches that provide benefits on a larger scale beyond what the current wastewater treatment and disposal model can achieve. Accordingly, aquifer replenishment can protect and enhance the region's groundwater supplies, as well as reducing the potential damage caused by discharge (of nutrients) to the lower James River and the Chesapeake Bay and maybe slowing or arresting relative sea level rise in the region.

GEOCHEMICAL CHALLENGES FACING SWIFT PROJECT

HRSD will recharge between 77 and 131 MGD to the PAS using over 60 injection wells with maximum capacities approaching 3.0 MGD per well, at seven WWTPs. Groundwater flow modeling revealed that the injection wells will screen combinations of two or all three PAS aquifers zones, according to the hydrologic conditions found at individual WWTPs. Both physical and geochemical challenges can emerge while recharging clean water into aquifers composed of reactive metal-bearing minerals, and potentially unstable clay minerals, while also containing brackish native groundwater, typical conditions found in the PAS.

Physical and chemical reactions are important relative both to the well facility operation and aquifer water quality. The potential damaging effects can come from

- Water to water mixing of injectate water with native groundwater—impacts are in the form of physical plugging pore spaces with solids or precipitated metals, reduced permeability local to the wellbore, and eventually lower injectivity.
- Interaction of the injectate/native groundwater mix with the aquifer matrix—impacts are in the form of damage to firewater sensitive clays, precipitation or dissolution of metal-bearing minerals, and potential release of metals troublesome to injection activities (iron and manganese), or water quantity issues (arsenic).

REDUCTION IN INJECTIVITY

The two following factors affect the injectivity of an injection well:

- Physical plugging
- Mineral precipitation

Physical Plugging

A well's injection rate per unit of head buildup (draw-up) in an injection well is known as its injectivity, express in gpm per foot (gpm/ft) of draw-up (Warner and Lehr, 1981). When injection begins, the water level in the well rises as a function of the transmitting properties of the receiving aquifer, and the well's efficiency. While the transmitting character of the aquifer should remain stable over the service life of an injection well, the available had for injection will decline as the injectate recharges the aquifer. This causes the static water level to rise toward the ground surface. The well's efficiency will decrease with time, depending on the quality of the injectate, particularly to its TSS content. The TSS content, which, in water, is commonly expressed as a concentration in terms of milligrams per liter and typically is described as the amount of filterable solids in a wastewater sample. More specifically, the term *solids* mean any material suspended or dissolved in water and wastewater. Although normal domestic wastewater contains a very small amount of solids (usually less than 0.1%), most treatment processes are designed specifically to remove or convert solids to a form that can be removed or discharged without causing environmental harm. However, 100% removal is unlikely. Even the most purified injectate can contain small amounts of TSS. If left to accumulate in the borehole environment (wellbore), solids can clog the screen, filter pack, and aquifer proximal to the well, which reduces the well's injectivity. Injectivity reduction increase draw-up and eventually lowers the well's injection capacity (Pyne, 2005). TSS can originate from scale or dirt in piping, treatment residuals, and reactions in the injectate that result in solids precipitation. One of the more common reactions occur when oxygen dissolved in the injectate reacts with dissolved iron or manganese, precipitating ferric or manganese oxides, turning a dissolved component of the injectate into a source of solids.

Mineral Precipitation

In addition to physical plugging, chemical reactions between the injectate and the native groundwater, or injectate and aquifer mineralogy can precipitate metal-bearing oxides and hydroxides. These reactions often come from injectate that contains dissolved oxygen (DO). Considering the relatively small surface areas around the wellbore, precipitating metal-bearing minerals can clog pore spaces, and reduce permeability and well injectivity. An important part of the research and planning involved with HRSD's SWIFT project is to determine precisely the type and composition of injectate from the RO, NF, or the BAC process to estimate its potential for plugging the wellbore.

GEOCHEMICAL CONCERNS

Beyond problems associated with physically plugging pore spaces around the borehole, several geochemical reactions can negatively affect injection well operations. These reactions include

- Clay mineral damage
- Metal precipitation
- Mineral dissolution

Damaging Clay Minerals

The term *clay* is applied to both materials having a particle size of less than 2 μm (25,400 μm = 1 inch) and a complex group of poorly defined hydrous silicate minerals that contain primarily aluminum, along with other cations (potassium and magnesium) according to the exact mineral species. Displaying a platy or tabular structure, clay minerals exhibit an extremely small grain size and typically adsorb water to their particle surfaces. In aquifer sand, clays occur in trace (less than 10%) amounts as components of the aquifer's interstitial spaces, coating framework particles like quartz grains, lining, or filling pore spaces, or as a weathering product of feldspars.

Damaging clays occur with the disruption of their mineral structure. The damage can arise when injecting water of significantly different ionic strengths than the native groundwater, a concern when injecting dilute fresh water into an aquifer containing brackish or saline native groundwater (Drever, 1988). The dilute water contains significantly less cations and a weaker charge than brackish native groundwater. When displacing the brackish water in the diffuse double layer between clay particles, the weaker charge can induce repulsive forces dispersing the particles, fragmenting the clay structure while mobilizing the fragments into flowing pore water. The particles can eventually accumulate in smaller pores physically plugging the pore space and reducing the permeability of the aquifer.

Damage can also arise when injectate displays differing cation chemistry than the native groundwater and the clay minerals (Langmuir, 1997). Exchanging cations can disrupt clay mineral structure particularly when their atomic radius exceeds the radius of the exchanged cation. The larger cation fragments the tabular structure, shearing off the edges of the mineral. Plate-like fragments break off the main mineral particle and migrate with flowing groundwater. Like the damage incurred by water of differing ionic strengths, migrating clay fragments will brush pile in pore spaces, physically plugging passageways and reducing aquifer permeability. Unlike the accumulation of TSS in the wellbore, formation damage by migrating clays develops in the aquifer away from the wellbore, making its removal difficult by back-flushing or even invasive rehabilitation techniques.

Mineral Precipitation

Metal-bearing minerals can precipitate in the aquifer away from the well. These reactions typically occur when the injectate contains DO at concentrations exceeding anoxic (DO less than 1.0 mg/L) levels but can also occur if the pH of the injectate exceed 9.0. As surface areas in the aquifer increase geometrically away from the well, mineral precipitation does not create as great a concern as the same reactions at the borehole wall.

Mineral Dissolution

Injectate reactions with minerals in the aquifer matrices can dissolve minerals leaching their elemental components (Stuyfzand, 1993). Injectate-containing dissolved oxygen above anoxic concentrations will react with common, reduced metal-bearing minerals like pyrite (FeS_2) and siderite ($FeCO_3$), to release iron and other metals like manganese that occupy sites in the mineral structure, iron and manganese can precipitate as oxide and hydroxide minerals if they contact injectate-containing DO.

Oxidation of arseniferous pyrite can release arsenic, creating a water quality concern in the migrating injectate (HRSD, 2016b).

WATER QUALITY AND AQUIFER MINERALOGY

Note: Because it would require an unwieldy number of permutations (63) to assess the injection of three injectate chemicals into three discreet aquifers at seven WWTPs, the chemical composition of three injectate types and native groundwater in the three PAS aquifer zones beneath the York River Treatment Plant was chosen for discussion here. The targeting of the York River WWTP for discussion that follows is not an issue and does not skew data because the York River Treatment Plant exhibits effluent and local aquifer characteristics typical of conditions across the HRSD.

Mass–balance relationships between raw water entering the plant and modeling of the AWTP were used to determine injectate chemistry. As no wells installed in the PAS aquifer zones currently exist at the York River WWTP, water quality data from the area around the site were obtained from the National Water Information System (NWIS) database (maintained by the USGS). The NWIS database provides samples collected by USGS personnel from local municipal, irrigation, and industrial supply wells, along with designated monitoring wells.

Two methods were used to identify potential minerals in the PAS aquifers that could react with the injectate. First, thermodynamic equilibrium models were applied to identifying the potential mineral suite in each aquifer. The models were run using water chemistry analyses obtained from the NWIS database for the PAS zones around the York River WWTP area. The models projects potential minerals that occur in equilibrium with water chemistry.

Second, mineralogical analysis of cores collected at the city of Chesapeake's ASR facility was examined to gain information on the mineralogy of the PAS. Cores were collected from the PAS zones in the city of Chesapeake. The composition of the PAS should remain fairly consistent across the HRSD service area. However, the grain size and sorting (texture) decline proceeding down the stratigraphic dip in the Virginia Coastal Plain (Treifke, 1973). Consistent with the changes in texture, the percentage of fines (texture) increases downdip. As the city of Chesapeake lies over 20 miles downdip from HRSD's York River WWTP, data from cores should portray more conservative aquifer properties than actually occur below York River.

INJECTATE WATER QUALITY CHEMISTRY

Injectate chemistry was estimated by modeling water quality entering water quality entering the York River Treatment Plant and through the AWTPs. Effluent chemistry was estimated for RO, NF, and biological activated carbon AWTPs.

Reverse Osmosis

Injectate modeled for treatment by the RO process featured a pH of approximately 7.8 after adjustment with lime ($CaOH_2$), dilute total dissolved solids (TDS) (46 mg/L), and, correspondingly, a low ionic strength (0.0015). Cations and anions in the influent were reduced following the treatment process, resulting in concentrations of cations like potassium (less than 1 mg/L), magnesium (less than 1 mg/L), and sodium (less than 10 mg/L) falling below their method detection limits (MDLs), with similarly lower concentrations of anions like phosphate (0.01 mg/L), chloride (less than 10 mg/L), and sulfate (less than 1 mg/L). At the York River Treatment Plant, RO-treated waste displayed a calcium bicarbonate water type. Metals such as iron, manganese, and arsenic exhibited low concentrations, and all near MDLs. RO has limited effect on DO concentrations in the injectate, which ranged around 5 mg/L (HRSD, 2016b).

Nanofiltration

Injectate derived from the NF process exhibited a pH of approximately 7.8, moderate TDS (262 mg/L), and corresponding ionic strength (0.005). Cations and anions are reduced following the treatment process, but RO displayed measurable concentrations of major cations like potassium (7.7 mg/L), magnesium (2.5 mg/L), and sodium (58 mg/L), and anions including phosphate (0.03 mg/L), chloride (125 mg/L), and sulfate (1.8 mg/L). NF injectate displayed sodium chloride chemistry. Concentrations of metals including iron, manganese, and arsenic fell below MDLs. NF injectate also displayed near-saturated concentrations of DO around 5 mg/L.

Biologically Activated Carbon

Injectate originating from BAC treatment displayed a pH of approximately 7.8, slightly brackish TDS (615 mg/L), and corresponding ionic strength (0.009). Cations and anions exhibited less reduction following the treatment process compared with NF and RO. Cation concentrations of potassium (13 mg/L), magnesium (10 mg/L), and sodium (103 mg/L) exceeded the concentrations yielded by the membrane (RO and NF) treatments. Concentrations of anions like phosphate (0.5 mg/L), chloride (212 mg/L), and sulfate (44 mg/L) also appeared correspondingly higher. Unlike RO and NF, DO concentrations fell to 1.3 mg/L after treatment using BAC. BAC injectate featured sodium chloride water chemistry. Iron and arsenic displayed concentrations higher than the membrane treatment options at 0.002 and 0.73 mg/L, respectively. Iron concentrations above 0.1 mg/L create significantly large amounts of TSS that can quickly clog an injection well.

NATIVE GROUNDWATER

Native groundwater quality from the UPA, MPA, and LPA zones was obtained from nested observation wells maintained by the USGS and NWIS, located 5 miles west of the York River WWTP. At this location, observation wells installed were screened in the UPA from 527 to 537 fbg, in the MPA from 820 to 830 fbg, and in the LPA from 1205 to 1215 fbg.

UPPER POTOMAC AQUIFER ZONE

The UPA featured a slightly alkaline pH (8.2), brackish TSD (1280 mg/L), anoxic water (DO less than 1.0 mg/L), displaying an ionic strength of 0.02. The groundwater exhibited low amounts of nutrients, with concentrations of ammonia, nitrates, and phosphate falling below 0.01 mg/L. Chloride concentrations approached 500 mg/L, while sodium concentrations appeared similarly elevated (516 mg/L). Concentrations of other cations including calcium (4.5 mg/L) and potassium (13.5 mg/L) were comparatively low. Iron concentrations fell around its method detection limit (0.01–0.04 mg/L), while manganese concentrations approached the drinking water maximum contaminant level of 0.05 mg/L. Water from the UPA displayed a sodium chloride chemistry.

MIDDLE POTOMAC AQUIFER ZONE

The MPA also featured a slightly alkaline pH (8.0), brackish TDS (2780 mg/L), and anoxic groundwater (DO less than 1.0 mg/L), with an ionic strength of 0.04. Similar to the UPA, concentrations of nutrients fell below 0.01 mg/L. Concentrations of anions, comprising chloride (1200 mg/L), alkalinity (370 mg/L), and sulfate 73 mg/L), exceeded concentrations encountered in the UPA. Sodium concentrations appeared similarity elevated at 870 mg/L, while concentrations of other cations, such as calcium, magnesium, and potassium, fell below 15 mg/L. Iron concentrations were near Safety Data Sheet limits (0.01–0.04 mg/L), while manganese concentrations appeared at 0.02 mg/L. Similar to the UPA, groundwater in the MPA exhibited sodium chloride chemistry.

LOWER POTOMAC AQUIFER ZONE

The LPA displayed a circum-neutral pH (7.7) brackish TDS (4580 mg/L) and anoxic groundwater (DO less than 1.0 mg/L). Similar to the other PAS aquifers, concentrations of nutrients fell below 0.01 mg/L, although phosphate concentrations, at 0.5 mg/L, appeared notably higher than in the other PAS aquifers. Concentrations of anions, comprising chloride (2950 mg/L) and sulfate (146 mg/L), exceeded concentrations encountered in the UPA and MPA. Sodium concentrations were similarly elevated (1700 mg/L). Concentrations of other cations such as potassium (25 mg/L) and calcium (51 mg/L) increased over the concentrations encountered in the other PAS zones. Yet, iron and manganese in the LPA mimicked groundwater from the other aquifers with concentrations at MDLs and 0.02 mg/L, respectively. Similar to the other PAS aquifers, groundwater from the LPA displayed sodium chloride chemistry.

GEOCHEMICAL ASSESSMENT OF INJECTATE AND GROUNDWATER CHEMISTRY

In this section, the discussion is about the chemical assessment of the injectate (that is, the effluent from the AWTPs, RO, NF, and BAC) that potentially is to be injected as injectate into native groundwater. Obviously, as pointed out with the example of the beakers of pure water and sea water mixing that adulterated the pure water into something not potable HRSD does not want a similar outcome with its injection of its treated wastewater into the Potomac Aquifer's native groundwater. The goal is to determine the appropriate injectate. Keeping the desired outcome and goal in mind, the evaluation of the chemistry of the injectate water and native groundwater from the PAS revealed the following:

- RO and NF displayed ionic strengths differing by over one order of magnitude from the native groundwater in the PAS.

- Biological activated carbon displayed ionic strengths within the same order of magnitude as the PAS.
- RO exhibits a differing cationic chemistry than the groundwater from the PAS.

Influence of Ionic Strength

The ionic strength of RO diluted, treated water appeared lower than groundwater in the three PAS zones by at least one order of magnitude. By comparison, the ionic strength displayed by NF differed from the LPA by over none order of magnitude. The ionic strength of biologically treated carbon, although lower than the PAS aquifers, fell within the same order of magnitude. The low ionic strength of RO compared to the PAS groundwater represents a concern for injection operations, particularly for RO' potential to disperse clay minerals. Clay dispersion represents an electro-kinetic process (Meade, 1964; Reed, 1972; Gray and Rex, 1966), where an electrostatic attraction between negatively charged clay particles is opposed by the tendency of ions to diffuse uniformly throughout an aqueous solution. One of the most important factors leading to the dispersion of clay minerals involves a change in the double-layer thick of a clay particle. A double layer of ions lies adjacent to the clay mineral surface or between the mineral's structural layers because a negative charge attracts cations toward the surface. As the fluid must maintain electrical neutrality, a more diffuse layer of anions surrounds the cations.

As in brackish water, when the concentration of ions is large, the double layer around the particle or between the clay's structure layers gets compressed to a smaller thickness. Compressing the double layer causes particles to coalesce, forming larger aggregates. This process is called clay flocculation. When the ionic concentration of a fluid invading the aquifer is significantly lower than the native groundwater, the diffuse double layer expands, forcing clay particles and the structural layers within clay minerals apart. The expansion prevents the clay particles from moving closer together and forming an aggregate. The tendency toward dispersion is measured in clay minerals by their zeta potential—i.e., where colloids with high zeta potential are electrically stabilized while colloids with low zeta potentials tend to coagulate or flocculate—according to the following relationship:

$$Z = 4\pi nq/D$$

(15.3)

where
   Z = zeta potential
   π = thickness of the zone of influence surrounding the charge particle
   q = charge on the clay particle before attaching cations
   D = dielectric constant of the liquid

For any solution and clay mineral, reducing the zeta potential involves lowering the thickness of the zone of influence. Substituting small, double- or triple-charged cations such as Ca^{+2} or Al^{+3}, respectively, in place of large singly charged and hydrated ions like Na^{+2}, lowers the zeta potential, permitting clay particles to coalesce. This behavior explains the tendency for sodium to cause clay dispersion, while calcium and aluminum induce its flocculation.

Aquifer sands containing more complex clay minerals like mixed-layer clays and smectite group clays that display small particle sizes, yet large surface charges usually exhibit the greatest sensitivity to fresh water (Brown and Silvey, 1977). As little as 0.4% smectite in a sand body has reduced the aquifer's hydraulic conductivity by 55% after exposure to fresher water (Hewitt, 1963). Mixed-layer clays and smectite encountered in cores from the UPA and MPA at the city of Chesapeake exhibited abundance equaling trace (less than 1%–4% of the whole rock composition of the sand).

In the 1970s, the USGS tested an ASR facility in Norfolk, Virginia; it exhibited greater than 50% reduction in injectivity after only 150 minutes of starting injection operations (Brown and Silvey, 1977). The ASR well was installed in the UPA, screening nearly 85 feet of sand in the unit. The USGS employed nuclear, electrical, and mechanical geophysical logging techniques to evaluate the origin of the injectivity losses and discriminate between the causes of physical plugging documented at other sites, like TSS loading. Injectivity losses caused by physical plugging from TSS loading typically occur at discrete zones through the well screen. In contrast, geophysical logging of the aquatic storage and recovery test well at Norfolk showed hydraulic conductivity losses distributed evenly across the entire screen. Also, in comparison to physical plugging by TSS which responds positively to mechanical and chemical rehabilitation techniques, the USGS was unable to restore even a fraction of the well's original injectivity.

USGS used calcium chloride ($CaCl_2$ greater than 10,000 mg/L) solution to treat the wellbore and proximal aquifer to arrest the declining injectivity. The double charged calcium cation forms a stronger particle and inter-layer bond than the monovalent cation, sodium. Using a concentrated solution ensures calcium exchanges for sodium at the maximum number of sites. After applying the treatment at Norfolk, the injectivity of the aquifer storage and recharge test remained stable over two more test cycles before the project ended. Concentrated solutions containing the trivalent aluminum proved to be effective in stabilizing clay minerals prone to dispersion in the presence of dilute injectate (Civan, 2000). Applying a calcium or aluminum chloride treatment to the PAS before initiating injection operations, offers a viable alternative for stabilizing clay minerals in situ, precluding formation damage, and injectivity loss should regulators select RO as the most viable method for protecting local water users. These treatments could also benefit injection operations using NF or BAC as the preferred injectate.

Cation Exchange

In addition to differing ionic strengths, RO, as calcium carbonate water, differs from the sodium chloride chemistry encountered in the UPA, MPA, and LPA. As previously described, the doubly charged calcium ion should benefit the long-term stability of clay minerals where calcium exchanges for sodium. However, calcium exhibits a large ionic radius that can damage clay mineral when entering the position left by the sodium, fragmenting the edges of the mineral, and mobilizing the fragments in the aquifer environment.

Iron and Manganese

None of the injectates or native groundwater from the PASs aquifers appears to exhibit problematic concentrations of iron or manganese. Iron concentrations in RO and NF effluent typically occurred below method detection limits (MDL)s. During injection operation, iron and manganese contained in the injectate or native groundwater can precipitate oxide and hydroxide minerals when exposed to DO. Formation of these minerals presents a problem if they precipitate close to the wellbore, which is a zone featuring small surface areas sensitive to physical plugging. Accordingly, the absence of iron and manganese in injectate or native groundwater benefits injection operations.

LITHOLOGY OF THE POTOMAC AQUIFER SYSTEM

The Lithology and minerals comprising the PAS aquifers is described in this section, stating with the general composition across the study area, and then focusing on cores collected from the UPA and MPA near the city of Chesapeake, Virginia.

LITHOLOGY

As previously mentioned, the PAS consists of three discrete aquifer zones (UPA, MPA, and LPA) named for their position in the section. Deposited in river (fluvial) and shallow marine environments, the aquifers consist of coarse to fine sands with occasional gravel, interbedded with thin gray to pale green clays (Treifke, 1973). The aquifers are separated by clay beds of thicknesses exceeding 20 feet. However, thinner clay beds transect the sand units in the MPA and the LPA. Because of the abundance of clay beds, the Middle Potomac and Lower Potomac often consist of multiple, stacked units requiring repeated screen and blank combinations for supply wells installed in these aquifers.

Sands are composed primarily of quartz (Meng and Harsh, 1988), often reaching amounts exceeding 90% by weight, forming the predominant framework mineral. Accessory minerals include orthoclase, muscovite, glauconite, and locally lignite. Trace minerals mostly occupy the interstitial spaces in the sands and comprise biotite, pyrite, siderite, magnetite, and clays.

CITY OF CHESAPEAKE AQUIFER STORAGE AND RECOVERY FACILITY CORE SAMPLES

In 1989, at the city of Chesapeake's ASR facility, ten core samples were collected from the UPA and MPA, at depths ranging from 560 to 835 fbg. The cores were submitted to Mineralogy Inc., a laboratory specializing in mineralogical assays, for the following analyses:

- Specific gravity
- Porosity
- Permeability
- X-ray diffraction
- Cation exchange capacity (CEC)
- Grain size distribution
- Energy dispersive chemical analysis
- Scanning electron microscopy

PAS sediments found at the city of Chesapeake locations, even though they are located approximately 35 miles apart, should display similar characteristics as those underlying the Yorktown WWTP. Because aquifer characteristics like grain size, sorting, textural maturity, porosity, and permeability decline moving downdip, PAS sediments at Yorktown WWTP should display characteristics better suited to injection operations than at the city of Chesapeake.

Core samples from the UPA and MPA were composed of coarse to very coarse-grained sands, in a medium-grained matrix. Aquifer sands appeared conglomeratic and unsorted. However, as unconsolidated sands they displayed open pore spaces yielding good porosity (21%–34%) and air permeability. Grain size diminished with depth. Samples from the deeper portions of the MPA exhibited a medium-grain size with a larger percentage of fine sands than shallower samples. Most clay minerals were found in interstitial spaces of the aquifers and showed a high degree of crystallinity, which suggests they formed after deposition and burial (i.e., were authigenic).

Sands from the UPA and MPA consisted of 84%–89% quartz with 8%–12% potassium and plagioclase feldspar, classifying the sands as subarkosic, or lithic arkosic. Trace (less than 10%) amounts of calcite and dolomite were detected in every sample of the aquifer sands. Clay minerals, comprising kaolinite, illite/mica, and smectite made up to 4% of the same samples. The iron carbonate mineral, siderite ($FeCO_3$) was encountered in a confining bed sample (595 fbg) at amount up to 19%. Siderite was also encountered in an aquifer core (685.2 fbg), at trace amounts.

Sands from the UPA and MPA consisted of 84%–89% quartz with 8%–12% potassium and plagioclase feldspar, classifying the sands as subarkosic or lithic arkosic. Trace (less than 10%) amounts of calcite and dolomite were detected in every sample of the aquifer sands (Table 15.2). Clay minerals, comprising kaolinite, illite/mica, and smectite made up to 4% of the same samples. The iron carbonate mineral siderite ($FeCO_3$) was encountered in a confined bed sample (595 fbg) at amounts up to 19%. Siderite was also encountered in an aquifer core (685.2 fbg) at amounts up to 19%. Siderite was also encountered in an aquifer core (685.2 fbg) at trace amounts.

Permeabilities in air (intrinsic permeability) ranged from 1280 to 5900 millidarcies. Generally, intrinsic permeability and porosity values declined with depth, so the greatest permeabilities were encountered in cores from the UPA. Intrinsic permeability displayed minimal anisotropy with horizontal and vertical values from the same core yielding near equal permeabilities. CEC refers to the number of exchangeable cations per dry weight that a soil can hold at a given pH and are available for exchange with the soil–water solution which is influenced by the amount and type of clay and the amount of organic matter (Drever, 1982). CEC serves as a measure of soil fertility, nutrient retention capacity, and the capacity to protect groundwater from cation contamination. The CEC of minerals contained in confining beds often controls the cation chemistry in the adjacent aquifers by exchanging cations across the contact between the units. For injection purposes, knowing the CEC of the aquifer and confining bed materials can help assess how these materials will react with recharged water displaying a specific cation ionic chemistry. CEC is expressed as milliequivalents of hydrogen per 100 g of dry soil (meq+/100 g). Table 15.3 presents the CEC values of some clay minerals.

All ten samples collected from the Chesapeake site were analyzed for CEC. Sodium represented the most dominant exchangeable cation followed by magnesium, calcium,

---

TABLE 15.3
Cation Exchange Capacities
of Some Clay Minerals
(meq/100g) (Drever, 1988)

Smectites	80–150
Vermiculites	120–200
Illites	10–40
Kaolinite	1–10
Chlorite	<10

---

and potassium. The confining bed sample at 685.2 fbg displayed the most elevated CEC at 12.5 meq/100 g of core. Aquifer sand samples from the UPA and MPA exhibited CECs for sodium ranging from 0.7 to 3.9 meq/100 g of core. Sodium, a monovalent ion in the exchange position of clays, will not benefit from injection operations.

Despite the dominance of sodium, CEC values from cores from city of Chesapeake were low, suggesting that the clays should display minimal tendency to exchange cations. In environments showing more elevated CECs, divalent ions like calcium or magnesium in the injectate can exchange with sodium temporarily disrupting the clay's atomic structure. Over a long term, replacing sodium with a divalent ion will strengthen the clay mineral's atomic structure eventually transitioning to a stable smectite.

MINERALOGY—GEOCHEMICAL MODELING

The thermodynamic equilibrium model PHREEQC (Parkhurst, 1995) was used to gain a greater understanding of the stability of the clay minerals in the PAS aquifers beneath the Yorktown WWTP, based on the native groundwater and injectate chemistries. As previously described, the stability of clay minerals can control the success of injection operations in sandy aquifers like the PAS.

Thermodynamic equilibrium models consist of computer programs using a relatively sophisticated set of equations (Davies, 1962; Truesdell and Jones, 1974; Debye and Huykel, 1923) to stimulate the chemical equilibrium of a solution under natural or laboratory conditions, and to simulate the effects of chemical reactions. These models perform the following types of calculations:

- Correct all equilibrium constants to the temperature of the specific sample.
- Calculates speciation: the distribution of chemical species by element by solving a matrix of equations.
- Calculations activity coefficients of each chemical species.
- Calculates the state of saturation for potential mineral species that occur in equilibrium with the samples water chemistry. These calculations identify potential mineral species, whether they will dissolve or precipitate under the changing conditions consistent with ASR operations.
- The models perform a wide variety of calculations related to oxidation–reduction processes.

Thermodynamic equilibrium computer models represent a power tool for predicting chemical behavior in a natural system. Manual manipulation of the same equations performed by these programs is time-consuming and prone to calculation errors.

STABILITY OF CLAY MINERALS

PHREEQC was employed for evaluating the stability of clays contained in the CaO$-$Al$_2$O$_3$$-SiO_2$$-H_2O, NaO-Al_2O_3SiO_2$$-H_2$O, and K$_2O-Al_2O_3$$-SiO_2$$-H_2$O mineral systems. Minerals contained in these systems represent clays and their weathering products (gibbsite, kaolinite) commonly found in sediments of the PAS. The simulations' objective involved determine how native groundwater chemistries fall into the stability fields of clay minerals and identifying potential; instabilities. Along with ambient clay stabilities, PHREEQC simulates how clays can evolve during the exchange of cations. However, the program does not address instability arising from introducing injectate of a differing ionic strength.

The chemistry of groundwater from the UPA, MPA, and LPA, along with potential injectate waters from the Yorktown Treatment Plant, was plotted on stability diagrams for three systems describing common clay minerals (CaO$-$Al$_2$O$_3$$-SiO_2$$-H_2O, NaO-Al_2O_3SiO_2$$-H_2$O, and K$_2O-Al_2O_3$$-SiO_2$$-H_2$O). Common clay minerals including smectite, beidellite, montmorillonite, illite, and gibbsite (weathering product) were over-saturated in recharge water samples, which suggests a tendency to precipitate over time.

When injecting waters of incompatible ionic strength or differing cations, damage to clay minerals can arise; however, this was not a concern during injection operations in the PAS because precipitation of clay minerals represented a relatively minimal matter regarding permeability loss. Moreover, the precipitation of clay minerals requires significant amounts of geologic time, rather than the relatively short service life of an injection facility.

THE BOTTOM LINE

Modeling and geochemical evaluation resulting from mixing native groundwater and injectate and the reactions between injectate and aquifer minerals in the PAS beneath HRSD's York River Treatment have generated several conclusions. These bottom line conclusions are from *Sustainable Water Recycling Initiative: Groundwater Injection Geochemical Compatibility Feasibility Evaluation. Report 20* by HRSD (2016a), Virginia Beach Virginia and its primary consultant CH2M Newport News, VA, and are summarized as follows:

- The chemistry of RO, a potential injectate, differed significantly from the chemistry of native groundwater exhibited by the UPA, MPA, and LPA.
  - RO displayed a dilute ionic strength that differed by over one order of magnitude from the chemistry encountered in the UPA and MPA, while approaching two orders of magnitude when compared against the Lower Potomac.
  - RO a calcium bicarbonate water chemistry, while groundwater from the three PAS aquifers uniformly exhibited sodium chloride chemistry.

- The low ionic strength of RO compared to groundwater from the PAS represents a concern for injection operations, particularly for its potential to disperse clay minerals. Once dispersed, clay particles migrate through connected pores in the aquifer until accumulating and blocking narrowed pores, reducing aquifer permeability and ultimately injection well capacity.
- A USGS-sponsored ASR facility tested at Norfolk in the 1970s used an injectate similar in ionic strength and cation chemistry to RO. The injection capacity of the ASR well declined by 50% after only 4 hours of operation, dropping 75% over several days. The USGS was not able to restore the capacity of the ASR well, despite applying several, for the time, state of the art rehabilitation techniques.
- Cores collected at the city of Chesapeake's Aquifer Storage rand recovery facility exhibited trace concentrations of smectitic clays dispersed throughout the interstices of every sample collected in aquifer sands. Smectites possess a complex lattice expanding structure vulnerable to dispersion or swelling when exposure to dilute water.
- In considering the varied cation chemistry between RO and groundwater in the PAS aquifers, the doubly charged calcium ion should benefit the long-term stability of clay minerals where calcium exchanges for sodium. However, calcium, when hydrated, exhibits a large ionic radius that can damage clay minerals upon entering the position vacated left by sodium, fragmenting the edges of the mineral, and mobilizing the fragments in the aquifer environment.
- Conversely, cores from the city of Chesapeake project, analyzed for CEC displayed little tendency to exchange, which is a benefit of injection operations.
- Geochemical modeling of potential clay minerals in the PASs aquifers produced a similar result with the stability of clay minerals improving over time during injection operations.
- Given the concerns with RO as a source of injectate, and the problems experienced at the City of Norfolk's ASR project, HRSD should consider eliminating RO from further evaluation on the SWIFT project.
- The ionic strength of NF and BAC injectate fell within one order of magnitude of the groundwater chemistries originating from the PAS aquifers. NF and biological activated carbon as a source of injectate represent significantly less of a concern of dispersing water-sensitive clays in the PAS aquifers during injection operations.
- Applying a calcium or aluminum chloride treatment to the PAS aquifers before initiating injection operations offers a viable alternative for stabilizing clay minerals in situ, precluding formation damage, and injectivity losses should regulators selected RO as the most viable method for protecting local water users. These treatments could also benefit injection operations using NF of BAC as the preferred injectate.
- BAC exhibits 0.7 mg/L iron, which presents a considerable source of TSS in the injectate, and a strong physical plugging agent in injection wells.

HRSD will need to remove iron from BAC effluent before employing it as an injectate.

- Geochemical modeling runs, simulating mixing between RO, NF, and BAC and groundwater from the PAS aquifers showed no evidence of deleterious reactions that might clog the injection wells or surrounding aquifer such as precipitating oxide, hydroxide, carbonate, or sulfate minerals, or the dissolution of silicate minerals.
- Geochemical modeling, simulating the mixing between the three groundwaters in an injection well screening the UPA, MPA, and LPA displayed no evidence of deleterious reactions that might clog the injection well or surrounding aquifers.
- Cores collected at the city of Chesapeake contained the iron carbonate mineral siderite at amounts ranging from 0.5% to 19% of the whole rock composition. In reactions with injectate-containing DO, siderite released up to 130 mg/L Fe II.
- Although not a concern for injection operations, dissolving siderite can compromise the quality of the disposed water, prompting attention from state and federal regulators.
- Adjusting the pH of the injectate water with a source of hydroxyl like sodium or potassium hydroxide can help lower Fe II concentrations. During model runs simulating reactions between injectate containing varying amounts of DO and pH of 8.5 with pyrite. Fe II concentrations fell below 10E−7 mg/L. Fe II oxidized to Fe III, which precipitated Fe III oxide and Fe III hydroxide minerals.

NOTES

1 Much of the material in this chapter is adapted from F. Spellman's SWIFT (2020). Lanham, Maryland: Bernan Press.
2 Much of the information in this section is based on HRSD's *Proposal to establish and extensometer station at the Nansemond wastewater treatment plant in Suffolk, Va* (2016). Proposed by USGS.
3 Much of the information in this section is based on information from HRSD (2016) and compiled by CH2M.
4 HRSD (2016b). *Sustainable Water Recycling Initiative: Groundwater Injection Geochemical Compatibility Feasibility Evaluation.* Virginia Beach, VA. Hampton Roads Sanitation District. Compiled by CH2M Newport News, VA.

REFERENCES

Anderman, E.R. and Hill, M.C. (2000). MODFLOW 2000. Water. USGS.gov/ogw/mudflow/MODFLOW.html.

Bahremand, A. and De Smedt, F. (2008). Distributed hydrological modeling and sensitivity analysis in Torysa Watershed, Slovakia. *Water Resources Management*, v. 22, no. 3, pp. 293–408.

Bair, E.S., Springer, A.E., and Roadcap, G.S. (1992). *CAPZONE.* Columbus: Ohio State University.

Brown, D.L. and Silvey, W.D. (1977). Artificial recharge to a freshwater-sensitive brackish-water sand aquifer, Norfolk, VA: U.S. Geological survey Professional Paper 939, 53 p.

CH2 M. (2016). *Sustainable Water Recycling Initiative: Groundwater Injection Geochemical Compatibility Feasibility Evaluation*, Report No. 1. Newport News, VA: CH2M.

Civan, F. (2000). *Reservoir Formation Damage: Fundamentals, Modeling, Assessment, and Mitigation*. Houston, TX: Gulf Pub. Co.

Davies, C.W. (1962). *Ion Association*. Washington, DC: Butterworths, 190 p.

Debye, P. and Huckel, E. (1923). On the theory of electrolytes, I. Freezing point depression and related phenomena. *Physikalische Zeitschrift*, v. 24, no. 9, pp. 185–206.

Drever, J.I. (1982).Geochemistry in Natural Waters, 2nd ed. New Jersey, Prentice Hall.

Drever, J.M. (1988). *The Geochemistry of Natural Waters*, 2nd ed. New York: Prentice-Hall Englewood Cliffs.

Eggleston, J. and Pope, J. (2013). Land subsidence and relative sea-level rise in the southern Chesapeake Bay region: U.S. Geological Survey Circular 1392, 30 p. Accessed from https://doi.org/10.3111/cir1392.

Eykhoff, P. (1974). *System Identification Parameter and State Estimation*. New York: Wiley & Sons.

Gray, D.H. and Rex, R.W. (1966). *Formation Damage in Sandstones Caused by Clay Dispersion and Migration*. La Habra, CA: Chevron Research Company.

Hamilton, P.A. and Larson, J.D. (1988). Hydrogeology and analysis of the ground-water-flow system in the coastal plan of southeastern Virginia. U.S. Geological Survey Water Resources Investigations Report 87-4240, Richmond, Virginia.

Hewitt, E.J. (1963). Mineral nutrition of plants in culture media. In *Plant Physiology*, Vol III. Ed. Steward, F.C. New York: Academic Press, pp. 97–133.

Heywood and Pope. (2009). Simulation of groundwater flow in the coastal plain aquifer system of Virginia: U.S. Geological Survey Scientific Investigation Report 2009-5039, Water.usgs.gov/ogw/seawat.

Hill, M., Kavetski, D., Clark, M., Ye, M., Arabic, M., Lu, D., Foglia, L., and Mehl, S. (2015). Practical use of computationally frugal model analysis methods. *Groundwater*, v. 54, no. 2, pp. 159–170.

Hill, M. and Tiedeman, C. (2007). *Effective Groundwater Model Calibration, with Analysis of Data, Sensitivities, Prediction, and Uncertainty*. New York: John Wiley & Sons.

Holdahl, S.R. and Morrison, N. (1974). Regional investigations of vertical crustal movements in the US using precise relevelings and macrograph data. *Tectonophysics*, v. 23, no. 4, pp. 373–390.

HRSD. (2016a). *Sustainable Water Recycling Initiative: Groundwater Injection Hydraulic Feasibility Evaluation. Report 1*. Virginia Beach, VA: Hampton Roads Sanitation District. Compiled by CH2M Newport News, VA.

HRSD. (2016b). *Sustainable Water Recycling Initiative: Groundwater Injection Geochemical Compatibility Feasibility Evaluation. Report 2*. Virginia Beach, VA. Hampton Roads Sanitation District. Compiled by CH2M Newport News, VA.

Laczniak, R.J. and Meng, III, A.A. (1988). Ground-water resources of the York-James Peninsula of Virginia. U.S. Geological Survey Water Resources Investigations Report 88-4059, Richmond, VA.

Langevin, C.D., Thorne, D.T., Jr., Dausman, A.M., Sukop, M.S., and Guo, W. (2008). SEWAT Version. U.S. Geological Survey. Accessed from https://pubs.usgs.gov/tm/tmbaza//.

Langmuir, D. (1997). *Aqueous Environmental Geochemistry*. New York: Prentice-Hall.

McFarland, E.R. (2013). Sediment distribution and hydrologic conditions of the potomac aquifer in Virginia and Parts of Maryland and North Caroline. United States Geological Survey Scientific Investigations Report 2013-5116, Reston, VA.

McFarland, E.R. and Bruce, T.S. (2006). The Virginian coastal plain hydrogeologic framework. United States Geologic Survey Professional paper 1731, Reston, VA.

McGill, K. and Lucas, M.C. (2009). Mitigating specific capacity losses in aquifer storage and recovery wells in the New Jersey Coastal Plain: New Jersey Chapter of American Water Works Association, Annual Conference, Atlantic City, NJ.

Meade, R.H. (1964). Removal of water and rearrangement of particles during the compaction of clayey sediments: Geological Survey Prof. Paper 497-B.

Meng, A.A. and Harsh, J.F. (1988). Hydrogeologic framework of the Virginia Coastal Plain. United States Geological Survey Professional Paper 1404-C, Washington, DC.

Parkhurst, D.D. (1995). User's guide to PHREEQC-A computer program for speciation, reaction-path, advective-transport, and inverse geochemical calculations: U.S. Geological Survey Water-Resources Investigations Report 95-4227, 143 p.

Parnell, D.J. (1997). Sensitivity analysis of normative economic models: Theoretical framework and practical strategies. *Agricultural Economics*, v. 16, pp. 139–152.

Pyne, D.G. (1995). *Groundwater Recharge and Wells*. Ann Arbor, MI: Lewis Publishers.

Pyne, D.G. (2005). *Aquifer Storage and Recovery: A Guide to Groundwater Recharge through Wells*. Gainesville, FL: ASR Press.

Reed, M.G. (1972). Stabilization of Formation Clays with Hydroxy-Aluminum Solutions. Society of Petroleum Engineers. Accessed from https://doi.org/10.2118/3694-PA.

Smith, B.S. (1999). The potential for saltwater intrusion in the Potomac aquifers of the York-James Peninsula. U.S. Geological survey Water Resources Investigations Report 98-4187, Richmond, Virginia.

Spellman, F.R. (1996). Process Safety Management. Lancaster, PA: Technomic Publishing Company.

Spellman, F.R. (2017). *The Science of Environmental Pollution*, 3rd ed. Boca Raton, FL: CRC Press.

Stuyfzand, P.J. (1993). Hydrochemistry and hydrology of the coastal dune area of the western Netherlands: Amsterdam, Netherlands, Free Univ., Ph.D. thesis, 366 p.

Theis, C.V. (1935). The relation between lowering of the piezometric surface and the rate of duration of discharge of a well groundwater storage. *Transactions of the America Geophysical Union*, v. 16, Part 2, pp. 211–214.

Treifke, R.H. (1973). Geologic studies Coastal Plain of Virginia, *Bulletin 83*. Richmond: Virginia Division of Mineral Resources.

Truesdell, A.H. and Jones, B.F. (1974). WATEQ, a computer program for calculating chemical equilibria of natural waters. *Journal of Research*, v. 2, pp. 233–274.

USEPA. (2016). Aquifer Recharge and Aquifer Storage and Recovery. Accessed from https://epa.gov/uic/aquifer-recharge-and-aquifer-storage-and-recovery.

USGS (2015). The Potomac Aquifer. Hampton Roads, VA.

Virginia Department of Environmental Quality (VADEQ). (2006a). *Status of Virginia's Water Resources*. Richmond, Commonwealth of Virgina, VA.

Virginia Department of Environmental Quality (VADEQ). (2006b). *Virginia Coastal Palin Model 2005 Withdrawals Simulation*. Richmond, Commonwealth of Virgina, VA.

Warner, D.L. and Lehr, J. (1981). *Subsurface Wastewater Injection, the Technology of Injecting Wastewater into Deep Wells for Disposal*. Berkeley, CA: Premier Press.

16 Advanced Water Purification

SWIFT PROJECT

It is common practice when treating wastewater to turn out water cleaner than the local waterways it ultimately outfalls into. It is ultimately sent on to the ocean with no downstream use—in other words, "one and done" usage. Why? Why waste such a valuable resource? Why not reuse it? Don't we already use it … in de facto water recycling (aka human engineering water cycle)?

Hampton Roads Sanitation District (HRSD) is faced with a variety of future challenges related to the treatment and disposition of wastewater in its region of responsibility; this region includes much of southern Chesapeake Bay, with its many major tributaries and surrounding communities. HRSD envisions that it can protect and enhance the region's groundwater supplies by reusing highly purified wastewater through advanced treatment and subsequent injection into the region's groundwater aquifers. For those of us who understand the natural water cycle, humankind's urban water cycle, the proper use of various advanced wastewater treatment processes to purify the water, and HRSD's commitment to absolute excellence to its ratepayers and all those who live in the Hampton Roads Region, as well as to restoring and sustaining Chesapeake Bay—we understand that HRSD's vision not only has merit but is also necessary. It is necessary because the groundwater supply within the Potomac Aquifer is dwindling; it is in danger of contamination from saltwater intrusion; treated wastewater from HRSD's wastewater treatment plants outfalls nutrients into the Bay, which contribute to dead zones and other environmental issues; and, finally, HRSD understands that as native groundwater is withdrawn from the underlying aquifers, this contributes to land subsidence in the region. Land subsidence plus global sea level rise is the main contributor to relative sea level rise, and if not abated, it will soon (in less than 150 years) inundate many of the Hampton Roads major cities and other low-lying areas in the region.

In the previous chapter, it was pointed out that HRSD and its contractor, along with the U.S. Geological Survey (USGS), are working in unison to implement steps and modeling to ensure the compatibility of treated injectate with native groundwater. It is also important to make sure the chemical match between the injectate and native groundwater is safe for consumption. This is where advanced water treatment (AWT) comes into play. Wastewater treated only to conventional standards is probably safe enough for pipe-to-pipe connection but suffers from the public perception of the "Yuck Factor". That is, the old "your toilet water at my tap, no way, Jose and Maria" syndrome. It should be pointed out that when HRSD's plans were published in the surrounding Hampton Roads area, the overwhelming majority of the populace voiced no objection to HRSD's plans. However, one of the few but

DOI: 10.1201/9781003461265-19

common complaints was from those with wells drawing water from the Potomac Aquifer; they were worried that the HRSD SWIFT project might contaminate their water source with toilet water. This is exactly what HRSD is working hard to prevent by implementing AWT. Understand that the AWT process is in addition to normal wastewater treatment and filtration. In this chapter, we describe the three treatment processes, reverse osmosis (RO), nanofiltration (NF), and biologically activated carbon (BAC), and the pilot studies that were used to determine which process or processes is best-suited to facilitate HRSD's goal of producing and injecting the safest water possible.

FUNCTIONING BY THIS BOOK

The goal of most public service entities is to perform their functions by the book. In most cases, this book is the written volume that contains the applicable regulations—the so-called laws of the land—that apply to their activities. For operations that can directly or indirectly affect the environment, the applicable regulations are generally federal-based and enforced (by the Environmental Protection Agency, for example). However, it is interesting to note that injection of reclaimed water into an aquifer that is used as a potable water supply is referred to as indirect potable reuse (IPR), and regulations have not been developed by the U.S. Environmental Protection Agency (EPA) for potable reuse projects; therefore, states in which IPR is being practiced (or is being actively considered) have developed state-specific potable reuse regulations. With regard to indirect potable water reuse compliance, it is the state regulator knocking at the door, not the Feds.

THOSE PLAYING BY THE BOOK IN INDIRECT POTABLE REUSE[1]

California and Florida have developed regulations governing the practice of IPR. Other states allow IPR but establish project-specific requirements on a case-by-case basis (e.g., Virginia, Texas). Because Virginia has not developed IPR regulations but does allow IPR, the state will likely look to successful full-scale IPR projects within Virginia (e.g., Upper Occoquan Service Authority (USOA), Loudon Water, and other states' IPR regulations) for guidance in regulating HRSD's proposed direct injection IPR project. Table 16.1 presents the treated water quality requirements for other IPR projects and associated regulations that the State of Virginia could reference with respect to an HRSD IPR project.
What does Table 16.1 indicate?

- *Total organic carbon (TOC)*: TOC is the amount of carbon found in an organic compound is often used as a non-specific indicator of water quality. California's strict TOC limit of 0.5 mg/L will most likely require the use of RO, which could increase HRSD's project costs significantly. Conversely, application of the TOC and chemical oxygen demand (COD) limits (used to measure the amount of organic compounds in water) used with other IPR facilities (approximately 2–3 mg/L TOC) may allow implementation of

TABLE 16.1

Example Treated Water Quality Requirements for Indirect Potable Reuse

Parameter Regulations	Virginia's Occoquan and Dulles Area Watershed Policies (Surface Water Augmentation)	TCEQ Policy for El Paso, TX (Direct Injection)	Florida IPR for Direct Injection	California's IPR Regulations for Direct Injection	EPA's IPR Guidelines for Direct Injection
Relevant IPR Projects	Upper Occoquan Service Authority, Centreville, VA	Hueco Bolson Recharge Project, El Paso, TX	N/A Broad Run WRF, Loudoun County, VA	West Basin Water Recycling Plant, Los Angeles, CA; Los Alamitos Seawater Intrusion Barrier, Long Beach, CA	N/A Groundwater Replenishment System, Orange County, CA
TOC	COD < 10 mg/L (approx. 3 mg/L TOC)	None	<3 mg/L, TOX <0.2 mg/L	0.5 mg/L	≤2 mg/L (of wastewater origin)
Pathogens	Multiple barriers required, *E. coli* < 2 cfu/100 mL	None, but multiple barriers required	Multiple barriers required; total coliform < 4 cfu/100 mL	LRV from raw wastewater to finished water: 12-log for enteric viruses; 10-log for Giardia and Cryptosporidium	Multiple barriers required; total coliform below detection limit
Nitrogen	TKN < 1 mg/L; TN < 4 mg/L (Broad Run WRF only)	NOx-N < 10 mg/L	TN < 10 mg/L	TN < 10 mg/L	None
TDS	None	<1000 mg/L	None	RO-treatment required	None
Misc.	TSS < 1 mg/L turbidity <0.5 ntu; TP <0.1 mg/L	Turbidity <1 NTU	Turbidity < 2–2.5 NTU	RO and AOP treatment required for CECs	Turbidity ≤ 2 NTU

Source: HRSD/CH2M (2016).

Note: Not all parameters are listed; for example, other requirements such as compliance with all drinking water MCLs, travel time, disinfection residual, and such are required in some states and locations.

more sustainable alternative treatment technologies. For example, the use of ozone and activated carbon operating in biological and adsorption modes has been studied for use in potable reuse projects and can often produce water with a TOC less than 3 mg/L.

- *Pathogens*: California's strict Log Reduction Values (LRVs) are more challenging to meet than requirements at other locations, which may require additional disinfection-based treatment technologies to achieve and increase project costs significantly if adopted by Virginia.
- *Nitrogen*: USEPA's maximum contaminant level (MCL)—i.e., the legal threshold limit on the amount of substance that is allowed in public water systems—for nitrate in drinking water is 10 mg/L [as nitrogen (N)]. Therefore, total nitrogen (TN) or nitrite/nitrate (if the predominant nitrogen species) is typically limited in IPR applications to 10 mg/L to achieve MCL compliance to meet the nitrate limit or to prevent the conversion of ammonia (NH_3) or nitrite or nitrate in the aquifer.
- *Total dissolved solids*: USEPA's secondary MCL for total dissolved solids (TDS)—i.e., the measure of the combined content of all inorganic and organic substances contained in a liquid in molecular or colloidal suspended form—is 500 mg/L. Secondary MCLs are not enforceable, and therefore compliance is not required, but drinking water customers may complain of objectionable taste when TDS levels exceed 500 mg/L, depending on the ionic makeup of the TDS. However, the TDS concentration in the Potomac Aquifer is suspected to be high (>750 mg/L, see Table 16.2), so TDS removal may not be necessary, although further investigation is warranted (HRSD, 2016).

TABLE 16.2
Estimated TDS Concentrations (mg/L) in the Potomac Aquifer at HRSD's WWTP Locations

WWTP	Upper Potomac	Middle Potomac	Lower Potomac
Boat Harbor	750	1000	3500
Army Base	1500	2500	15,000
Virginia Initiative Plant	1000	2600	15,000
Nansemond	750	800	5000
James River	1000	1500	10,000
York River	5000	5000	10,000
Williamsburg	1000	1500	3000

Source: TDS data from Focazio, M.J., Speiran, G.K., and Rowan, M.E. (1993). Quality of groundwater in the Coastal Plain physiographic Province of Virginia: U.S. Geological Survey Water-Resources Investigations Report 92-4175, 20 p.

ADVANCED WATER TREATMENT PROCESSES

Advanced treatment provided at IPR plants varies but is typically focused on providing multiple barriers for the removal of pathogens and organics. Nitrogen and TDS removal is provided at some locations where necessary. Water extracted from direct injection and surface spreading projects that recharge groundwater isn't typically treated again prior to distribution into the potable water system; however, water from surface augmentation projects is typically treated again at water treatment plants because of water treatment requirements stipulated by USEPA's Surface Water Treatment Rule. For example, Fairfax County's Griffith Water Treatment Plant provides coagulation, sedimentation, ozone oxidation, BAC filtration, and chlorine disinfection for water extracted from that Occoquan Reservoir that is augmented by the Upper Occoquan Service Authority's (UOSA) IPR plant.

Treatment provided for IPR projects is typically a combination of multiple barriers for the removal of pathogens and organics. Multiple barriers for pathogens are typically provided through a combination of coagulation, flocculation, sediment, lime clarification, filtration (granular or membrane), and disinfection (chlorine, ultraviolet [UV], or ozone). Multiple barriers for organics removal are typically provided through a combination of advanced treatment processes (e.g., RO, granular-activated carbon [GAC], ozone in combination with GAC [BAC]), although conventional treatment processes (e.g., coagulation and softening) also provide removal at some locations. All potable reuse plants and/or processes discussed in this presentation include a robust organics removal process of either GAC, RO, or either GAC, RO, or soil aquifer treatment (SAT), which are effective barriers to bulk and trace organics and represent the backbone of the potable treatment process. SAT land treatment is the controlled application of wastewater to earthen basins in permeable soils at a rate typically measured in terms of meters of liquid per week. The purpose of a SAT system is to provide a receiver aquifer capable of accepting liquid intended to recharge shallow groundwater. System design and operating criteria are developed to achieve that goal. However, there are several alternatives with respect to the utilization or final fate of the treated water (USEPA, 2006):

- Groundwater recharge.
- Recovery of treated water for subsequent reuse or discharge.
- Recharge of adjacent surface streams.
- Seasonal storage of treated water beneath the site with seasonal recovery for agriculture.

The SAT process typically includes application of the reclaimed water using spreading basins and subsequent percolation through the vadose zone. SAT provides significant removal of both pathogens and organics through biological activity and natural filtration. However, because the Potomac Aquifer is confined, providing SAT for treatment through the vadose zone to recharge the aquifer is not possible for the HRSD SWIFT project. On the other hand, movement of reclaimed water through

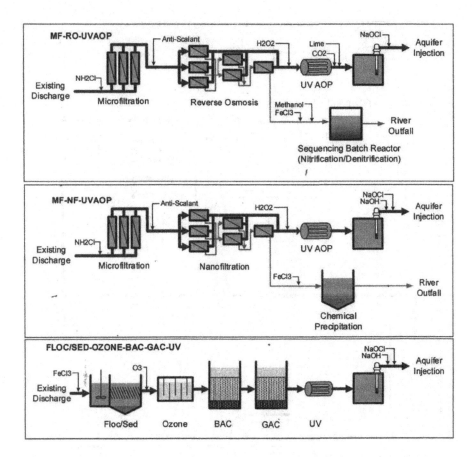

FIGURE 16.1 Process flow diagrams for three advanced wastewater treatment processes. Source: HRSD/CH2M (2016). CH2M Report No. 3.

the aquifer after direct injection will provide significant treatment benefits, including excellent removal of pathogens, and RO- and GAC-based AWT plants are often provided at locations where SAT treatment through the vadose zone is not feasible because these processes can be implemented at most locations.

RO- and GAC-based advanced treatment trains were developed for the HRSD groundwater recharge project using the historical WWTP effluent water quality data and the preliminary aquifer recharge water quality goals, as discussed previously. Three treatment trains were developed from this analysis and include an RO-based train, a NF-based train, and a GAC-based train (Figure 16.1). Figure 16.2 shows the actual treatment train used by HRSD. Consideration for each of these treatment trains includes the following (HRSD, 2016):

- *RO-based train*: RO has become common for potable water reuse projects in California and many international locations (e.g., Singapore and Australia) because of its effective removal of TDS, TOC, and TOs. California

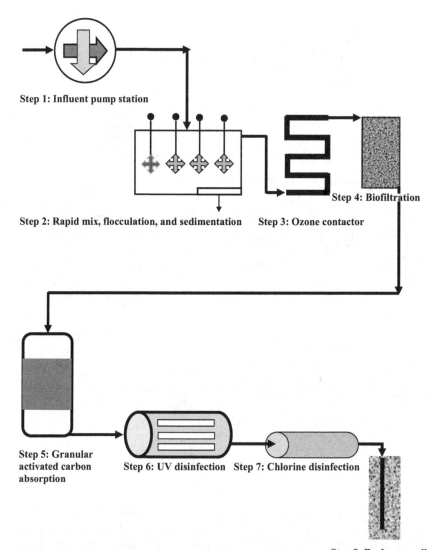

Step 1: Influent pump station

Step 2: Rapid mix, flocculation, and sedimentation   **Step 3: Ozone contactor**

Step 4: Biofiltration

Step 5: Granular activated carbon absorption   **Step 6: UV disinfection**   **Step 7: Chlorine disinfection**

Step 8: Recharge well

FIGURE 16.2 Process flow diagram for SWIFT facility.

regulations require the use of RO for direct injection reuse projects or a comparable alternative with regulatory approval. RO creates a waste (concentrate) stream that can be difficult and costly to dispose of, especially at inland locations. Most locations where RO has been implemented are located near the ocean, where disposal of RO concentrate is convenient and much less costly than at inland locations.

• *NF-based train*: The NF-based train is similar to the RO-based train but operates at significantly lower pressure and generates a less saline concentrate, which results in significant cost savings. This process does not meet California's IPR regulatory requirements but provides excellent treatment with significant removal of pathogens and organics. This train provides partial TDS removal by providing a high level of removal of divalent ions (e.g., calcium and magnesium) and moderate removal of monovalent ions (e.g., sodium chloride). NH_3 and NOx-N removal is much lower with NF compared to RO, which results in a lower TN concentration in the concentrate.

- *GAC-based train*: This train is a modernized version of full-scale operational IPR plants that have successfully been in operation for decades in Virginia (1978, 2008), Texas (1985), and more recently, Georgia (2000). GAC adsorption is used as the backbone process for organics removal, and other treatments have been added for multiple barriers to pathogens and organics. Flocculation and sedimentation provide removal of solids, pathogens, organics, and phosphorous. Ozone provides disinfection of pathogens and oxidation of organics, including oxidation of CECs and high-molecular-weight organic matter to smaller organic fractions that can be assimilated by biological activity present on GAC media, which is referred to as BAC filtration. This treatment train does not provide any TDS removal and, therefore, does not generate a TDS-enriched waste that might require further treatment prior to discharge.

CURRENT STATUS OF SWIFT PROJECT

On November 19, 2023, I contacted Jay Bernas, General Manager of HRSD, and he provided me with a personal communication on the status of the SWIFT project. In a water quality report dated November 9, 2023, he made it clear that the SWIFT Project is showing signs of success. He stated that, upon full implementation of the project, more than 100 million gallons a day will be delivered to the Potomac Aquifer. Eventually, HRSD's SWIFT Project will be one of the world's largest groundwater recharge programs. In addition to augmenting Virginia's groundwater supply, SWIFT supports Chesapeake Bay restoration efforts, protects the aquifer from saltwater intrusion, and shows signs of slowing the impact of observed sea level rise by decreasing the rate of land subsidence.

THE BOTTOM LINE

Based on my observation of the SWIFT Project from start-up through operation and its continuous growth and progress, I firmly believe that we can mitigate many of the effects of groundwater withdrawal and subsequent land subsidence through far thinking technology. Instead of waiting for someone else to solve our land subsidence problems, let's figure out how we can work together at state and local levels to develop practical fixes. And it is HRSD's SWIFT Project that is showing the way.

NOTE

1   Much of the material in this section is from HRSD (2016). *Sustainable Water Recycling Initiative: Advanced Water Purification Process Feasibility Evaluation. Report 3.* Hampton Roads Sanitation District. Compiled by CH2M Newport News, VA.

REFERENCES

HRSD/CH2M. (2016). *Sustainable Water Recycling Initiative: Groundwater Injection Hydraulic Feasibility Evaluation. Report* 1. Virginia Beach, VA: Hampton Roads Sanitation District. Compiled by CH2M Newport News, VA. *Sustainable Water Recycling Initiative: Groundwater Injection Geochemical Compatibility Feasibility Evaluation. Report 2.* Virginia Beach, VA. Hampton Roads Sanitation District. Compiled by CH2M Newport News, VA. *Sustainable Water* Recycling Initiative: *Advance Water Purification Process Feasibility Evaluation, Report 3.* Hampton Roads Sanitation District. Compiled by CH2M Newport News, VA.

USEPA (2006). *UV light toolkit.* Washington, DC: United Stated Environmental Protection Agency.

Conversion Factors and SI Units

The units most commonly used by environmental engineering professionals and environmental practitioners in all areas of interest are based on the complicated English system of weights and measures. However, bench work is usually based on the metric system or the International System of Units (SI) due to the convenient relationship between milliliters (mL), cubic centimeters (cm^3), and grams (g).

The SI is a modernized version of the metric system established by international agreement. The metric system of measurement was developed during the French Revolution and was first promoted in the United States in 1866. In 1902, proposed congressional legislation requiring the US government to use the metric system exclusively was defeated by a single vote. Although we use both systems in this text, SI provides a logical and interconnected framework for all measurements in engineering, science, industry, and commerce. The metric system is much simpler to use than the existing English system since all its units of measurement are divisible by 10.

Before listing the various conversion factors commonly used in environmental engineering, it is important to describe the prefixes commonly used in the SI system. These prefixes are based on the power 10. For example, a "kilo" means 1000 g, and a "centimeter" means one-hundredth of 1 m. The 20 SI prefixes used to form decimal multiples and submultiples of SI units are given in Table 1.

Note that kilogram is the only SI unit with a prefix as part of its name and symbol. Because multiple prefixes may not be used, in the case of the kilogram, the prefix names of Table 1 are used with the unit name "gram" and the prefix symbols are used with the unit symbol "g." With this exception, any SI prefix may be used with any SI unit, including the degree Celsius and its symbol °C.

TABLE 1
SI Prefixes

Factor	Name	Symbol
10^{24}	Yotta	Y
10^{21}	Zetta	Z
10^{18}	Exa	E
10^{15}	Peta	P
10^{12}	Tera	T
10^9	Giga	G
10^6	Mega	M
10^3	Kilo	k

(Continued)

TABLE 1 (*Continued*)
SI Prefixes

Factor	Name	Symbol
10^2	Hecto	h
10^1	Deka	da
10^{-1}	Deci	d
10^{-2}	Centi	c
10^{-3}	Milli	m
10^{-6}	Micro	μ
10^{-9}	Nano	n
10^{-12}	Pico	p
10^{-15}	Femto	f
10^{-18}	Atto	a
10^{-21}	Zepto	z
10^{-24}	Yocto	y

Example 1

10^{-6} kg = 1 mg (one milligram), but not 10^{-6} kg = 1 μkg (one micro-kilogram)

DID YOU KNOW?

The Fibonacci sequence is the following sequence of numbers:

1, 1, 2, 3, 5, 8, 13, 21, 34, 55, 89, 144, …

Or, alternatively,

0.1,1, 2, 3, 5, 8, 13, 21, 34, 55, 89, 144, …

Two important points arise: The first one is each term from the third onward is *the sum of the previous two.* Another point to notice is that if you divide each number in the sequence by the next number, beginning with the first, an interesting thing appears to be happening:

$1/1 = 1$, $1/2 = 0.5$, $2/3 = 0.66666$ …, $3/5 = 0.6$, $5/8 = 0.625$, $8/13 = 0.61538$ …, $13/21 = 0.61904$ …, (the first of these ratios appear to be converging to a number just a bit larger than 0.6).

CONVERSION FACTORS

Conversion factors are given below in the alphabetical order (Table 2) and in unit category:

Example 2

Consider the height of the Washington Monument. We may write $h_w = 169,000$ mm $= 16,900$ cm $= 169$ m $= 0.169$ km using the millimeter (SI prefix "milli," symbol "m"), centimeter (SI prefix "centi," symbol "c"), or kilometer (SI prefix "kilo," symbol "k").

Example 3

Problem: Find degree Celsius of water at 72°F.

Solution:

$$°C = (F - 32) \times 5/9 = (72 - 32) \times 5/9 = 22.2$$

A table of conversion factors is provided in Tables 2, and 3 lists conversion factors by unit category.

TABLE 2
Alphabetical List of Conversion Factors

Factor	Metric (SI) or English Conversions
1 atm (atmosphere) =	1.013 bars
	10.133 N/cm² (newtons/square centimeter)
	33.90 ft. of H_2O (feet of water)
	101.325 kp (kilopascals)
	1013.25 mg (millibars)
	13.70 psia (pounds/square inch—absolute)
760 torr	
760 mm Hg (millimeters of mercury)	
1 bar =	0.987 atm (atmospheres)
	1×10^6 dynes/cm² (dynes/square centimeter)
	33.45 ft. of H_2O (feet of water)
	1×10^5 pascals [N/m²] (newtons/square meter)
	750.06 torr
	750.06 mm Hg (millimeters of mercury)
1 Bq (becquerel) =	1 radioactive disintegration/second
	2.7×10^{-11} Ci (curie)
	2.7×10^{-8} mCi (millicurie)
1 BTU (British thermal unit) =	252 cal (calories)
	1055.06 J (joules)
	10.41 L-atmosphere
	0.293 watt-hours
1 cal (calories) =	3.97×10^{-3} BTUs (British thermal units)
	4.18 J (joules)

(Continued)

TABLE 2 (*Continued*)
Alphabetical List of Conversion Factors

Factor	Metric (SI) or English Conversions
	0.0413 L-atmospheres
	1.163×10^{-3} watt-hours
1 cm (centimeters) =	0.0328 ft (feet)
	0.394 in (inches)
	10,000 microns (micrometers)
	100,000,000 Å $= 10^8$ Å (Ångstroms)
1 cc (cubic centimeter) =	3.53×10^{-5} ft^3 (cubic feet)
	0.061 in^3 (cubic inches)
	2.64×10^{-4} gal (gallons)
	52.18 L (liters)
	52.19 mL (milliliters)
1 ft^3 (cubic foot) =	28.317 cc (cubic centimeters)
	1728 in^3 (cubic inches)
	0.0283 m^3 (cubic meters)
	7.48 gal (gallons)
	28.32 L (liters)
	29.92 qts (quarts)
1 in^3	16.39 cc (cubic centimeters)
	16.39 mL (milliliters)
	5.79×10^{-4} ft^3 (cubic feet)
	1.64×10^{-5} m^3 (cubic meters)
	4.33×10^{-3} gal (gallons)
	0.0164 L (liters)
	0.55 fl oz (fluid ounces)
1 m^3 (cubic meter) =	1,000,000 cc $= 10^6$ cc (cubic centimeters)
	33.32 ft^3 (cubic feet)
	61,023 in^3 (cubic inches)
	264.17 gal (gallons)
	1000 L (liters)
1 yd^3 (cubic yard) =	201.97 gal (gallons)
	764.55 L (liters)
1 Ci (curie) =	3.7×10^{10} radioactive disintegrations/second
	3.7×10^{10} Bq (becquerel)
	1000 mCi (millicurie)
1 day =	24 h (hours)
	1440 min (minutes)
	86,400 s (seconds)

(*Continued*)

TABLE 2 (*Continued*)
Alphabetical List of Conversion Factors

Factor	Metric (SI) or English Conversions
	0.143 weeks
	2.738×10^{-3} yrs (years)
1°C (expressed as an interval) =	1.8°F = [9/5]°F (degrees Fahrenheit)
	1.8°R (degrees Rankine)
	1.0 K (degrees Kelvin)
°C (degree Celsius) =	[(5/9)(°F − 32°)]
1°F (expressed as an interval) =	0.556°C = [5/9]°C (degrees Celsius)
	1.0°R (degrees Rankine)
	0.556 K (degrees Kelvin)
°F (degree Fahrenheit) =	[(9/5)(°C) + 32°
1 dyne =	1×10^{-5} N (newton)
1 eV (electron volt) =	1.602×10^{-12} ergs
	1.602×10^{-19} J (joules)
1 erg =	1 dyne-centimeters
	1×10^{-7} J (joules)
	2.78×10^{-11} watt-hours
1 fps (feet/second) =	1.097 kmph (kilometers/hour)
	0.305 mps (meters/second)
	0.01136 mph (miles/hour)
1 ft (foot) =	30.48 cm (centimeters)
	12 in (inches)
	0.3048 m (meters)
	1.65×10^{-4} nt (nautical miles)
	1.89×10^{-4} mi (statute miles)
1 gal (gallon) =	3,785 cc (cubic centimeters)
	0.134 ft^3 (cubic feet)
	231 in^3 (cubic inches)
	3.785 L (liters)
1 gm (gram)	0.001 kg (kilogram)
	1000 mg (milligrams)
	1,000,000 ng = 10^6 ng (nanograms)
	2.205×10^{-3} lbs (pounds)
1 gm/cc (grams/cubic cent.) =	62.43 lbs/ft^3 (pounds/cubic foot)
	0.0361 lbs/in^3 (pounds/cubic inch)
	8.345 lbs/gal (pounds/gallon)
1 Gy (gray) =	1 J/kg (joules/kilogram)
	100 rad

(*Continued*)

TABLE 2 (*Continued*)
Alphabetical List of Conversion Factors

Factor	Metric (SI) or English Conversions
	1 Sv (sievert) – [unless modified through division by an appropriate factor, such as Q and/or N]
1 hp (horsepower) =	745.7 J/sec (joules/sec)
1 hr (hour) =	0.0417 days
	60 min (minutes)
	3600 sec (seconds)
	5.95×10^{-3} weeks
	1.14×10^{-4} yrs (years)
1 in (inch) =	2.54 cm (centimeters)
	1000 mils
I inch of water =	1.86 mm Hg (millimeters of mercury)
	249.09 pascals
	0.0361 psi (lbs/in^2)
1 J (joule) =	9.48×10^{-4} BTUs (British thermal units)
	0.239 cal (calories)
	10,000,000 ergs $= 1 \times 10^7$ ergs
	***9.87×10^{-3} L-atmospheres
	1.0 N-m (newton-meters)
1 kcal (kilocalories) =	3.97 BTUs (British thermal units)
	1000 cal (calories)
	4186.8 J (joules)
1 kg (kilogram) =	1000 g (grams)
	2205 lbs (pounds)
1 km (kilometer) =	3280 ft (feet)
	0.54 nt (nautical miles)
	0.6214 mi (statute miles)
1 kw (kilowatt) =	56.87 BTU/min (British thermal units)
	1.341 hp (horsepower)
	1000 J/sec (kilocalories)
1 kw-hr (kilowatt-hour) =	3412.14 BTU (British thermal units)
	3.6×10^6 J (joules)
	859.8 kcal (kilocalories)
1 L (liter) =	1000 cc (cubic centimeters)
	1 dm^3 (cubic decimeters)
	0.0353 ft^3 (cubic feet)
	61.02 in^3 (cubic inches)
	0.264 gal (gallons)
	1000 mL (milliliters)

(*Continued*)

TABLE 2 (*Continued*)
Alphabetical List of Conversion Factors

Factor	Metric (SI) or English Conversions
	1.057 qts (quarts)
1 m (meter) =	1×10^{10} Å (Ångstroms)
	100 cm (centimeters)
	3.28 ft (feet)
	39.37 in (inches)
	1×10^{-3} km (kilometers)
	1000 mm (millimeters)
	1,000,000 $\mu = 1 \times 10^6$ μ (micrometers)
	1×10^9 nm (nanometers)
1 mps (meters/second) =	196.9 fpm (feet/minute)
	3.6 kmph (kilometers/hour)
	2.237 mph (miles/hour)
1 mph (mile/hour) =	88 fpm (feet/minute)
	1.61 kmph (kilometers/hour)
	0.447 mps (meters/second)
1 kt (nautical mile) =	6076.1 ft (feet)
	1.852 km (kilometers)
	1.15 mi (statute miles)
	2025.4 yds (yards)
1 mi (statute mile) =	5280 ft (feet)
	1.609 km (kilometers)
	1609.3 m (meters)
	0.869 nt (nautical miles)
	1760 yds (yards)
1 miCi (millicurie) =	0.001 Ci (curie)
	3.7×10^{10} radioactive disintegrations/second
	3.7×10^{10} Bq (becquerel)
1 mm Hg (mm of mercury) =	1.316×10^{-3} atm (atmosphere)
	0.535 in H_2O (inches of water)
	1.33 mb (millibars)
	133.32 pascals
	1 torr
	0.0193 psia (pounds/square inch – absolute)
1 min (minute) =	6.94×10^{-4} days
	0.0167 h (hours)
	60 s (seconds)
	9.92×10^{-5} weeks

(*Continued*)

TABLE 2 (*Continued*)
Alphabetical List of Conversion Factors

Factor	Metric (SI) or English Conversions
	1.90×10^{-6} yrs (years)
1 N (newton) =	1×10^5 dynes
1 N-m (newton-meter) =	1.00 J (joules)
	2.78×10^{-4} watt-hours
1 ppm (parts/million-volume) =	1.00 mL/m³ (milliliters/cubic meter)
1 ppm [wt] (parts/million-weight) =	1.00 mg/kg (milligrams/kilograms)
1 pascal =	9.87×10^{-6} atm (atmospheres)
	4.015×10^{-3} in H_2O (inches of water)
	0.01 mb (millibars)
	7.5×10^{-3} mm Hg (milliliters of mercury)
1 lb (pound) =	453.59 g (grams)
	16 oz (ounces)
1 lbs/ft³ (pounds/cubic foot) =	16.02 g/L (grams/liter)
1 lbs/ft³ (pounds/cubic inch) =	27.68 g/cc (grams/cubic centimeter)
	1728 lbs/ft³ (pounds/cubic feet)
1 psi (pounds/square inch)=	0.068 atm (atmospheres)
	27.67 in H_2O (inches or water)
	68.85 mb (millibars)
	51.71 mm Hg (millimeters of mercury)
	6894.76 pascals
1 qt (quart) =	946.4 cc (cubic centimeters)
	57.75 in³ (cubic inches)
	0.946 L (liters)
1 rad =	100 ergs/gm (ergs/gram)
	0.01 Gy (gray)
	1 rem [unless modified through division by an appropriate factor, such as Q and/or N]
1 rem	1 rad [unless modified through division by an appropriate factor, such as Q and/or N]
1 Sv (sievert) =	1 Gy (gray) [unless modified through division by an appropriate factor, such as Q and/or N]
1 cm² (square centimeter)=	1.076×10^{-3} ft² (square feet)
	0.155 in² (square inches)
1×10^{-4} m² (square meters)	
1 ft² (square foot) =	2.296×10^{-5} acres
	9.296 cm² (square centimeters)
	144 in² (square inches)
	0.0929 m² (square meters)
1 m² (square meter) =	10.76 ft² (square feet)

(*Continued*)

TABLE 2 (*Continued*)
Alphabetical List of Conversion Factors

Factor	Metric (SI) or English Conversions
	1550 in^2 (square inches)
1 mi^2 (square mile) =	640 acres
	2.79 × 10^7 ft^2 (square feet)
	2.59 × 10^6 m^2 (square meters)
1 torr =	1.33 mb (millibars)
1 watt =	3.41 BTI/hr (British thermal units/hour)
	1.341 × 10^{-3} hp (horsepower)
	52.18 J/sec (joules/second)
1 watt-hour =	3.412 BTUs (British thermal units)
	859.8 cal (calories)
	3600 J (joules)
	35.53 L-atmosphere
1 week =	7 days
	168 h (hours)
	10,080 min (minutes)
	6.048 × 10^5 s (seconds)
	0.0192 yrs (years)
1 yr (year) =	365.25 days
	8766 h (hours)
	5.26 × 10^5 min (minutes)
	3.16 × 10^7 s (seconds)
	52.18 weeks

TABLE 3
Conversion Factors by Unit Category

	Units of Length
1 cm (centimeter) =	0.0328 ft (feet)
	0.394 in (inches)
	10,000 μm (micrometers)
	100,000,000 Å = 10^8 Å (Ångstroms)
1 ft (foot) =	30.48 cm (centimeters)
	12 in (inches)
	0.3048 m (meters)
	1.65 × 10^{-4} nt (nautical miles)
	1.89 × 10^{-4} mi (statute miles)
1 in (inch) =	2.54 cm (centimeters)
	1000 mils

(*Continued*)

TABLE 3 (*Continued*)
Conversion Factors by Unit Category

Units of Length

1 km (kilometer) =	3280.8 ft (feet)
	0.54 nt (nautical miles)
	0.6214 mi (statute miles)
1 m (meter) =	1×10^{10} Å (Ångstroms)
	100 cm (centimeters)
	3.28 ft (feet)
	39.37 in (inches)
	1×10^{-3} km (kilometers)
	1000 mm (millimeters)
	1,000,000 $\mu = 1 \times 10^6 \mu$ (micrometers)
	1×10^9 nm (nanometers)
1 kt (nautical mile) =	6076.1 ft (feet)
	1.852 km (kilometers)
	1.15 km (statute miles)
	2.025.4 yds (yards)
1 mi (statute mile) =	5280 ft (feet)
	1.609 km (kilometers)
	1.690.3 m (meters)
	0.869 nt (nautical miles)
	1760 yds (yards)

Units of Area

1 cm² (square centimeter) =	1.076×10^{-3} ft² (square feet)
	0.155 in² (square inches)
	1×10^{-4} m² (square meters)
1 ft² (square foot) =	2.296×10^{-5} acres
	929.03 cm² (square centimeters)
	144 in² (square inches)
	0.0929 m² (square meters)
1 m² (square meter) =	10.76 ft² (square feet)
	1550 in² (square inches)
1 mi² (square mile) =	640 acres
	2.79×10^7 ft² (square feet)
	2.59×10^6 m² (square meters)

Units of Volume

1 cc (cubic centimeter) =	3.53×10^{-5} ft³ (cubic feet)
	0.061 in³ (cubic inches)
	2.64×10^{-4} gal (gallons)
	0.001 L (liters)
	1.00 mL (milliliters)

(*Continued*)

TABLE 3 (*Continued*)
Conversion Factors by Unit Category

Units of Volume

1 ft³ (cubic foot) =	28,317 cc (cubic centimeters)
	1728 in³ (cubic inches)
	0.0283 m³ (cubic meters)
	7.48 gal (gallons)
	28.32 L (liters)
	29.92 qts (quarts)
1 in³ (cubic inch) =	16.39 cc (cubic centimeters)
	16.39 mL (milliliters)
	5.79×10^{-4} ft³ (cubic feet)
	1.64×10^{-5} m³ (cubic meters)
	4.33×10^{-3} gal (gallons)
	0.0164 L (liters)
	0.55 fl oz (fluid ounces)
1 m³ (cubic meter) =	$1,000,000$ cc $= 10^6$ cc (cubic centimeters)
	35.31 ft³ (cubic feet)
	61,023 in³ (cubic inches)
	264.17 gal (gallons)
	1000 L (liters)
1 yd³ (cubic yards) =	201.97 gal (gallons)
	764.55 L (liters)
1 gal (gallon) =	3785 cc (cubic centimeters)
	0.134 ft³ (cubic feet)
	231 in³ (cubic inches)
	3.785 L (liters)
1 L (liter) =	1000 cc (cubic centimeters)
	1 dm³ (cubic decimeters)
	0.0353 ft³ (cubic feet)
	61.02 in³ (cubic inches)
	0.264 gal (gallons)
	1000 mL (milliliters)
	1.057 qts (quarts)
1 qt (quart) =	946.4 cc (cubic centimeters)
	57.75 in³ (cubic inches)
	0.946 L (liters)

Units of Mass

1 g (grams) =	0.001 kg (kilograms)
	1000 mg (milligrams)

(*Continued*)

TABLE 3 *(Continued)*
Conversion Factors by Unit Category

	Units of Mass
	$1,000,000$ mg $= 10^6$ ng (nanograms)
	2.205×10^{-3} lbs (pounds)
1 kg (kilogram) =	1000 g (grams)
	2.205 lbs (pounds)
1 lbs (pound) =	453.59 g (grams)
	16 oz (ounces)

	Units of Time
1 day =	24 h (hours)
	1440 min (minutes)
	86,400 s (seconds)
	0.143 weeks
	2.738×10^{-3} yrs (years)
1 h (hours) =	0.0417 days
	60 min (minutes)
	3600 sec (seconds)
	5.95×10^{-3} yrs (years)
1 h (hour) =	0.0417 days
	60 min (minutes)
	3600 s (seconds)
	5.95×10^{-3} weeks
	1.14×10^{-4} yrs (years)
1 min (minutes) =	6.94×10^{-4} days
	0.0167 h (hours)
	60 s (seconds)
	9.92×10^{-5} weeks
	1.90×10^{-6} yrs (years)
1 week =	7 days
	168 h (hours)
	10,080 min (minutes)
	6.048×10^5 s (seconds)
	0.0192 yrs (years)
1 yr (year) =	365.25 days
	8766 h (hours)
	5.26×10^5 min (minutes)
	3.16×10^7 s (seconds)
	52.18 weeks

(Continued)

TABLE 3 (*Continued*)
Conversion Factors by Unit Category

Units of the Measure of Temperature

°C (degrees Celsius) =	$[(5/9)(°F - 32°)]$
1°C (expressed as an interval) =	$1.8°F = [9/5]°F$ (degrees Fahrenheit)
	1.8°R (degrees Rankine)
	1.0 K (degrees Kelvin)
°F (degree Fahrenheit) =	$[(9/5)(°C) + 32°]$
1°F (expressed as an interval) =	$0.556°C = [5/9]°C$ (degrees Celsius)
	1.0°R (degrees Rankine)
	0.556 K (degrees Kelvin)

Units of Force

1 dyne =	1×10^{-5} N (newtons)
1 N (newton) =	1×10^5 dynes

Units of Work or Energy

1 BTU (British thermal unit) =	252 cal (calories)
	1055.06 J (joules)
	10.41 L-atmospheres
	0.293 watt-hours
1 cal (calories) =	3.97×10^{-3} BTUs (British thermal units)
	4.18 J (joules)
	0.0413 L-atmospheres
	1.163×10^{-3} watt-hours
1 eV (electron volt) =	1.602×10^{-12} ergs
	1.602×10^{-19} J (joules)
1 erg =	1 dyne-centimeter
1×10^{-7} J (joules)	
	2.78×10^{-11} watt-hours
1 J (joule) =	9.48×10^{-4} BTUs (British thermal units)
	0.239 cal (calories)
	$10,000,000$ ergs $= 1 \times 10^7$ ergs
	9.87×10^{-3} L-atmospheres
	1.00 N-m (newton-meters)
1 kcal (kilocalorie) =	3.97 BTUs (British thermal units)
	1000 cal (calories)
	4186.8 J (joules)
1 kw-hr (kilowatt-hour) =	3412.14 BTU (British thermal units)
	3.6×10^6 J (joules)
	859.8 kcal (kilocalories)

(Continued)

TABLE 3 (*Continued*)
Conversion Factors by Unit Category

Units of Work or Energy

1 N-m (newton-meter) =
    1.00 J (joules)
    2.78×10^{-4} watt-hours

1 watt-hour =
    3.412 BTUs (British thermal units)
    859.8 cal (calories)
    3600 J (joules)
    35.53 L-atmospheres

Units of Power

1 hp (horsepower) =
    745.7 J/s (joules/sec)

1 kw (kilowatt) =
    56.87 BTU/min (British thermal units/minute)
    1.341 hp (horsepower)
    1000 J/sec (joules/sec)

1 watt =
    3.41 BTU/hr (British thermal units/hour)
    1.341×10^{-3} hp (horsepower)
    1.00 J/sec (joules/second)

Units of Pressure

1 atm (atmosphere) =
    1.013 bars
    10.133 N/cm^2 (newtons/square centimeters)
    33.90 ft. of H_2O (feet of water)
    101.325 kp (kilopascals)
    14.70 psia (pounds/square inch – absolute)
    760 torr
    760 mm Hg (millimeters of mercury)

1 bar =
    0.987 atm (atmospheres)
    1×10^6 dynes/cm^2 (dynes/square centimeter)
    33.45 ft. of H_2O (feet of water)
    1×10^5 pascals [N/m^2] (newtons/square meter)
    750.06 torr
    750.06 mm Hg (millimeters of mercury)

1 inch of water =
    1.86 mm Hg (millimeters of mercury)
    249.09 pascals
    0.0361 psi (lbs/in^2)

1 mm Hg (millimeter of merc.) =
    1.316×10^{-3} atm (atmospheres)
    0.535 in H_2O (inches of water)
    1.33 mb (millibars)
    133.32 pascals
    1 torr
    0.0193 psia (pounds/square inch – absolute)

1 pascal =
    9.87×10^{-6} atm (atmospheres)
    4.015×10^{-3} in H_2O (inches of water)

(*Continued*)

TABLE 3 (*Continued*)
Conversion Factors by Unit Category

Units of Pressure

	0.01 mb (millibars)
	7.5×10^{-3} mm Hg (millimeters of mercury)
1 psi (pounds/square inch) =	0.068 atm (atmospheres)
	27.67 in H_2O (inches of water)
	68.85 mb (millibars)
	51.71 mm Hg (millimeters of mercury)
	6894.76 pascals
1 torr =	1.33 mb (millibars)

Units of Velocity or Speed

1 fps (feet/second) =	1.097 kmph (kilometers/hour)
	0.305 mps (meters/second)
	0.01136 mph (miles/hours)
1 mps (meters/second) =	196.9 fpm (feet/minute)
	3.6 kmph (kilometers/hour)
	2.237 mph (miles/hour)
1 mph (mile/hour) =	88 fpm (feet/minute)
	1.61 kmph (kilometers/hour)
	0.447 mps (meters/second)

Units of Density

1 gm/cc (grams/cubic cent) =	62.43 lbs/ft³ (pounds/cubic foot)
	0.0361 lbs/in³ (pounds/cubic inch)
	8.345 lbs/gal (pounds/gallon)
1 lbs/ft³ (pounds/cubic foot) =	16.02 g/L (grams/liter)
1 lbs/in² (pounds/cubic inch) =	27.68 g/cc (grams/cubic centimeter)
	1.728 lbs/ft³ (pounds/cubic foot)

Units of Concentration

1 ppm (parts/million-volume) =	1.00 mL/m³ (milliliters/cubic meter)
1 ppm (wt) =	1.00 mg/kg (milligrams/kilograms)

Radiation and Dose-Related Units

1 Bq (becquerel) =	1 radioactive disintegration/second
	2.7×10^{-11} Ci (curie)
	2.7×10^{-8} (millicurie)
1 Ci (curie) =	3.7×10^{10} radioactive disintegration/second
	3.7×10^{10} Bq (becquerel)
	1000 mCi (millicurie)
1 Gy (gray) =	1 J/kg (joule/kilogram)
	100 rad
	1 Sv (sievert) – [unless modified through division by an appropriate factor, such as Q and/or N]

(Continued)

TABLE 3 (*Continued*)

Conversion Factors by Unit Category

Radiation and Dose-Related Units

1 mCi (millicurie) =	0.001 Ci (curie)
	3.7×10^{10} radioactive disintegrations/second
	3.7×10^{10} Bq (becquerel)
1 rad =	100 ergs/gm (ergs/gm)
	0.01 Gy (gray)
	1 rem – [unless modified through division by an appropriate factor, such as Q and/or N]
1 rem =	1 rad – [unless modified through division by an appropriate factor, such as Q and/or N]
1 Sv (sievert) =	1 Gy (gray) – [unless modified through division by an appropriate factor, such as Q and/or N]

DID YOU KNOW?

Units and dimensions are not the same concepts. Dimensions are concepts like time, mass, length, and weight. Units are specific cases of dimensions, like hour, gram, meter, and lb. You can *multiply* and *divide* quantities with different units: 4 ft × 8 lb = 32 ft-lb; but you can *add* and *subtract* terms only if they have the same units: 5 lb + 8 kg = **NO WAY!!!**

Conversion Factors
Practical Examples

Sometimes we have to convert between different units. Suppose that a 60-inch piece of pipe is attached to an existing 6-foot piece of pipe. Joined together, how long are they? Obviously, we cannot find the answer to this question by adding 60 to 6. Why? Because two lengths are given in different units. Before we can add the two lengths, we must convert one of them to the unit of the other. Then, when we have two lengths in the same unit, we can add them.

To perform this conversion, we need a *conversion factor*. That is, in this case, we have to know how many inches make up a foot—that is, 12 inches is 1 foot. Knowing this, we can perform the calculation in two steps:

1. 60 inches is really $60/12 = 5$ ft
2. 5 ft + 6 ft = 11 ft

From the example above, it can be seen that a conversion factor changes known quantities in one unit of measure to an equivalent quantity in another unit of measure.

In making the conversion from one unit to another, we must know two things:

1. The exact number that relates the two units
2. Whether to multiply or divide by that number

When making conversions, confusion over whether to multiply or divide is common; on the other hand, the number that relates the two units is usually known and, thus, is not a problem. Understanding the proper methodology—the "mechanics"—to use for various operations requires practice and common sense.

Along with using the proper "mechanics" (and practice and common sense) in making conversions, probably the easiest and fastest method of converting units is to use a conversion table. The simplest conversion requires that the measurement be multiplied or divided by a constant value. For instance, if the depth of wet cement in a form is 0.85 feet, multiplying by 12 inches per foot converts the measured depth to inches (10.2 inches). Likewise, if the depth of the cement in the form is measured as 16 inches, dividing by 12 inches per feet converts the depth measurement to feet (1.33 feet).

WEIGHT, CONCENTRATION, AND FLOW

Using Table 4 to convert from one unit expression to another and vice versa is good practice. However, in making conversions to solve process computations in water treatment operations, for example, we must be familiar with conversion calculations based upon a relationship between weight, flow or volume, and concentration. The basic relationship is

$$\text{Weight} = \text{Concentration} \times \text{Flow or Volume} \times \text{Factor} \tag{1}$$

TABLE 4
Conversion Table

To Convert	Multiply by	To Get
Feet	12	Inches
Yards	3	Feet
Yards	36	Inches
Inches	2.54	Centimeters
Meters	3.3	Feet
Meters	100	Centimeters
Meters	1000	Millimeters
Square yards	9	Square feet
Square feet	144	Square inches
Acres	43,560	Square feet
Cubic yards	27	Cubic feet
Cubic feet	1728	Cubic inches
Cubic feet (water)	7.48	Gallons
Cubic feet (water)	62.4	Pounds
Acre-feet	43,560	Cubic feet
Gallons (water)	8.34	Pounds
Gallons (water)	3.785	Liters
Gallons (water)	3785	Milliliters
Gallons (water)	3785	Cubic centimeters
Gallons (water)	3785	Grams
Liters	1000	Milliliters
Days	24	Hours
Days	1440	Minutes
Days	86,400	Seconds
Million gallons/day	1,000,000	Gallons/day
Million gallons/day	1.55	Cubic feet/second
Million gallons/day	3.069	Acre-feet/day
Million gallons/day	36.8	Acre-inches/day
Million gallons/day	3785	Cubic meters/day
Gallons/minute	1440	Gallons/day
Gallons/minute	63.08	Liters/minute
Pounds	454	Grams
Grams	1000	Milligrams
Pressure, psi	2.31	Head, ft (water)
Horsepower	33,000	Foot-pounds/minute
Horsepower	0.746	Kilowatts

TABLE 5
Weight, Volume, and Concentration Calculations

To Calculate	Formula
Pounds	Concentration, mg/L × Tank Vol. MG × 8.34 lb/MG/mg/L
Pounds/day	Concentration, mg/L × Flow, MGD × 8.34 lb/MG/mg/L
Million gallons/day	$\dfrac{\text{Quantity, lb/day}}{\left(\text{Conc., mg/L} \times 8.34\ \text{lb/mg/L/MG}\right)}$
Milligrams/liter	$\dfrac{\text{Quantity, lb}}{\left(\text{Tank volume, MG} \times 8.34\ \text{lb/mg/L/MG}\right)}$
Kilograms/liter	Conc., mg/L × Volume, MG × 3.785 lb/MG/mg/L
Kilograms/day	Conc., mg/L × Flow, MGD × 3.785 lb/MG/mg/L
Pounds/dry ton	Conc. mg/kg × 0.002 lb/d.t./mg/kg

Table 5 summarizes weight, volume, and concentration calculations. With practice, many of these calculations become second nature to users.

Note: The following conversion factors are used extensively in *The Science of Series* and come into play or usage in engineering involved with land subsidence work—this is especially the case when dealing with depletion of groundwater-caused subsidence, which is the primary cause of land subsidence. Based on years of study, I have determined that about 80% of land subsidence has something to do with water—usually its withdrawal from groundwater. This information is covered in this text. Also, conversions, mathematics, are a critical part of science—any science—and therefore are presented with examples given on their application.

- 7.48 gallons per ft^3
- 3.785 L per gallon
- 454 g per pound
- 1000 mL per liter
- 1000 mg per gram
- 1 ft^3/sec (cfs) = 0.6465 MGD

Key point: Density (also called specific weight) is mass per unit volume and may be registered as lb/cu ft, lb/gal, grams/mL, and grams/cu meter. If we take a fixed volume container, fill it with a fluid, and weigh it, we can determine density of the fluid (after subtracting the weight of the container).

- 8.34 pounds per gallon (water)—(density = 8.34 lb/gal)
- one milliliter of water weighs 1 g—(density = 1 g/mL)
- 62.4 pounds per ft^3 (water)—(density = 8.34 lb/gal)
- 8.34 lb/gal = mg/L (converts dosage in mg/L into lb/day/MGD)

Example: 1 mg/L × 10 MGD × 8.3 = 83.4 lb/day

- 1 psi = 2.31 feet of water (head)
- 1 foot head = 0.433 psi
- $°F = 9/5(°C + 32)$
- $°C = 5/9(°F - 32)$
- average water usage: 100 gallons/capita/day (gpcd)
- persons per single family residence: 3.7

Example 1

Convert million gallons per day (MGD) to gallons per day (gpd).

$$Flow = Flow, MGD × 1,000,000 \ gal/MG$$

Sample Problem

The industrial water 1 meter reads 28.8 MGD. What is the current flow rate in gpd?

$$Flow = 28.8 \ MGD × 1,000,000 \ gal/MG = 28,800,000 \ gpd$$

Example 2

Convert MGD to cfs.

$$Flow, cfs = Flow, MGD × 1.55 \ cfs/MGD$$

Sample Problem

The flow rate entering industrial operation is 2.89 MGD. What is the flow rate in cfs?

$$Flow = 2.89 \ MGD × 1.55 \ cfs/MGD = 4.48 \ cfs$$

Example 3

Convert gallons per minute (gpm) to MGD.

$$Flow, MGD = \frac{Flow, gpm × 1440 \ min/day}{1,000,000 \ gal/MG}$$

Sample Problem

The industrial flow meter indicates that the current flow rate is 1469 gpm. What is the flow rate in MGD?

$$\text{Flow, MGD} = \frac{1469 \text{ gpm} \times 1440 \text{ min/day}}{1,000,000 \text{ gal/MG}} = 2.12 \text{ MGD (rounded)}$$

Example 4

Convert gpd to MGD.

$$\text{Flow, MGD} = \frac{\text{Flow, gal/day}}{1,000,000 \text{ gal/MG}}$$

Sample Problem

The industrial totalizing flow meter indicates that 33,444,950 gallons of wastewater have left the plant in the past 24 hours. What is the flow rate in MGD?

$$\text{Flow} = \frac{33,444,950 \text{ gal/day}}{1,000,000 \text{ gal/MG}} = 33.44 \text{ MGD}$$

Example 5

Convert flow in cfs to MGD.

$$\text{Flow, MGD} = \frac{\text{Flow, cfs}}{1.55 \text{ cfs/MG}}$$

Sample Problem

The flow in a channel is determined to be 3.89 cfs. What is the flow rate in MGD?

$$\text{Flow, MGD} = \frac{3.89 \text{ cfs}}{1.55 \text{ cfs/MG}} = 2.5 \text{ MGD}$$

Example 6

Problem

The water in a tank weighs 675 pounds. How many gallons does it hold?

Solution

Water weighs 8.34 lbs/gal. Therefore,

$$\frac{675 \text{ lb}}{8.34 \text{ lb/gal}} = 80.9 \text{ gallons}$$

Example 7

Problem

A liquid chemical weighs 62-lb/cu ft. How much does a five-gallon can of it weigh?

Solution

Solve for specific gravity (SG); get lb/gal; multiply by 5.

$$\text{Specific gravity} = \frac{\text{wt. channel}}{\text{wt. water}}$$

$$\frac{62 \text{ lb/cu ft}}{62.4 \text{ lb/cu ft}} = 0.99$$

$$\text{Specific gravity} = \frac{\text{wt. chemical}}{\text{wt. water}}$$

$$99 = \frac{\text{wt. chemical}}{8.34 \text{ lb/gal}}$$

$$8.26 \text{ lb/gal} = \text{wt. chemical}$$

$$8.26 \text{ lb/gal} \times 5 \text{ gal} = 41.3 \text{ lb}$$

Example 8

Problem

A wooden piling with a diameter of 16 inches and a length of 16 feet weighs 50-lb/cu ft. If it is inserted vertically into a body of water, what vertical force is required to hold it below the water surface?

Solution

If this piling had the same weight as water, it would rest just barely submerged. Find the difference between its weight and that of the same volume of water. That is the weight needed to keep it down.

$$\frac{62.4 \text{ lb/cu ft (water)} - 50.0 \text{ lb/cu ft (piling)}}{12.4 \text{ lb/cu ft difference}}$$

$$\text{Volume of piling} = 0.785 \times 1.33^2 \times 16 \text{ ft} = 22.21 \text{ cu ft}$$

$$12.4 \text{ lb/cu ft} \times 22.21 \text{ cu ft} = 275.4 \text{ lb (needed to hold it below water surface)}$$

Example 9

Problem

A liquid chemical with a SG of 1.22 is pumped at a rate of 40 gpm. How many pounds per day are being delivered by the pump?

Solution

Solve for pounds pumped per minute; change to lb/day.

$$8.34 \text{ lb/gal water} \times 1.22 \text{ SG liquid chemical} = 10.2\text{lb/gal liquid}$$

$$40 \text{ gal/min} \times 10.2 \text{ lb/gal} = 408 \text{ lb/min}$$

$$408 \text{ lb/min} \times 1440 \text{ min/day} = 587{,}520 \text{ lb/day}.$$

Example 10

Problem

A cinder block weighs 70 pounds in air. When immersed in water, it weighs 40 pounds. What is the volume and SG of the cinder block?

Solution

The cinder block displaces 30 pounds of water; solve for cu ft of water displaced (equivalent to volume of cinder block).

$$\frac{30\text{-lb water displaced}}{62.4 \text{ lb/cu ft}} = 48 \text{ cu ft water displaced}$$

Cinder block volume = .48 cu ft; this weighs 70 lb.

$$\frac{70 \text{ lb}}{48 \text{ cu ft}} = 145.8 \text{ lb/cu ft density of cinder block}$$

$$\text{Specific gravity} = \frac{\text{density of cinder block}}{\text{density of water}}$$

$$= \frac{\text{density of cinder block}}{\text{density of water}} = 2.34$$

TEMPERATURE CONVERSIONS

Two commonly used methods used to make temperature conversions. We have already demonstrated the following method:

$$°C = 5/9 \ (°F - 32)$$

$$°F = 9/5 \ (°C) + 32$$

Example 11

Problem

At a temperature of 4°C, water is at its greatest density. What is the degree of Fahrenheit?
Solution

$$°F = (°C) = 9/5 + 32$$

$$= 4 \times 9/5 + 32$$

$$= 7.2 + 32$$

$$= 39.2$$

However, the difficulty arises when one tries to recall these formulas from memory. Probably the easiest way to recall these important formulae is to remember three basic steps for both Fahrenheit and Celsius conversions:

1. Add 40°
2. Multiply by the appropriate fraction (5/9 or 9/5)
3. Subtract 40°

Obviously, the only variable in this method is the choice of 5/9 or 9/5 in the multiplication step. To make the proper choice, you must be familiar with the two scales. The freezing point of water is 32° on the Fahrenheit scale and 0° on the Celsius scale. The boiling point of water is 212° on the Fahrenheit scale and 100° on the Celsius scale.

Key point: Note, for example, that at the same temperature, higher numbers are associated with the Fahrenheit scale and lower numbers with the Celsius scale. This important relationship helps you decide whether to multiply by 5/9 or 9/5.

Now look at a few conversion problems to see how the three-step process works.

Example 12

Suppose that we wish to convert 240°F to Celsius. Using the three-step process, we proceed as follows:

1. Step 1: add 40°

$$240° + 40° = 280°$$

2. Step 2: 280° must be multiplied by either 5/9 or 9/5. Because the conversion is to the Celsius scale, we will be moving to a number *smaller* than 280. Through reason and observation, obviously, if 280 were multiplied

by 9/5, the result would be almost the same as multiplying by 2, which would double 280 rather than make it smaller. If we multiply by 5/9, the result will be about the same as multiplying by ½, which would cut 280 in half. Because in this problem, we wish to move to a smaller number, we should multiply by 5/9:

$$(5/9)(280°)=156.0°C$$

3. Step 3: Now subtract 40°

$$156.0°C - 40.0°C = 116.0°C$$

Therefore, $240°F = 116.0°C$

Example 13

Convert 22°C to Fahrenheit.

1. Step 1: add 40°

$$22° + 40° = 62°$$

1. Because we are converting from Celsius to Fahrenheit, we are moving from a smaller to a larger number, and 9/5 should be used in the multiplications:

2. Step 2:

$$(9/5)(62°) = 112°$$

3. Step 3: Subtract 40

$$112° - 40° = 72°$$

Thus, $22°C = 72°F$

Obviously, knowing how to make these temperature conversion calculations is useful. However, in practical *in situ* or non-*in situ* operations, you may wish to use a temperature conversion table.

OTHER PERTINENT CONVERSION FACTORS

Tables 6 and 7 illustrate the conversion for various volumes to attain one part per million and also to illustrate conversion for parts per million (ppm) in proportion and percent.

TABLE 6
Conversion for Various Volumes to Attain One Part per Million

Amount Active Ingredient	Unit of Volume	Parts per Million
2.71 pounds	Acre-foot	1
1.235 g	Acre-foot	1
1.24 kg	Acre-foot	1
0.0283 g	Cubic foot	1
1 mg	Liter	1
8.34 pounds	Million gallons	1
1-g	Cubic meter	1
0.0038 g	Gallon	1
3.8 g	Thousand gallons	1

TABLE 7
Conversion for Parts per Million (PPM) in Proportion and Percent

Parts per Million	Proportion	Percent
0.1	1:10,000,000	0.00001
0.5	1:2,000,000	0.00005
1.0	1:1,000,000	0.0001
2.0	1:500,000	0.0002
3.0	1:333,333	0.0003
5.0	1:200,000	0.0005
7.0	1:142,857	0.0007
10.0	1:100,000	0.001
15.0	1:66,667	0.0015
25.0	1:40,000	0.0025
50.0	1:20,000	0.005
100.0	1:10,000	0.01
200.0	1:5000	0.02
250.0	1:4000	0.025
500.0	1:2000	0.05
1550.0	1:645	0.155
5000.0	1:200	0.5
10,000.0	1:100	1.0

Note: The expression ppm is without dimensions, that is, no units of weight or volume are specifically designed. Using the format of other units, the expression may be written as follows:

$$\frac{\text{parts}}{\text{million parts}}$$

"Parts" are not defined. If cubic centimeters replace parts, we obtain

$$\frac{\text{cubic centimeters}}{\text{million cubic centimeters}}$$

Similarly, we might write pounds per million pounds, tons per million tons, or liters per million liters. In each expression, identical units of weight of volume appear in both the numerator and denominator and may be canceled out, leaving a dimensionless term. An analog of ppm is the more familiar term "percent." Percent can be written as

$$\frac{\text{parts}}{\text{hundred parts}}$$

Note: Because of water's tendency to move slowly in groundwaters and fast in waterfalls, rivers, and some currents, it is important to include velocity conversions (Tables 8 and 9).

TABLE 8
Velocity

To Convert from	To	Multiply by
Meters/sec	Kilometers/hr	3.6
	Feet/sec	3.281
	Miles/hr	2.237
Kilometers/hr	Meters/sec	0.2778
	Feet/sec	0.9113
	Miles/hr	0.6241
Feet/hr	Meters/sec	0.3048
	Kilometers/hr	1.0973
	Miles/hr	0.6818
Miles/hr	Meters/sec	0.4470
	Kilometers/hr	1.6093
	Feet/sec	1.4667

Because in land subsidence, we are dealing with movement of soil.

TABLE 9
Concentration

To Convert from	To	Multiply by
Grams/cu m	Milligrams/cu m	1000.0
	Grams/cu ft	0.02832
	Micrograms/cu ft	1.0×10^6
	Pounds/1000 cu ft	0.06243
Micrograms/cu m	Milligrams/cu m	0.001
	Grams/cu ft	28.43×10^{-9}
	Grams/cu m	1.0×10^{-6}
	Micrograms/cu ft	0.02832
	Pounds/1000 cu ft	62.43×10^{-9}
Micrograms/cu ft	Milligrams/cu m	35.314×10^{-3}
	Grams/cu ft	1.0×10^{-6}
	Grams/cu m	35.314×10^{-6}
	Micrograms	35.314
	Pounds/1000 cu ft	2.2046×10^{-6}
Pounds/1000 cu ft	Milligrams/cu m	16.018×10^3
	Grams/cu ft	0.35314
	Micrograms/cu m	16.018×10^6
	Grams/cu m	16.018
	Micrograms/cu ft	353.14×10^2

SOIL TEST RESULTS CONVERSION FACTORS

Again, in this book, we are dealing with the movement of soils; thus, soil test results can be important and converted from ppm to pounds per acre by multiplying ppm by a conversion factor based on the depth to which the soil was sampled. Because a slice of soil 1 acre in area and 3 inches deep weighs approximately 1 million pounds, the conversion factors given in Table 10 can be used.

TABLE 10
Soil Test Conversion Factors

Soil Sample Depth Inches	Multiply ppm by
3	1
6	2
7	2.33
8	2.66
9	3
10	3.33
12	4

STANDARD CONVERSIONS FOR MANUAL CALCULATIONS

Volume	Weight	Length
1 gal. = 3.78 L	1 lb. = 453 g or 0.453 kg	1 inch = 2.54 cm
1 L = 0.26 gal	1 kg = 2.2 lbs	1 cm = 0.39 in.
1 tsp. = 5	3.28 ft = 1 m	

Acronyms and Abbreviations

°C	Degrees Celsius
μg/L	Micrograms per liter
AB	Army base
AFT	Alternate filtration technology
ALCR	Air-liquid conversion ratio
AOP	Advanced oxidation process
ASTM	American Society for Testing and Materials
AWT	Advance water treatment
AWTP	Advanced water treatment plant
BAC	Biologically active granular activated carbon
BH	Boat Harbor
BNR	Biological nitrogen removal
BOD	Biochemical oxygen demand
BOD$_5$	5-day biochemical oxygen demand
BV	Bed volume
BVs	Bed volumes
CaCO$_3$	Calcium carbonate
CA	Cellulose acetate
CAS	Conventional activated sludge
CBOD	Carbonaceous biochemical oxygen demand
CEC	Contaminant of emerging concern
CFU	Coliform forming unit
Cl$_2$	Chlorine disinfection
COD	Chemical oxygen demand
CPES	Cost estimating program
CWA	Clean Water Act
DBP	Disinfection by-product
DO	dissolved oxygen
DOC	dissolved organic carbon
DPR	direct potable reuse
EDC	endocrine disrupting compounds
EPA	US Environmental Protection Agency
FBG	Feet below grade
FCV	flow control valve
FLOC	Flocculation
GAC	Granular activated carbon
GMF	Granular media filtration
GPM/SF	Gallons per minute per square foot
H$_2$O$_2$	Hydrogen peroxide
HFF	Hollow fine fiber
HP	Horsepower

HPC	Heterotrophic plate count
HRSD	Hampton Roads Sanitation District
IMS	Integrated membrane system
IPR	Indirect potable reuse
JR	James River
kWh	Kilowatt hour
LRV	Log reduction value
MBR	Membrane bioreactor
MCC	Motor Control Center
MCL	Maximum contaminant level
MCR	Membrane cartridge filtration
MF	Microfiltration
mg/L	Milligrams per liter
MGD	Million gallons per day
MJ/CM²	Milli-joules per centimeter squared
mL	Milliliter
N	N
N/A	Not applicable
ND	No data
NDMA	N-nitrosodimethylamine
NF	Nanofiltration
NG/L	Nanogram per liter
NH$_3$	Ammonia
NH$_3$-N	Ammonia nitrogen
Nit/Denit	Nitrification/denitrification
NOM	Natural organic matter (humic and fulvic acids)
NOx-N	Nitrate/nitrite-nitrogen
NP	Nansemond
NPDES	National Pollutant Discharge Elimination System
NTU	Nephelometric turbidity unit
O&M	Operations and maintenance
PAS	Potomac Aquifer System
PCV	Pressure control valve
PPCP	Pharmaceuticals and personal care products
PSI	Pounds per square inch
RO	Reverse osmosis
SAT	Soil aquifer treatment
SDI	Silt density index
SDWA	Safe Drinking Water Act
SED	Sedimentation
SWTR	Surface Water Treatment Rule
TCEP	Tris (2-carboxyethyl)phosphine
TCEQ	Texas Commission on Environmental Quality
TDS	Total dissolved solids
THM	Trihalomethanes
TKN	Total Kjeldahl nitrogen

TMDL	Total maximum daily load
TN	Total nitrogen
TOC	Total organic carbon
TOX	Total organic halides
TP	Total phosphorus
TSS	Total suspended solids
TTHM	Total trihalomethanes
UF	Ultrafiltration
UOSA	Upper Occoquan Service Authority
USDA	US Department of Agriculture
UV	Ultraviolet
UVAOP	Ultraviolet advanced oxidation process
VIP	Virginia Initiative Plant
VFD	Variable frequency drive
WB	Williamsburg
WWTP	Wastewater treatment plant
YR	York River

VERTICAL DATUM

In this book, "sea level" refers to the National Geodetic Vertical Datum of 1929—a geodetic datum derived from a general adjustment of the first-order level nets of both the United States and Canada, formerly called "Sea Level datum of 1929." "Mean sea level" is not used with reference to any particular vertical datum; where used, the phrase means the average surface of the ocean as determined by calibration of measurements at tidal stations.

RELATIVE SEA LEVEL

In this book, "relative sea level" is defined as the sea level that is observed with respect to a land-based reference frame. The important point is (at least it is in the book) relative sea level can change by changes on land such as subsidence.

Glossary

A

Absorption: any process by which one substance penetrates the interior of another substance.

Acid: has a pH of water less than 5.5; pH modifier used in the US Fish and Wildlife Service wetland classification system; in common usage, acidic water has a pH less than 7.

Acidic deposition: the transfer of acidic or acidifying substances from the atmosphere to the surface of the Earth or to objects on its surface. Transfer can be either by wet deposition processes (rain, snow, dew, fog, frost, or hail) or by dry deposition (gases, aerosols, or fine to coarse particles).

Acid rain: precipitation with higher-than-normal acidity, caused primarily by sulfur and nitrogen dioxide air pollution.

Acre-foot (acre-ft.): the volume of water needed to cover an acre of land to a depth of 1 foot; equivalent to 43,560 cubic feet or 325,851.4 gallons.

Activated carbon: a very porous material that after being subjected to intense heat to drive off impurities can be used to adsorb pollutants from water.

Adsorption: the process by which one substance is attracted to and adheres to the surface of another substance, without actually penetrating its internal structure.

Aeration: a physical treatment method that promotes biological degradation of organic matter. The process may be passive (when waste is exposed to air) or active (when a mixing or bubbling device introduces the air).

Aerobic bacteria: a type of bacteria that require free oxygen to carry out metabolic function.

Algae: chlorophyll-bearing nonvascular, primarily aquatic species that have no true roots, stems, or leaves; most algae are microscopic, but some species can be as large as vascular plants.

Algal bloom: the rapid proliferation of passively floating, simple plant life, such as blue–green algae, in and on a body of water.

Alkaline: has a pH greater than 7; pH modifier in the US Fish and Wildlife Service wetland classification system; in common usage, a pH of water greater than 7.4.

Alluvial aquifer: a water-bearing deposit of unconsolidated material (sand and gravel) left behind by a river or other flowing water.

Alluvium: general term for sediments of gravel, sand, silt clay, or other particulate rock material deposited by flowing water, usually in the beds of rivers and streams, on a flood plain, on a delta, or at the base of a mountain.

Alpine snow glade: a marshy clearing between slopes above the timberline in mountains.

Amalgamation: the dissolving or blending of a metal (commonly gold and silver) in mercury to separate it from its parent material.

Ammonia: a compound of nitrogen and hydrogen (NH_3) that is a common by-product of animal waste. Ammonia readily converts to nitrate in soils and streams.

Anaerobic: pertaining to, taking place in, or caused by the absence of oxygen.

Anomalies: as related to fish, externally visible skin or subcutaneous disorders, including deformities, eroded fins, lesions, and tumors.

Anthropogenic: having to do with or caused by humans.

Anticline: a fold in the Earth's crust, convex upward, whose core contains stratigraphically older rocks.

Aquaculture: the science of farming organisms that live in water, such as fish, shellfish, and algae.

Aquatic: living or growing in or on water.

Aquatic guidelines: specific levels of water quality which, if reached, may adversely affect aquatic life. These are non-enforceable guidelines issued by a governmental agency or other institution.

Aquifer: a geologic formation, group of formations, or part of a formation that contains sufficient saturated permeable material to yield significant quantities of water to springs and wells.

Aquitard: A saturated, but poorly permeable, geologic unit that impedes groundwater movement and does not yield water freely to wells, but which may transmit appreciable water to and from adjacent aquifers and, where sufficiently thick, may constitute an important groundwater storage unit. Really extensive aquitards may function regionally as confined units within aquifer systems.

Arroyo: a small, deep, flat-floored channel or gully of an ephemeral or intermittent stream, usually with nearly vertical banks cut, into unconsolidated material.

Artesian: an adjective referring to confined aquifers. Sometimes the rem artesian is used to denote a portion of a confined aquifer where the altitude of the potentiometric surface is above the land surface (flowing wells and artesian wells are synonymous in this usage). But more generally, the term indicates that the altitudes of the potentiometric surface are above the altitude of the base or the confining unit (artesian wells and flowing wells are not synonymous in this case).

Artificial recharge: augmentation of natural replenishment of groundwater storage by some method of construction, spreading of water, or by pumping water directly into an aquifer.

Atmospheric deposition: the transfer of substances from the air to the surface of the Earth, either in a wet form (rain, fog, snow, dew, frost, hail) or in a dry form (gases, aerosols, particles).

Atmospheric pressure: the pressure exerted by the atmosphere on any surface beneath or within it, which is equal to 14.7 pounds per square inch at sea level.

Average discharge: as used by the US Geological Survey, the arithmetic average of all complete water years of record of surface water discharge whether consecutive or not. The term "average" generally is reserved for average of record, and "mean" is used for averages of shorter periods, namely, daily, monthly, or annual mean discharges.

B

Background concentration: a concentration of a substance in a particular environment that is indicative of minimal influence by human (anthropogenic) sources.

Backwater: a body of water in which the flow is slowed or turned back by an obstruction such as a bridge or dam, an opposing current, or the movement of the tide.

Bacteria: single-celled microscopic organisms.

Bank: the sloping ground that borders a stream and confines the water in the natural channel when the water level, or flow, is normal.

Bank storage: the change in the amount of water stored in an aquifer adjacent to a surface–water body resulting from a change in the stage of the surface-water body.

Barrier bar: an elongate offshore ridge, submerged at least at high tide, built up by the action of waves or currents.

Base flow: the sustained low flow of a stream, usually groundwater inflow to the stream channel.

Basic: the opposite of acidic; water that has a pH greater than 7.

Basin and range physiography: a region characterized by a series of generally north-trending mountain ranges separated by alluvial valleys.

Bed material: sediment comprising the streambed.

Bed sediment: the material that temporarily s stationary in the bottom of a stream or other watercourse.

Bedload: sediment that moves on or near the streambed and is in almost continuous contact with the bed.

Bedrock: a general term used for solid rock that underlies soils or other unconsolidated material.

Benthic invertebrates: insects, mollusk, crustaceans, worms, and other organisms without a backbone that live in, on, or near the bottom of lakes, streams, or oceans.

Benthic organism: a form of aquatic life that lives on or near the bottom of streams, lakes, or oceans.

Bioaccumulation: the biological sequestering of a substance at higher concentrations than concentrations at which it occurs in the surrounding environment or medium. Also, the process whereby a substance enters organisms through the gills, epithelial tissues, dietary, or other sources.

Bioavailability: the capacity of a chemical constituent to be taken up by living organisms either through physical contact or by ingestion.

Biochemical: chemical processes that occur inside or are mediated by living organisms.

Biochemical process: a process characterized by, produced by, or involving chemical reactions in living organism.

Biochemical oxygen demand: the amount of oxygen required by bacteria to stabilize decomposable organic matter under aerobic conditions.

Biodegradation: transformation of a substance into new compounds through biochemical reactions or the actions of microorganisms such as bacteria.

Biological treatment: a process that uses living organisms to bring about chemical changes.

Biomass: the amount of living matter, in the form of organisms, present in a particular habitat, usually expressed as weight per unit area.

Biota: all living organisms of an area.

Blue hole: a subsurface void, usually a solution sinkhole, developed in carbonate rocks that are open to the Earth's surface and contains tidally influenced waters of fresh, marine, or mixed chemistry.

Bog: a nutrient-poor, acidic wetland dominated by a waterlogged, spongy mat of sphagnum moss that ultimately forms a thick layer of acidic peat, generally has no inflow or outflow, and fed primarily by rainwater.

Brackish water: water with a salinity intermediate between seawater and freshwater (containing from 1000 to 10,000 mg/L of dissolved solids).

Breakdown product: a compound derived by chemical, biological, or physical action upon a pesticide. The breakdown is a natural process that may result in a more toxic or a less toxic compound and a more persistent or less persistent compound.

Breakpoint chlorination: the addition of chlorine to water until the chlorine demand has been satisfied and free chlorine residual is available for disinfection.

C

Calcareous: a rocks or substance formed of calcium carbonate or magnesium carbonate by biological deposition or inorganic precipitation or containing those minerals in sufficient quantities to effervesce when treated with cold hydrochloric acid.

Capillary fringe: the zone above the water table in which water is held by surface tension. Water in the capillary fringe is under a pressure less than atmospheric.

Carbonate rocks: rocks (such as limestone or dolostone) that are composed primarily of minerals (such as calcite and dolomite) containing a carbonate ion.

Cenote: steep-walled natural well that extends below the water table; generally caused by collapse of a cave roof; term reserved for features found in the Yucatan Peninsula of Mexico.

Center pivot irrigation: an automated sprinkler system involving a rotating pipe or boom that supplies water to a circular area of an agricultural field through sprinkler heads or nozzles.

Channel scour: erosion by flowing water and sediment on a stream channel; results in removal of mud, silt, and sand on the outside curve of a stream bend and the bed material of a stream channel.

Channelization: the straightening and deepening of a stream channel to permit the water to move faster or to drain a wet area for farming.

Chemical treatment: a process that results in the formation of a new substance or substances. The most common chemical water treatment processes include coagulation, disinfection, water softening, and filtration.

Chlordane: octachlor-4,7-methanotetrahydroindane. An organochlorine insecticide no longer registered for use in the US Technical chlordane is a mixture in which the primary components are cis- and trans-chlordane, cis- and trans-nonachlor, and heptachlor.

Chlorinated solvent: a volatile organic compound containing chlorine. Some common solvents are trichloroethylene, tetrachloroethylene, and carbon tetrachloride.

Chlorofluorocarbons: a class of volatile compounds consisting of carbon, chlorine, and fluorine. Commonly called freons, which have been in refrigeration mechanisms, as blowing agents in the fabrication of flexible and rigid foams, and, until banned from use several years ago, as propellants in spray cans.

Chlorination: the process of adding chlorine to water to kill disease-causing organisms or to act as an oxidizing agent.

Chlorine demand: a measure of the amount of chlorine that will combine with impurities and is therefore unavailable to act as a disinfectant.

Cienaga: a marshy area where the ground is wet due to the presence of seepage from springs.

Clean Water Act (CWA): federal law dating to 1972 (with several amendments) with the objective to restore and maintain the chemical, physical, and biological integrity of the nation's waters. Its long-range goal is to eliminate the discharge of pollutants into navigable waters and to make national waters fishable and swimmable.

Climate: the sum total of the meteorological elements that characterize the average and extreme conditions of the atmosphere over a long period of time at any one place or region of the Earth's surface.

Coagulants: chemicals that cause small particles to stick together to form larger particles.

Coagulation: a chemical water treatment method that causes very small, suspended particles to attract one another and form larger particles. This is accomplished by the addition of a coagulant that neutralizes the electrostatic charges that cause particles to repel each other.

Coliform bacteria: a group of bacteria predominantly inhabiting the intestines of humans or animals, but also occasionally found elsewhere. The presence of the bacteria in water is used as an indication of fecal contamination (contamination by animal or human wastes).

Color: a physical characteristic of water. Color is most commonly tan or brown from oxidized iron, but contaminants may cause other colors, such as green or blue. Color differs from turbidity, which is water's cloudiness.

Combined sewer overflow: a discharge of untreated sewage and stormwater to a stream when the capacity of a combined storm/sanitary sewer system is exceeded by storm runoff.

Communicable diseases: usually caused by *microbes*—microscopic organisms including bacteria, protozoa, and viruses. Most microbes are essential components of our environment and do not cause disease. Those that do are called pathogenic organisms, or simply *pathogens*.

Community: in ecology, the species that interact in a common area.

Community water system: a public water system that serves at least 15 service connections used by year-round residents, or regularly serves at least 25 year-round residents.

Compaction: in this book, compaction is used in its geologic sense and refers to the inelastic compression of the aquifer system. Compaction of the aquifer system reflects the rearrangement of the mineral grain pore structure and largely nonrecoverable reduction of the porosity under stress greater than the preconsolidation stress. Compaction, as used here, is synonymous with the term "virgin consolidation" used by soils engineers. The term refers to both the process and the measured change in thickness. As a practical matter, a very small amount (1%–5%) of the compaction is recoverable as a slight elastic rebound of the compacted material if stresses are reduced.

Compaction, residual: is compaction that would ultimately occur if a given increase in applied stress were maintained until steady-state pore pressures were achieved. Residual compaction may also be defined as the difference between (1) the amount of compaction that will occur ultimately for a given increase in applied stress and (2) that which has occurred at a specified time.

Composite sample: a series of individual or grab samples taken at different times from the same sampling point and mixed together.

Compression: in this book, compression refers to the decrease in thickness of sediments, as a result of increase in vertical compressive stress. Compression may be elastic (fully recoverable) or inelastic (nonrecoverable).

Concentration: the ratio of the quantity of any substance present in a sample of a given volume or a given weight compared to the volume or weight of the sample.

Cone of depression: the depression of heads around a pumping well caused by withdrawal of water.

Confined aquifer (artesian aquifer): an aquifer that is completely filled with water under pressure and that is overlain by material that restricts the movement of water.

Confining bed: a layer of rock having very low hydraulic conductivity that hampers the movement of water into and out of an aquifer.

Confining layer: a body of impermeable or distinctly less permeable material stratigraphically adjacent to one or more aquifers that restricts the movement of water into and out of the aquifers.

Confining unit: a saturated, relatively low-permeability geologic unit is really extensive and serves to confine an adjacent artesian aquifer of aquifers. Leaky confining units may transmit appreciable water to and from adjacent aquifers.

Confluence: the flowing together of two or more streams; the place where a tributary joins the mainstream.

Conglomerate: a coarse-grained sedimentary rock composed of fragments larger than 2 mm in diameter.

Consolidation: in soil mechanics, consolidation is the adjustment of a saturated soil in response to increased load, involving the squeezing of water from the pores and decrease in the void ratio or porosity of the soul. In this book, the geologic term "compaction" is used in preference to consolidation.

Constituent: a chemical or biological substance in water, sediment, or biota that can be measured by an analytical method.

Consumptive use: the quantity of water that is not available for immediate rescue because it has been evaporated, transpired, or incorporated into products, plant tissue, or animal tissue.

Contact recreation: recreational activities, such as swimming and kayaking, in which contact with water is prolonged or intimate and in which there is a likelihood of ingesting water.

Contaminant: a toxic material found as an unwanted residue in or on a substance.

Contamination: degradation of water quality compared to original or natural conditions due to human activity.

Contributing area: the area in a drainage basin that contributes water to streamflow or recharge to an aquifer.

Core sample: a sample of rock, soil, or other material obtained by driving a hollow tube into the undisturbed medium and withdrawing it with its contained sample.

Criterion: a standard rule or test on which a judgment or decision can be based.

Cross-connection: any connection between safe drinking water and a non-potable water or fluid.

CxT value: the product of the residual disinfectant concentration **C,** in milligrams per liter, and the corresponding disinfectant contact time **T,** in minutes. Minimum **CxT** values are specified by the Surface Water Treatment Rule, as a means of ensuring adequate kill or inactivation of pathogenic microorganisms in water.

D

Datum plane: a horizontal plane to which ground elevations or water surface elevations are referenced.

Deepwater habitat: permanently flooded lands lying below the deepwater boundary of wetlands.

Degradation products: compounds resulting from transformation of an organic substance through chemical, photochemical, and/or biochemical reactions.

Denitrification: a process by which oxidized forms of nitrogen such as nitrate are reduced to form nitrites, nitrogen oxides, ammonia, or fee nitrogen commonly brought about by the action of denitrifying bacteria and usually resulting in the escape of nitrogen to the air.

Detection limit: the concentration of a constituent or analyte below which a particular analytical method cannot determine, with a high degree of certainty, the concentration.

Diatoms: single-celled, colonial, or filamentous algae with siliceous cell walls constructed of two overlapping parts.

Direct runoff: the runoff entering stream channels promptly after rainfall or snowmelt.

Discharge: the volume of fluid passing a point per unit of time, commonly expressed in cubic feet per second, million gallons per day, gallons per minute, or seconds per minute per day.

Discharge area (groundwater): area where subsurface water is discharged to the land surface, to surface water, or to the atmosphere.

Disinfectants-disinfection by-products: a term used in connection with state and federal regulations designed to protect public health by limiting the concentration of either disinfectants or the by-products formed by the reaction of disinfectants with other substances in the water (such as trihalomethanes—THMs).

Disinfection: a chemical treatment method. The addition of a substance (e.g., chlorine, ozone, or hydrogen peroxide), which destroys or inactivates harmful microorganisms, or inhibits their activity.

Dispersion: the extent to which a liquid substance introduced into a groundwater system spreads as it moves through the system.

Dissociate: the process of ion separation that occurs when an ionic solid is dissolved in water.

Dissolved constituent: operationally defined as a constituent that passes through a 0.45-μm filter.

Dissolved oxygen: the oxygen dissolved in water usually expressed in milligrams per liter, parts per million, or percent of saturation.

Dissolved solids: any material that can dissolve in water and be recovered by evaporating the water after filtering the suspended material.

Diversion: a turning aside or alteration of the natural course of a flow of water, normally considered physically to leave the natural channel. In some States, this can be a consumptive use direct from another stream, such as by livestock watering. In other States, a diversion must consist of such actions as taking water through a canal, pipe, or conduit.

Dolomite: a sedimentary rock consisting chiefly of magnesium carbonate.

Domestic withdrawals: water used for normal household purposes, such as drinking, food preparation, bathing, washing clothes and dishes, flushing toilets, and watering lawns and gardens. The water may be obtained from a public supplier or may be self-supplied. Also called residential water use.

Drainage area: the drainage area of a stream at a specified location is that area, measured in a horizontal plane, which is enclosed by a drainage divide.

Drainage basin: the land area drained by a river or stream.

Drainage divide: boundary between adjoining drainage basins.

Drawdown: the difference between the water level in a well before pumping and the water level in the well during pumping. Also, for flowing wells, the reduction of the pressure head as a result of the discharge of water.

Drinking water standards: water quality standards measured in terms of suspended solids, unpleasant taste, and microbes harmful to human health. Drinking water standards are included in state water quality rules.

Drinking water supply: any raw or finished water source that is or may be used as a public water system or as drinking water by one or more individuals.

Drip irrigation: an irrigation system in which water is applied directly to the root zone of plants by means of applicators (orifices, emitters, porous tubing, or perforate pipe) operated under low pressure. The applicators can be placed on or below the surface of the ground or can be suspended from supports.

Drought: a prolonged period of less-than-normal precipitation such that the lack of water causes a serious hydrologic imbalance.

E

Ecoregion: an area of similar climate, landform, soil, potential natural vegetation, hydrology, or other ecologically relevant variables.

Ecosystem: a community of organisms considered together with the nonliving factors of its environment.

Effluent: outflow from a particular source, such as a stream that flows from a lake or liquid waste that flows form a factory or sewage treatment plant.

Effluent limitations: standards developed by the EPA to define the levels of pollutants that could be discharged into surface waters.

Electrodialysis: the process of separating substances in a solution by dialysis, using an electric field as the driving force.

Electronegativity: the tendency for atoms that do not have a complete octet of electrons in their outer shell to become negatively charged.

Ellipsoid, Earth: a mathematically determined three-dimensional surface obtained by rotating an ellipse about its semi-minor axis. In the case of the Earth, the ellipsoid is the modeled shape of its surface, which is relatively flattened in the polar axis.

Ellipsoid, height: the distance of a point above the ellipsoid measured perpendicular to the surface of the ellipsoid.

Emergent plants: erect, rooted, herbaceous plants that may be temporarily or permanently flooded at the base but do not tolerate prolonged inundation of the entire plant.

Enhanced Surface Water Treatment Rule: a revision of the original Surface Water Treatment Rule that includes new technology and requirements to deal with newly identified problems.

Environment: the sum of all conditions and influences affecting the life of organisms.

Environmental sample: a water sample collected from an aquifer or stream for the purpose of chemical, physical, or biological characterization of the sampled resource.

Environmental setting: land area characterized by a unique combination of natural and human-related factors, such as row-crop cultivation or glacial-till soils.

Ephemeral stream: a stream or part of a stream that flows only in direct response to precipitation; it receives little or no water from springs, melting snow, or other sources; its channel is at all times above the water table.

EPT richness index: an index based on the sum of the number of taxa in three insect orders, Ephemeroptera (mayflies), Plecoptera (stoneflies), and Trichoptera (caddisflies), that are composed primarily of species considered to be relatively intolerant to environmental alterations.

Equipotential line: a line on a map or cross-section along which total heads are the
    same.

Erosion: the process whereby materials of the Earth's crust are loosened, dissolved,
    or worn away and simultaneously moved from one place to another.

Eutrophication: the process by which water becomes enriched with plant nutrients,
    most commonly phosphorus and nitrogen.

Evaporite minerals (deposits): minerals or deposits of minerals formed by evapora-
    tion of water containing salts. These deposits are common in arid climates.

Evaporites: a class of sedimentary rocks composed primarily of minerals precipi-
    tated from a saline solution as a result of extensive or total evaporation of
    water.

Evapotranspiration: the process by which water is discharged to the atmosphere as
    a result of evaporation from the soil and surface–water bodies and transpira-
    tion by plants.

Exfoliation: the process by which concentric scales, plates, or shells or rock, from
    less than centimeter to several meters in thickness, are stripped from the
    bare surface of a large rock mass.

F

Facultative bacteria: a type of anaerobic bacteria that can metabolize its food either
    aerobically or anaerobically.

Fall line: imaginary line marking the boundary between the ancient, resistant crys-
    talline rocks of the Piedmont province of the Appalachian Mountains, and
    the younger, softer sediments of the Atlantic Coastal Plain province in the
    Eastern United States. Along rivers, this line commonly is reflected by
    waterfalls.

Fecal bacteria: microscopic single-celled organisms (primarily fecal coliforms and
    fecal streptococci) found in the wastes of warm-blooded animals. Their
    presence in water is used to assess the sanitary quality of water for body
    contact recreation or for consumption. Their presence indicates contamina-
    tion by the wastes of warm-blooded animals and the possible present of
    pathogenic (disease producing) organisms.

Federal Water Pollution Control Act (1972): The Act outlines the objective "to
    restore and maintain the chemical, physical, and biological integrity of the
    nation's waters." This 1972 act and subsequent Clean Water Act amend-
    ments are the most far-reaching water pollution control legislation ever
    enacted. They provided comprehensive programs for water pollution con-
    trol, uniform laws, and interstate cooperation. They provided grants for
    research, investigations, training, and information on national programs on
    surveillance, the effects of pollutants, pollution control, and the identifi-
    cation and measurement of pollutants. Additionally, they allot grants and
    loans for the construction of treatment works. The Act established national
    discharge standards with enforcement provisions.

The Federal Water Pollution Control Act established several milestone achievement
    dates. It required secondary treatment of domestic waste by publicly owned

treatment works (POTWs), and application of "best practicable" water pollution control technology by industry by 1977. Virtually all industrial sources have achieved compliance (because of economic difficulties and cumbersome federal requirements, certain POTWs obtained an extension to July 1, 1988, for compliance). The Act also called for new levels of technology to be imposed during the 1980s and 1990s, particularly for controlling toxic pollutants.

The Act mandates a strong pretreatment program to control toxic pollutants discharged by industry into POTWs. The 1987 amendments require that stormwater from industrial activity must be regulated.

Fertilizer: any of a large number of natural or synthetic materials, including manure and nitrogen, phosphorus, and potassium compound, spread on or worked into soil to increase its fertility.

Filtrate: liquid that has been passed through a filter.

Filtration: a physical treatment method for removing solid (particulate) matter from water by passing the water through porous media such as sand or a man-made filter.

Flocculation: the water treatment process following coagulation, it uses gentle stirring to bring suspended particles together so that they will form larger, more settleable clumps called floc.

Flood: any relatively high streamflow that overflows the natural or artificial banks of a stream.

Flood attenuation: a weakening or reduction in the force or intensity of a flood.

Flood irrigation: the application of irrigation water whereby the entire surface of the soil is covered by pond water.

Flood plain: a strip of relatively flat land bordering a stream channel that is inundated at times of high water.

Flow line: the idealized path followed by particles of water.

Flow net: The grid pattern formed by a network of flow lines and equipotential lines.

Flowpath: an underground route for groundwater movement, extending from a recharge (intake) zone to a discharge (output) zone such as a shallow stream.

Fluvial: pertaining to a river or stream.

Freshwater: water that contains less than 1000 mg/L of dissolved solids.

Freshwater chronic criteria: the highest concentration of a contaminant that freshwater aquatic organisms can be exposed to for an extended period of time (4 days) without adverse effects.

G

Geodetic datum: a set of constants specifying the coordinate system used for geodetic control, for example, for calculating the coordinates of points on the Earth.

Geoid, Earth: The sea level equipotential surface or figure of the Earth. If the earth were completely covered by a shallow sea, the surface of this sea would conform to the geoid shaped by the hydrodynamic equilibrium of the water subject to gravitational and rotational forces. Mountains and valleys are departures from the reference geoid.

Grab sample: a single water sample collected at one time from a single point.

Groundwater: the fresh water found under the Earth's surface, usually in aquifers. Groundwater is a major source of drinking water and a source of a growing concern in areas where leaching agricultural or industrial pollutants, or substances from leaking underground storage tanks are contaminating groundwater.

H

Habitat: the part of the physical environment in which a plant or animal lives.

Hardness: a characteristic of water caused primarily by the salts of calcium and magnesium. It causes deposition of scale in boilers, damage in some industrial processes, and sometimes objectionable taste. It may also decrease soup's effectiveness.

Head: the height above a datum plane of a column of water. In a groundwater system, it is composed of elevation head and pressure head.

Headwaters: the source and upper part of a stream.

Hydraulic conductivity: the capacity of a rock to transmit water. It is expressed as the volume of water at the existing kinematic viscosity that will move in unit time under a unit hydraulic gradient through a unit area measured at right angles to the direction of flow.

Hydraulic gradient: the change of hydraulic head per unit of distance in a given direction.

Hydrocompaction: the process of volume decrease and density increase that occurs when certain moisture-deficient deposits compact as they are wetted for the first time since burial. The vertical downward movement of the land surface that results from this process has also been termed "shallow subsidence" and "near-surface subsidence."

Hydrogen bonding: the term used to describe the weak but effective attraction that occurs between polar covalent molecules.

Hydrograph: graph showing variation of water elevation, velocity, streamflow, or other property of water with respect to time.

Hydrologic cycle: literally the water-earth cycle. The movement of water in all three physical forms through the various environmental mediums (air, water, biota, and soil).

Hydrology: the science that deals with water as it occurs in the atmosphere, on the surface of the ground, and underground.

Hydrostatic pressure: the pressure exerted by the water at any given point in a body of water at rest.

Hygroscopic: a substance that readily absorbs moisture.

I

Impermeability: the incapacity of a rock to transmit a fluid.

Index of Biotic Integrity: an aggregated number, or index, based on several attributes or metrics of a fish community that provides an assessment of biological conditions.

Indicator sites: stream sampling sites located at outlets of drainage basins with relatively homogeneous land use and physiographic conditions; most indicator-site basins have drainage areas ranging from 20 to 200 square miles.

Infiltration: the downward movement of water from the atmosphere into soil or porous rock.

Influent: Water flowing into a reservoir, basin, or treatment plant.

Inorganic: containing no carbon; matter other than plant or animal.

Inorganic chemical: a chemical substance of mineral origin not having carbon in its molecular structure.

Inorganic soil: soil with less than 20% organic matter in the upper 16 inches.

Instantaneous discharge: the volume of water that passes a point at a particular instant of time.

Ionic Bond: the attractive forces between oppositely charged ions—for example, the forces between the sodium and chloride ions in a sodium chloride crystal.

Instream use: water use takes place within the stream channel for such purposes as hydroelectric power generation, navigation, water quality improvement, fish propagation, and recreation. Sometimes it is called non-withdrawal use or in-channel use.

Intermittent stream: a stream that flows only when it receives water from rainfall runoff or springs, or from some surface source such as melting snow.

Internal drainage: surface drainage whereby the water does not reach the ocean, such as drainage toward the lowermost or central part of an interior basin or closed depression.

Intertidal: alternately flooded and exposed by tides.

Intolerant organisms: organisms that are not adaptable to human alterations to the environment and thus decline in numbers where alterations occur.

Invertebrate: an animal having no backbone or spinal column.

Ion: a positively or negatively charged atom or group of atoms.

Irrigation: controlled application of water to arable land to supply requirements of crops not satisfied by rainfall.

Irrigation return flow: the part of irrigation applied to the surface that is not consumed by evapotranspiration or up take by plants and that migrates to an aquifer or surface–water body.

Irrigation withdrawals: withdrawals of water for application on land to assist in the growing of crops and pastures or to maintain recreational lands.

K

Karst: a type of topography that is formed on limestone, dolomite, gypsum, and other rocks, primarily by dissolution, and that is characterized by sinkholes, caves, and subterranean drainage.

Karst mantled: a terrane of karst features, usually subdued, and covered by soil or a thin alluvium.

Karst topography: type of topography that is formed in limestone, gypsum, and other similar type rock by dissolution and is characterized by sinkholes, caves, and rapid underground water movement.

Karstification: action by water, mainly chemical but also mechanical, that produces features of a karst topography.

Kill: Dutch term for stream or creek.

L

Lacustrine: pertaining to, produced by, or formed in a lake.

Leachate: a liquid that has percolated through soil containing soluble substances and that contains certain amounts of these substances in solution.

Leaching: the removal of materials in solution from soil or rock; also refers to movement of pesticides or nutrients form the land surface to groundwater.

Limnetic: the deepwater zone (greater than 2 m deep).

Littoral: the shallow water zone (less than 2 m deep).

Load: material that is moved or carried by streams, reported as weight of material transported during a specified time period, such as tons per year.

M

Main stem: the principal trunk of a river or a stream.

Marsh: a water-saturated, poor drained area, intermittently or permanently water covered, having aquatic and grasslike vegetation.

Maturity (stream): the stage in the development of a stream at which it has reached its maximum efficiency, when velocity is just sufficient to carry the sediment delivered to it by tributaries; characterized by a broad, open, flat-floored valley having a moderate gradient and gentle slope.

Maximum Contaminant Level (MCL): a primary standard, whereas an MCLG is a maximum concentration goal for a drinking water contaminant, which would be desirable bade don human health concerns and assuming all feasibility issues such as cost and technological capacity are not considered. Stated differently, MCL is the maximum allowable concentration of a contaminant in drinking water, as established by state and/or federal regulations. Primary MCLs are health related and mandatory. Secondary MCLs are related to the aesthetics of the water and are highly recommended, but not required.

Mean discharge: the arithmetic means of individual daily mean discharges of a stream during a specific period, usually daily, monthly, or annually.

Membrane filter method: a laboratory method used for coliform testing. The procedure uses an ultra-thin filter with a uniform pore size smaller than bacteria (less than a micron). After water is forced through the filter, the filter is incubated in a special media that promotes the growth of coliform bacteria. Bacterial colonies with a green-gold sheen indicate the presence of coliform bacteria.

Method detection limit: the minimum concentration of a substance that can be accurately identified and measured with current lab technologies.

Midge: a small fly in the family Chironomidae. The larval (juvenile) life stages are aquatic.

Minimum reporting level (MRL): the smallest measured concentration of a constituent that may be reliably reported using a given analytical method. In many cases, the MRL is used when documentation for the method detection limit is not available.

Mitigation: actions taken to avoid, reduce, or compensate for the effects of human-induced environmental damage.

Modes of transmission of disease: the ways in which diseases spread from one person to another.

Monitoring: repeated observation, measurement, or sampling at a site, on a scheduled or event basis, for a particular purpose.

Monitoring well: a well-designed for measuring water levels and testing groundwater quality.

Multiple-tube fermentation method: a laboratory method used for coliform testing, which uses a nutrient broth placed in a culture tubes. Gas production indicates the presence of coliform bacteria.

N

National Pollutant Discharge Elimination System: a requirement of the CWA that discharges meet certain requirements prior to discharging waste to any water body. It sets the highest permissible effluent limits, by permit, prior to making any discharge.

National Primary Drinking Water Regulations: regulations developed under the Safe Drinking Water Act (SDWA), which establish MCLs, monitoring requirements, and reporting procedures for contaminants in drinking water that endanger human health.

Near coastal water initiative: this initiative was developed in 1985 to provide for management of specific problems in waters near coastlines that are not dealt with in other programs.

Nitrate: an ion consisting of nitrogen and oxygen (NO_3). Nitrate is a plant nutrient and is very mobile in soils.

Nonbiodegradable: substances that do not break down easily in the environment.

Nonpoint source: a source (of any water-carried material) from a broad area, rather than from discrete points.

Nonpoint-source contaminant: a substance that pollutes or degrades water that comes from lawn or cropland runoff, the atmosphere, roadways, and other diffuse sources.

Nonpoint-source water pollution: water contamination that originates from a broad area (such as leaching of agricultural chemicals from crop lad) and enters the water resource diffusely over a large area.

Nonpolar covalently bonded: a molecule composed of atoms that share their electrons equally, resulting in a molecule that does not have polarity.

Nutrient: any inorganic or organic compound needed to sustain plant life.

O

Organic: containing carbon, but possibly also containing hydrogen, oxygen, chlorine, nitrogen, and other elements.

Organic chemical: a chemical substance of animal or vegetable origin having carbon in its molecular structure.

Organic detritus: any loose organic material in streams—such as leaves, bark, or twigs—removed and transported by mechanical means, such as disintegration or abrasion.

Organic soil: soil that contains more than 20% organic matter in the upper 16 inches.

Organochlorine compound: synthetic organic compounds containing chlorine. As generally used, term refers to compounds containing mostly or exclusively carbon, hydrogen, and chlorine.

Outwash: soil material washed down a hillside by rainwater and deposited upon more gently sloping land.

Overdraft: any withdrawal of groundwater in excess of the *Safe Yield*.

Overland flow: the flow of rainwater or snowbelt over the land surface toward stream channels.

Oxidation: when a substance either gains oxygen or loses hydrogen or electrons in a chemical reaction. One of the chemical treatment methods.

Oxidizer: a substance that oxidizes another substance.

P

Paleokarst: a karstified area that has been buried by later deposition of sediments.

Parts per million (ppm): the number of weight or volume units of a constituent present with each one million units of the solution or mixture. Formerly used to express the results of most water and wastewater analyses, **ppm** is being replaced by milligrams per liter **M/L**. For drinking water analyses, concentration in parts per million and milligrams per liter are equivalent. A single ppm can be compared to a shot glass full of water inside a swimming pool.

Pathogens: types of microorganisms that can cause disease.

Perched groundwater: unconfined groundwater separated from an underlying main body of groundwater by an unsaturated zone.

Percolation: the movement, under hydrostatic pressure, of water through interstices of a rock or soil (except the movement through large openings such as caves).

Perennial stream: a stream that normally has water in its channel at all times.

Periphyton: microorganisms that coat rocks, plants, and other surfaces on lake bottoms.

Permeability: the capacity of a rock for transmitting a fluid; a measure of the relative ease with which a porous medium can transmit a liquid. The quality of the soil enables water to move downward through the soil profile. Permeability is measured as the number of inches per hour that water moves downward through the saturated soil. Terms describing permeability are as follows:

Very slow	Less than 0.06 inches/hour
Slow	0.06–0.2 inches/hour
Moderately slow	0.2–0.6 inches/hour
Moderate	0.6–2.0 inches/hour
Moderately rapid	2.0–6.0 inches/hour
Rapid	6.0–20 inches//hour
Very rapid	More than 20 inches/hour

pH: a measure of the acidity (less than 7) or alkalinity (greater than 7) of a solution; a pH of 7 is considered neutral.

Phosphorus: a nutrient essential for growth that can play a key role in stimulating aquatic growth in lakes and streams.

Photosynthesis: the synthesis of compounds with the aid of light.

Physical treatment: any process that does not produce a new substance (e.g., screening, adsorption, aeration, sedimentation, and filtration).

Plutonic: a loosely defined term with a number of current usages. We use it to describe igneous rock bodies that crystallized at great depth or, more generally, any intrusive igneous rock.

Point source: originating at any discrete source.

Polar covalent bond: the shared pair of electrons between two atoms are not equally held. Thus, one of the atoms becomes slightly positively charged and the other atom becomes slightly negatively charged.

Polar covalent molecule: (water) one or more polar covalent bonds result in a molecule that is polar covalent. Polar covalent molecules exhibit partial positive and negative poles, causing them to behave like tiny magnets. Water is the most common polar covalent substance.

Pollutant: any substance introduced into the environment that adversely affects the usefulness of the resource.

Pollution: the presence of matter or energy whose nature, location, or quantity produces undesired environmental effects. Under the Clean Water Act, for example, the term is defined as a man-made or man-induced alteration of the physical, biological, and radiological integrity of water.

Polychlorinated biphenyls (PCBs): a mixture of chlorinated derivatives of biphenyl, marketed under the trade name Aroclor with a number designating the chlorine content (such as Aroclor 1260). PCBs were used in transformers and capacitors for insulating purposes and in gas pipeline systems as a lubricant. Further sale or new use was banned by law in 1979.

Polycyclic aromatic hydrocarbon (PAH): a class of organic compounds with a fused-ring aromatic structure. PAH results from incomplete combustion of organic carbon (including wood), municipal solid waste, and fossil fuels, as well as from natural or anthropogenic introduction of uncombusted coal and oil. PAHs included benzo(a)pyrene, fluoranthene, and pyrene.

Population: a collection of individuals of one species or mixed species making up the residents of a prescribed area.

Porosity: (1) the ratio of the aggregate volume of pore spaces in rock or soil to its total volume, usually stated as a percent. (2) a measure of the water-bearing capacity of subsurface rock. With respect to water movement, it is not just the total magnitude of porosity that is important, but the size of the voids and the extent to which they are interconnected, as the pores in a formation may be open, or interconnected, or closed and isolated. For example, clay may have a very high porosity with respect to potential water content, but it constitutes a poor medium as an aquifer because the pores are usually so small.

Potable water: water that is safe and palatable for human consumption.

Potentiometric surface: a surface that represents the total head in an aquifer; that is, it represents the height above a datum plane at which the water level stands in tightly cased wells that penetrated the aquifer.

Precipitation: any or all forms of water particles that fall from the atmosphere, such as rain, snow, hail, and sleet. The act or process of producing a solid phase within a liquid medium.

Pretreatment: any physical, chemical, or mechanical process used before the main water treatment processes. It can include screening, presedimentation, and chemical addition.

Primary drinking water standards: regulations on drinking water quality (under SWDA) considered essential for preservation of public health.

Primary treatment: the first step of treatment at a municipal wastewater treatment plant. It typically involves screening and sedimentation to remove materials that float or settle.

Public-supply withdrawals: water withdrawn by public and private water suppliers for use within a general community. Water is used for a variety of purposes such as domestic, commercial, industrial, and public water use.

Public water system: as defined by the SDWA, any system, publicly or privately owned, that serves at least 15 service connections 60 days out of the year or serves an average of 25 people at least 60 days out of the year.

Publicly Owned Treatment Works (POTW): a waste treatment works owned by a state, local government unit or Indian tribe, usually designed to treat domestic wastewaters.

R

Rain shadow: a dry region on the lee side of a topographic obstacle, usually a mountain range, where rainfall is noticeably less than on the windward side.

Reach: a continuous part of a stream between two specified points.

Reaeration: the replenishment of oxygen in water from which oxygen has been removed.

Receiving water: a river, lake, ocean, stream, or other water source into which wastewater or treated effluent is discharged.

Recharge: the process by which water is added to a zone of saturation, by percolation from the soil surface or by artificial injection.

Recharge area (groundwater): an area within which water infiltrates the ground and reaches the zone of saturation.

Reference dose (RfD): an estimate of the amount of a chemical that a person can be exposed to on a daily basis that is not anticipated to cause adverse systemic health effects over the person's lifetime.

Representative sample: a sample containing all the constituents present in the water from which it was taken.

Return flow: that part of irrigation water that is not consumed by evapotranspiration and that returns to its source or another body of water.

Reverse osmosis (RO): solutions of differing ion concentration are separated by a semipermeable membrane. Typically, water flows from the chamber with lesser ion concentration into the chamber with the greater ion concentration, resulting in hydrostatic or osmotic pressure. In RO, enough external pressure is applied to overcome this hydrostatic pressure, thus reversing the flow of water. This results in the water on the other side of the membrane becoming depleted in ions and demineralized.

Riffle: a shallow part of the stream where water flows swiftly over completely or partially submerged obstructions to produce surface agitation.

Riparian: pertaining to or situated on the bank of a natural body of flowing water.

Riparian rights: a concept of water law under which authorization to use water in a stream is based on ownership of the land adjacent to the stream.

Riparian zone: pertaining to or located on the bank of a body of water, especially a stream.

Rock: any naturally formed, consolidated, or unconsolidated material (but not soil) consisting of two or more minerals

Runoff: a part of precipitation or snowmelt that appears in streams or surface–water bodies.

Rural withdrawals: water used in suburban or farm areas for domestic and livestock needs. The water generally is self-supplied and includes domestic use, drinking water for livestock, and other uses such as dairy sanitation, evaporation from stock-watering ponds, and cleaning and waste disposal.

S

Safe Drinking Water Act (SDWA): a federal law passed in 1974 with the goal of establishing federal standards for drinking water quality, protecting underground sources of water, and setting up a system of state and federal cooperation to assure compliance with the law.

Saline water: water that is considered unsuitable for human consumption or for irrigation because of its high content of dissolved solids; generally expressed as milligrams per liter (mg/L) of dissolved solids; seawater is generally considered to contain more than 35,000 mg/L of dissolved solids. A general salinity scale is:

	Concentration of Dissolved Solids in mg/L
Slightly saline	1000–3000
Moderately saline	3000–10,000
Very saline	10,000–35,000
Brine	More than 35,000

Saturated zone: a subsurface zone in which all the interstices or voids are filled with water under pressure greater than that of the atmosphere.

Screening: a pretreatment method that uses coarse screens to remove large debris from the water to prevent clogging of pipes or channels to the treatment plant.

Secondary drinking water standards: regulations developed under the SDWA that established maximum levels of substances affecting the aesthetic characteristics (taste, color, or odor) of drinking water.

Secondary maximum contaminant level (SMCL): the maximum level of a contaminant or undesirable constituent in public water systems that, in the judgment of US EPA, is required to protect the public welfare. SMCLs are secondary (non-enforceable) drinking water regulations established by the US EPA for contaminants that may adversely affect the odor or appearance of such water.

Secondary treatment: the second step of treatment at a municipal wastewater treatment plant. This step uses growing numbers of microorganisms to digest organic matter and reduce the amount of organic waste. Water leaving this process is chlorinated to destroy any disease-causing microorganisms before its release.

Sedimentation: a physical treatment method that involves reducing the velocity of water in basins so that the suspended material can settle out by gravity.

Seep: a small area where water percolates slowly to the land surface.

Seiche: a sudden oscillation of the water in a moderate-size boy of water, caused by wind.

Sinkhole: a depression in a karst area. At land surface its shape is generally circular and its size measured in meters to tens of meters; underground it is commonly funnel-shaped and associated with subterranean drainage.

Sinuosity: the ratio of the channel length between two points on a channel to the straight-line distance between the same two points; a measure of meandering.

Soil: the layer of material at the eland surface that supports plant growth.

Soil horizon: a layer of soil that is distinguishable from adjacent layers by characteristic physical and chemical properties.

Soil moisture: water occurs in the pore spaces between the soil particles in the unsaturated zone from which water is discharged by the transpiration of plants or by evaporation from the soil.

Solution: formed when a solid, gas, or another liquid in contact with a liquid becomes dispersed homogeneously throughout the liquid. The substance, called a solute is said to dissolve. The liquid is called the solvent.

Solvated: when either a positive or negative ion becomes completely surrounded by polar solvent molecules.

Sorb: to take up and hold either by absorption or adsorption.

Sorption: general term for the interaction (binding or association) of a solute ion or molecule with a solid.

Spall: a chip or fragment removed from a rock surface by withering; especially by the process of exfoliation.

Specific capacity: the yield of a well per unit of drawdown.

Specific retention: the ratio of the volume of water retained in a rock after gravity drainage to the volume of the work.

Specific storage: the volume of water that an aquifer system releases or takes into storage per unit volume per unit change in head. The specific storage is equivalent to the *Storage Coefficient* divided by the thickness of the aquifer system.

Specific yield: the ratio of the volume of water that will drain under the influence of gravity to the volume of saturated rock.

Spring: place where any natural discharge of groundwater flows at the ground surface.

Storage: the capacity of an aquifer, aquitard, or aquifer system to release or accept water into groundwater storage, per unit change in hydraulic head.

Storage coefficient: the volume of water released from storage in a unit prism of an aquifer when the head is lowered a unit distance.

Strain: Deformation that results from a stress. Expressed in terms of the amount of deformation per inch.

Stratification: the layered structure of sedimentary rocks.

Stress and strain: in materials, stress is a measure of the deforming force applied to a body. Strain (which is often erroneously used as a synonym for stress) is really the resulting change in its shape (deformation). For perfectly elastic material, stress is proportional to stain. This relationship is explained by Hooke's Law, which states that the deformation of a body is proportional to the magnitude of the deforming force, provided that the body's elastic limit is not exceeded. If the elastic limit is not reached, the body will return to its original size once the force is removed. For example, if a spring is stretched by 2 cm by a weight of 1 N, it will be stretched by 4 cm by a weight of 2 N, and so on; however, once the load exceeds the elastic limit for the spring, Hooke's law will no longer be obeyed, and each successive increase in weight will result in a greater extension until the spring finally breaks.

Stress forces are categorized in three ways:

1. Tension (or tensile stress), in which equal and opposite forces that act away from each other are applied to a body; tends to elongate a body.
2. Compression stress, in which equal and opposite forces that act toward each other are applied to a body; tends to shorten it.
3. Shear stress, in which equal and opposite forces that do not act along the same line of action or plane are applied to a body; tends to change its shape without changing its volume.

Stress, geostatic (lithostatic): the total weight (per unit area) of sediments and water above some plane of reference. Geostatic stress normal to any horizontal plane of reference in a saturated deposit may also be defined as the sum of the effective stress and the fluid pressure at that depth.

Stress, preconsolidation: the maximum antecedent effective stress to which a deposit has been subjected and which it can stand without undergoing additional permanent deformation. Stress changes in the range less than the preconsolidation stress produce elastic deformations of small magnitude. In fine-grained materials, stress increases beyond the preconsolidation stress produced much larger deformations that are principally inelastic (nonrecoverable). Synonymous with "virgin stress."

Stress, seepage: force (per unit area) transferred from the water to the medium by viscous friction when water flows through a porous medium. The forced transferred to the medium is equal to the loss of hydraulic head and is termed the seepage force exerted in the friction of flow.

Subsidence: a dropping of the land surface as a result of ground water being pumped. Cracks and fissures can appear in the land. Some state that subsidence is virtually an irreversible process. Others, like the author of this book, state that the jury is still out on the validity of this statement; HRSD's SWIFT project may prove that land subsidence can be reversed.

Surface runoff: runoff that travels over the land surface to the nearest stream channel.

Surface tension: the attractive forces exerted by the molecules below the surface upon those at the surface, resulting in them crowding together and forming a higher density.

Surface water: all water naturally open to the atmosphere, and all springs, wells, or other collectors that are directly influenced by surface water.

Surface Water Treatment Rule (SWTR): a federal regulation established by the USEPA under the SDWA that imposes specific monitoring and treatment requirements on all public drinking water systems that draw water from a surface water source.

Suspended sediment: sediment that is transported in suspension by a stream.

Suspended solids: different from suspended sediment only in the way that the sample is collected and analyzed.

Synthetic organic chemicals (SOCs): generally applied to manufactured chemicals that are not as volatile as volatile organic chemicals. Included are herbicides, pesticides, and chemicals widely used in industries.

T

Total head: the height above a datum plane of a column of water. In a groundwater system, it is composed of elevation head and pressure head.

Total suspended solids (TSS): solids present in wastewater.

Transmissivity (groundwater): the capacity of a rock to transmit water under pressure. The coefficient of transmissibility is the rate of flow of water, at the prevailing water temperature, in gallons per day, through a vertical strip of the aquifer one foot wide, extending the full saturated height of the aquifer under a hydraulic gradient of 100%. A hydraulic gradient of 100% means a one-foot drop in head in one foot of flow distance.

Transpiration: the process by which water passes through living organisms, primarily plants, into the atmosphere.

Trihalomethanes (THMs): a group of compounds formed when natural organic compounds from decaying vegetation and soil (such as humic and fulvic acids) react with chlorine.

Turbidity: a measure of the cloudiness of water caused by the presence of suspended matter, which shelters harmful microorganisms and reduces the effectiveness of disinfecting compounds.

U

Unconfined aquifer: an aquifer whose upper surface is a water table free to fluctuate under atmospheric pressure.

Unsaturated zone: a subsurface zone above the water table in which the pore spaces may contain a combination of air and water.

V

Vehicle of disease transmission: any nonliving object or substance contaminated with pathogens.

Vernal pool: a small lake or pond that is filled with water for only a short time during the spring.

Vug: a small cavity or chamber in rock that may be lined with crystals.

W

Wastewater: the spent or used water from individual homes, a community, a farm, or an industry that contains dissolved or suspended matter.

Water budget: an accounting of the inflow to, outflow from, and storage changes of water in a hydrologic unit.

Water column: an imaginary column extending through a water body from its floor to its surface.

Water demand: water requirements for a particular purpose, such as irrigation, power, municipal supply, plant transpiration, or storage.

Water softening: a chemical treatment method that uses either chemicals to precipitate or a zeolite to remove those metal ions (typically Ca^{2+}, Mg^{2+}, Fe^{3+}) responsible for hard water.

Water table: the top water surface of an unconfined aquifer at atmospheric pressure.

Waterborne disease: water is a potential vehicle of disease transmission, and waterborne disease is possibly one of the most preventable types of communicable illness. The application of basic sanitary principles and technology have virtually eliminated serious outbreaks of waterborne diseases in developed countries. The most prevalent waterborne diseases include *typhoid fever, dysentery, cholera, infectious hepatitis,* and *gastroenteritis.*

Watershed: the land area that drains into a river, river system, or other body of water.

Wellhead protection: the protection of the surface and subsurface areas surrounding a water well or well field supplying a public water system from contamination by human activity.

Y

Yield: the mass of material or constituent transported by a river in a specified period of time divided by the drainage area of the river basin.

Yield, optimal: an optimal amount of groundwater, by virtue of its use, that should be withdrawn from an aquifer system or groundwater basin each year. It is a dynamic quantity that must be determined from a set of alternative groundwater management decisions subject to goals, objectives, and constraints of the management plan.

Yield, perennial: the amount of usable water from an aquifer than can be economically consumed each year for an indefinite period of time. It is a specified amount that is commonly specified equal to the mean annual recharge to the aquifer system, which thereby limits the amount of groundwater that can be pumped for beneficial use.

Yield, safe: the amount of groundwater that can be safely withdrawn from a groundwater basin annually, without producing an undesirable result. Undesirable results include but are not limited to depletion of groundwater storage, the intrusion of water of undesirable quality, the contraventions of existing water rights, the deterioration of the economic advantages of pumping (such as excessively lower water level and the attendant increased pumping lifts and associated energy costs), excessive depletion of stream flow by induced infiltration, and land subsidence.

Z

Zone of aeration: the zone above the water table. Water in the zone of aeration does not flow into a well.

Zone of capillarity: the area above a water table where some or all of the interstices (pores) are filled with water that is held by capillarity.

Index

Note: **Bold** page numbers refer to tables and *italic* page numbers refer to figures.

Printed in the United States
by Baker & Taylor Publisher Services